H. Ward Silver, NØAX

ARRL's HANDS-ON RADIO EXPERIMENTS

Volume 2

60 electronic experiments from the pages of QST magazine

Compiled by Steve Ford, WB8IMY
Production: Jodi Morin, KA1JPA
Cover Design: Sue Fagan, KB1OKW

Published by:
ARRL The national association for **AMATEUR RADIO**®
225 Main Street ■ Newington, CT 06111-1494

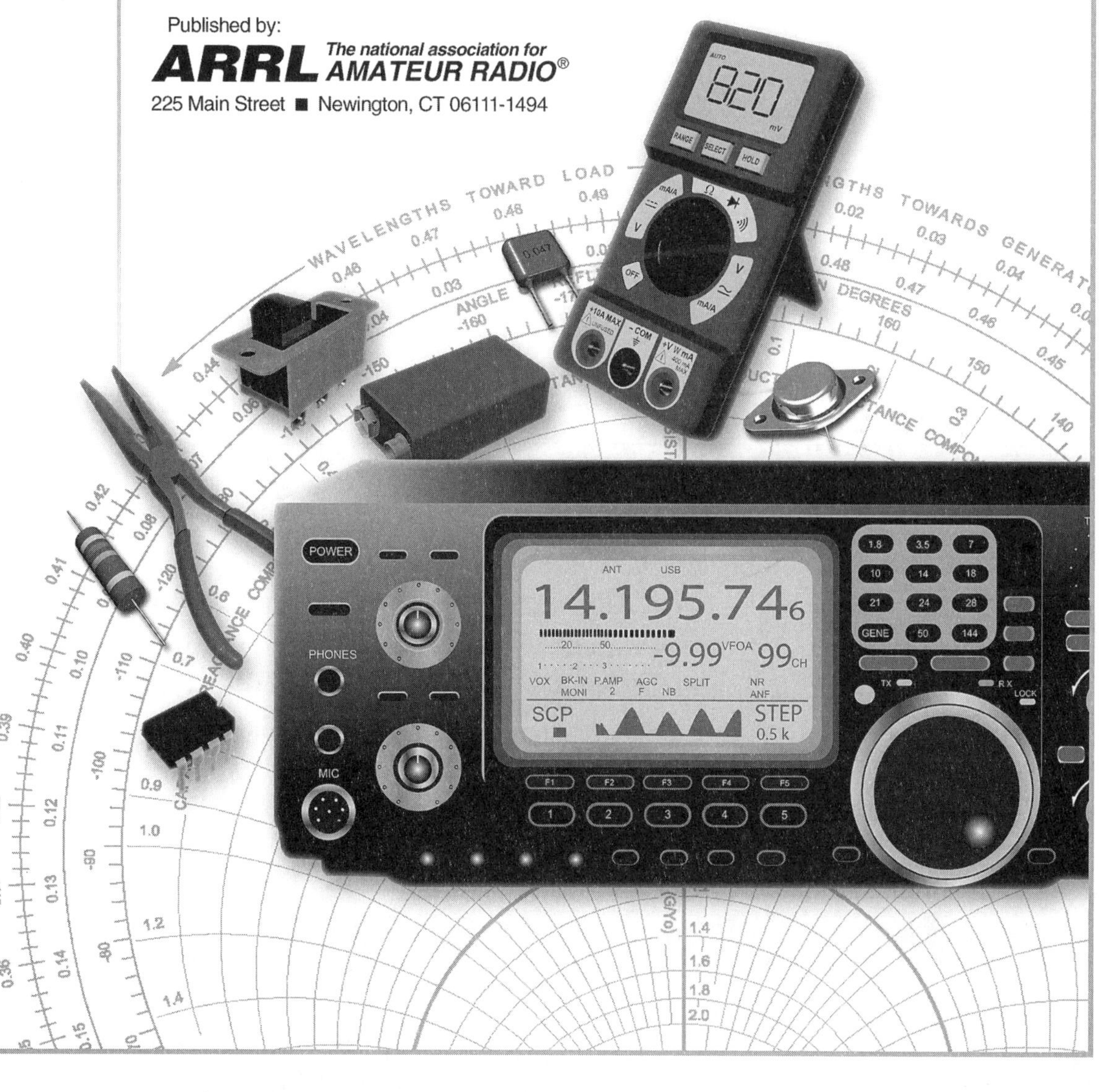

Copyright © 2013 by

The American Radio Relay League

Copyright secured under the Pan-American Convention

International Copyright secured.
No part of this work may be reproduced in any form except by written permission of the publisher. All rights of translation are reserved.

Printed in USA

Quedan reservados todos los derechos

ISBN: 0-87259-341-1

First Edition
First Printing

Updates

There is often additional information regarding the Hands-On Radio experiments — you can find it on the Hands-On Radio Web page at **www.arrl.org/hands-on-radio**. The contents include links to reference articles and sources of supplies and equipment. There is also a Frequently Asked Questions section that provides explanations about many of the experiments. Readers have contributed their observations and sometimes tools and links that relate to the topic in the experiment. You may find the information helpful as you perform the experiments in this anthology.

www.arrl.org/tis/hands-on-radio

Contents

Electronic Fundamentals............................7
Experiment #64 – Waveforms & Harmonics
Experiment #65 – Spectrum Modification
Experiment #117 – Laying Down the Laws
Experiment #118 – The Laws at Work
Experiment #95 – Watt's In A Waveform?
Experiment #75 – Series-Parallel Conversion
Experiment #74 – Resonant Circuits
Experiment #77 – Load Lines
Experiment #105 – Gain-Bandwidth Product
Experiment #106 – Effects of Gain-Bandwidth Product
Experiment #72 – Return Loss & S-Parameters
Experiment #104 – Words to Watch For
Experiment #100 – Hands-On Hundred

Simulation..33
Experiment #83 – Circuit Simulation, Part 1
Experiment #84 – Circuit Simulation, Build and Test
Experiment #85 – Circuit Simulation, Complex Parts
Experiment #86 – Viewing Waveforms in LTSpice
Experiment #87 – ELSIE Filter Design, Part 1
Experiment #88 – ELSIE Filter Design, Part 2
Experiment #107 – PCB Layout, Part 1
Experiment #108 – PCB Layout, Part 2
Experiment #109 – PCB Layout, Part 3
Experiment #110 – PCB Layout, Part 4

Antennas and Transmission Lines...........53
Experiment #82 – Antenna Height
Experiment #113 – Radiation Patterns
Experiment #92 – The 468 Factor
Experiment #94 – SWR and Transmission Line Loss
Experiment #81 – Synchronous Transformer
Experiment #111 – Coiled-Coax Chokes
Experiment #96 – Open-Wire Transmission Lines
Experiment #116 – The Quarter-Three-Quarter
   Wave (Q3Q) Balun
Experiment #119 – The Q3Q Balun Redux

Electronic Circuits and Components 71
Experiment #62 – About Resistors
Experiment #63 – About Capacitors
Experiment #78 – Bridge Circuits
Experiment #115 – All About Tapers
Experiment #101 – Rotary Encoders
Experiment #76 – Diode Junctions
Experiment #99 – Cascode Amplifier
Experiment #73 – Choosing an Op Amp
Experiment #70 – Three-Terminal Regulator
Experiment #98 – Linear Supply Design
Experiment #97 – Programmable Frequency Reference

RF Techniques 93
Experiment #91 – Common-Mode Chokes
Experiment #112 – RFI Hunt
Experiment #114 – Recording Signals
Experiment #93 – An LED AM Modulator
Experiment #102 – Detecting RF, Part 1
Experiment #103 – Detecting RF, Part 2
Experiment #66 – Mixer Basics
Experiment #68 – Phase-Locked Loop, Part 1
Experiment #69 – Phase-Locked Loop, Part 2
Experiment #79 – Pi and T Networks

Practical Construction............................113
Experiment #120 – Power Protection
Experiment #121 – Transient Protection
Experiment #90 – Construction Techniques
Experiment #89 – Overvoltage Protection
Experiment #80 – Battery Capacity
Experiment #67 – The Return of the Kit
Experiment #71 – Circuit Layout

Parts List..127

Foreword

The Hands-On Radio series of columns in QST magazine just keeps chugging away with no shortage of topics – many suggested by readers, others discovered in reference books, and still more result from the author's journey of personal education. As they say, if you really want to learn something, teach it! The response to the columns has been very gratifying, to say the least, and notes are continuously received as readers work through even the very first columns.

This second volume begins with some fundamental ideas, covering basic electricity and the discovery of its relationship to magnetism which led to Maxwell's equations and the beginning of what we call "wireless" today. At the same time, there is a section of experiments that deal with the use of computers to design, simulate, and construct electronic circuits. That's quite a span of time and of technical understanding! In the next five years, robotic circuit construction may become a topic, too.

In between, hams continue to learn about all manner of circuits and components. Even with radio circuits disappearing into the equations used to create digital gate arrays, the antenna and transmission line are as useful as ever, so there is a collection of material for these subjects, too. Traditional analog circuits are well-represented, as always.

In short, there is no danger of running out of things to write about or experiment with in the world of Amateur Radio. We continue using everything from the venerable 2N2222 transistor to the latest high-speed analog-to-digital flash converter to generate and receive signals. We use sophisticated antenna modeling software along with rules-of-thumb and cut-and-try to build our antennas. As long as we continue to put our hands on radios, there will be things to learn and techniques to try. I hope to be there with you along the way.

Thanks for reading and happy experimenting!

73, Ward, NØAX
February 2013

Other Books by Ward Silver, NØAX

Other books by the author include three popular titles in the …for Dummies series by Wiley Publishing; *Ham Radio for Dummies, Two-Way Radios and Scanners for Dummies*, and *Circuitbuilding Do-It-Yourself for Dummies*. He has just released a ham radio detective novel, *Ray Tracy: Zone of Iniquity* and an a beginner's guide to antenna modeling software, *Antenna Modeling for Beginners*. He is also the author of the three ARRL license guides and is the Lead Editor for the *ARRL Handbook* and the *ARRL Antenna Book*.

About the ARRL

The seed for Amateur Radio was planted in the 1890s, when Guglielmo Marconi began his experiments in wireless telegraphy. Soon he was joined by dozens, then hundreds, of others who were enthusiastic about sending and receiving messages through the air—some with a commercial interest, but others solely out of a love for this new communications medium. The United States government began licensing Amateur Radio operators in 1912.

By 1914, there were thousands of Amateur Radio operators—hams—in the United States. Hiram Percy Maxim, a leading Hartford, Connecticut inventor and industrialist, saw the need for an organization to band together this fledgling group of radio experimenters. In May 1914 he founded the American Radio Relay League (ARRL) to meet that need.

Today ARRL, with more than 158,000 members, is the largest organization of radio amateurs in the United States. The ARRL is a not-for-profit organization that:

- promotes interest in Amateur Radio communications and experimentation
- represents US radio amateurs in legislative matters, and
- maintains fraternalism and a high standard of conduct among Amateur Radio operators.

At ARRL headquarters in the Hartford suburb of Newington, the staff helps serve the needs of members. ARRL is also International Secretariat for the International Amateur Radio Union, which is made up of similar societies in 150 countries around the world.

ARRL publishes the monthly journal *QST*, as well as newsletters and many publications covering all aspects of Amateur Radio. Its headquarters station, W1AW, transmits bulletins of interest to radio amateurs and Morse code practice sessions. The ARRL also coordinates an extensive field organization, which includes volunteers who provide technical information and other support services for radio amateurs as well as communications for public-service activities. In addition, ARRL represents US amateurs with the Federal Communications Commission and other government agencies in the US and abroad.

Membership in ARRL means much more than receiving *QST* each month. In addition to the services already described, ARRL offers membership services on a personal level, such as the ARRL Volunteer Examiner Coordinator Program and a QSL bureau.

Full ARRL membership (available only to licensed radio amateurs) gives you a voice in how the affairs of the organization are governed. ARRL policy is set by a Board of Directors (one from each of 15 Divisions) elected by the membership. The day-to-day operation of ARRL HQ is managed by a Chief Executive Officer.

No matter what aspect of Amateur Radio attracts you, ARRL membership is relevant and important. There would be no Amateur Radio as we know it today were it not for the ARRL. We would be happy to welcome you as a member! (An Amateur Radio license is not required for Associate Membership.) For more information about ARRL and answers to any questions you may have about Amateur Radio, write or call:

ARRL — the national association for Amateur Radio®

225 Main Street
Newington CT 06111-1494
Voice: 860-594-0200
Fax: 860-594-0259
E-mail: **hq@arrl.org**
Internet: **www.arrl.org/**

Prospective new amateurs call (toll-free):
800-32-NEW HAM (800-326-3942)
You can also contact us via e-mail at **newham@arrl.org**
or check out ARRLWeb at **http://www.arrl.org/**

Electronic Fundamentals

Experiment #64 — Waveforms and Harmonics

This month I'm going to introduce three new tools; two are combinations of software and a sound card and one is a combination of software and external hardware. Even better — two are free and the third is a great bargain! We'll make plenty of use of these tools in the months to come.

Tool number 1 is the free *Function Generator* software that was included in the 2008 edition of *The ARRL Handbook*.[1] It uses a PC sound card to produce a variety of audio waveforms from 20 Hz to 20 kHz. Tool number 2 is the also-free *Audio Spectrum Analyzer* software also included with *The ARRL Handbook*. It will display audio from a sound card input or from WAV files in three forms — frequency spectrum, oscilloscope or waterfall format. Both packages were written by Grant Connell, WD6CNF.

Tool number 3 is a new kind of oscilloscope — the Parallax USB Oscilloscope (**www.arrl.org/catalog/?item=1259#top**). This USB-driven oscilloscope converts signals to digital format in a small external pod and sends the data to a host PC for display on the screen where all the controls are simulated graphically. This saves a lot of expense because there is very little in the way of stand-alone electronics. The pod connects to two signal probes and one trigger probe. Power is supplied by the USB host (the PC). While the specifications [**tinyurl.com/2fwkek**] are limited compared to an expensive hardware version, the USB 'scope is perfect for a beginning workbench and projects involving signals below 200 kHz. It is also an excellent introduction to the oscilloscope, in general. I'll be using this 'scope in numerous experiments for this column.

[1]*The ARRL Handbook for Radio Communications*, 2008 Edition. Available from your ARRL dealer or the ARRL Bookstore, ARRL order no. 1018. Telephone 860-594-0355, or toll-free in the US 888-277-5289; **www.arrl.org/shop/**; **pubsales@arrl.org**.

Waveform Terms

A signal can be broadly defined as any *waveform* or group of waveforms intentionally produced by a single piece of equipment for the purpose of conveying energy or information. A waveform is any defined sequence of voltages or currents, usually *periodic,* or repeating. The simplest signal is a sine wave of a single frequency. A *complex* signal consists of sine waves of more than one frequency. AM broadcast signals are complex because they consist of a carrier signal and two sideband signals.

If a complex signal consists of sine waves at integer multiples of some frequency, the lowest frequency sine wave is called the *fundamental* and the higher frequency sine waves are its *harmonics*. There is no such thing as the "first harmonic" — there is a fundamental and a second, third, fourth, fifth, etc. harmonic.

Components are sine waves that make up a complex signal. Components need not be harmonics of a fundamental, such as the carrier and sidebands mentioned above. Make it a habit to say "component" unless you are referring specifically to a harmonic. "Harmonic" can also mean a replica of the complex signal, such as the second harmonic of a 20 meter SSB signal appearing on 10 meters. A *spectrum* is a set of components — almost always more than one. *Spectra* refers to a set of more than one spectrum.

Meet *Mssr* Fourier

One of the great advances in physics and mathematics was the discovery that periodic waveforms of any sort (electrical or not) could be decomposed into a set of sine waves of different frequencies and phases. While the fundamental math was envisioned by Euler (*Oy*-ler), French physicist and mathematician Joseph Fourier (*For*-ee-yay) led the way in creating comprehensive methods of analysis to work with these sets of sine waves. These methods are known as *Fourier analysis*; it is key to electrical and all sorts of other kinds of engineering.

The average radio amateur can put Fourier's discoveries to work by learning how the shape of a waveform is determined by its spectrum. This is useful in understanding the bandwidth of analog and digital signals. Knowing how a

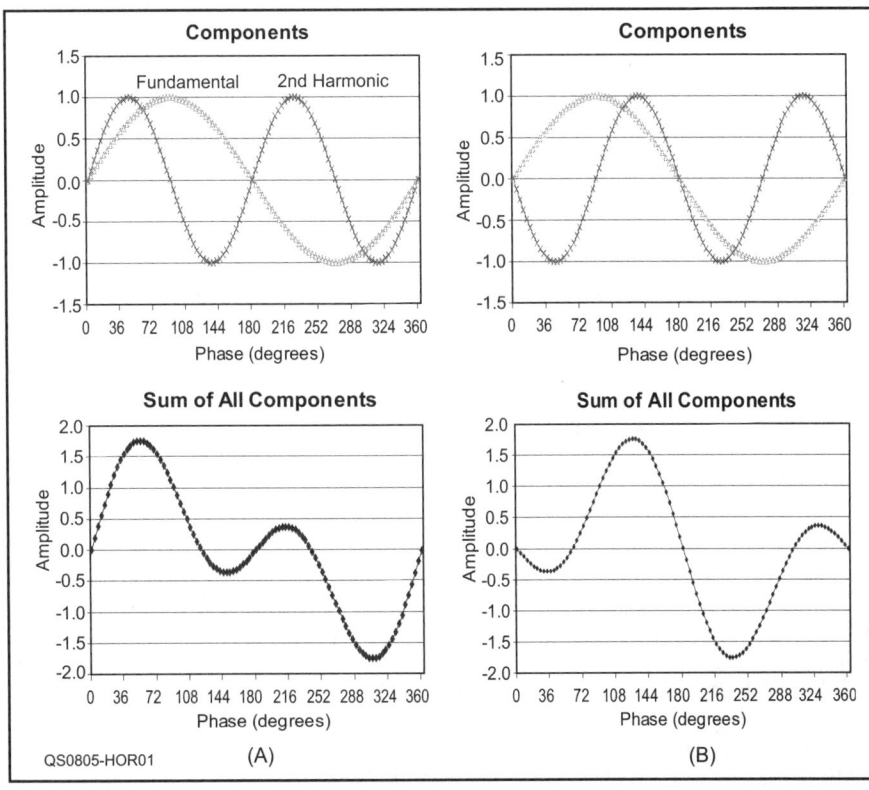

Figure 1 — The results of combining a fundamental and its second harmonic. In A both signals begin in phase and in B, out of phase. The spreadsheet to generate these and other waveforms is available on the Hands-On Radio Web site.

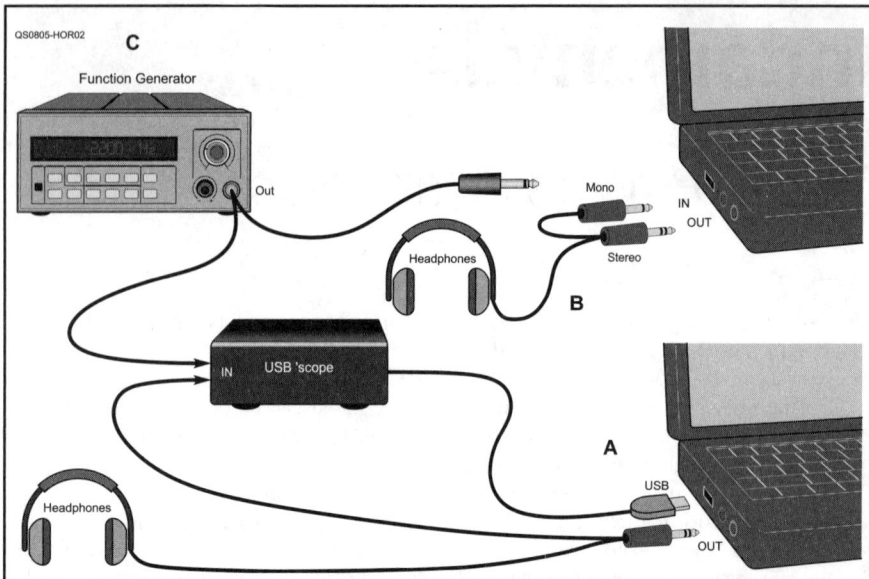

Figure 2 — Three methods of viewing the spectrum of a complex waveform. At A, a USB 'scope is used along with the PC sound card function generator. At B, the sound card's output is connected directly to its input and analyzed in the PC. C shows how to use a stand-alone function generator, instead.

complex waveform can be deconstructed into its individual components provides a better understanding of the effects of filters and amplifiers. For example, the source of key clicks becomes clear as the higher harmonics making a sharply keyed CW signal's envelope. The cure also becomes clear: slow down the rise and fall time of the envelope and those harmonics are greatly reduced!

Along with each component's frequency, its amplitude and phase also affect the resulting waveform. Imagine a sine wave and its second harmonic, both with the same amplitude and starting phase. What would the sum of these two waveforms look like? That waveform would rise quickly, since both waves are rising at the same time, but the second harmonic peaks 45° before the fundamental and begins dropping. The sum waveform turns out to peak a little later and then drops. The entire waveform is shown in Figure 1A. If the second harmonic is initially out of phase with the fundamental, decreasing while the fundamental increases, the waveform takes on the different shape in Figure 1B.

Getting Started

In the following sequence of experiments, you'll create some complex signals of your own, by creating and viewing the various components. I've put a spreadsheet on the Hands-On Radio Web site (**www.arrl.org/hands-on-radio**) with which you can experiment to make your own waveforms. The actual signals can be generated by either a stand-alone function generator or the *Function Generator* software mentioned above. You can view and analyze the signals either by using the USB 'scope or *Audio Spectrum Analyzer* software. The Hands-On Radio

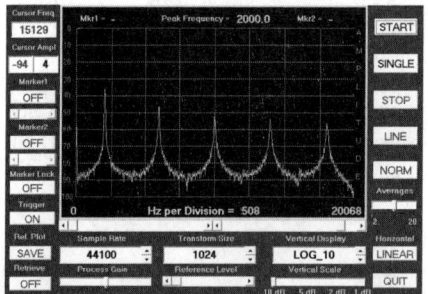

Figure 3 — The spectrum of a 2000 Hz square wave includes all odd harmonics of the fundamental at gradually decreasing amplitudes.

Web site lists the configuration for the two software packages and the USB 'scope.

Figure 2 shows three methods of hooking up your equipment and PC to generate signals and observe the resulting spectra. At (A), the PC sound card generates the waveforms while both the USB 'scope and a pair of headphones listen. To let both the 'scope and headphones "hear" the output signals, use a stereo splitter for the headphones and a stereo plug with its contacts exposed for the 'scope clips. At (B), the sound card output is routed directly into the sound card input and spectrum analysis software by using a cable that is stereo on one end and mono on the other (or some equivalent series of adapters). At (C), you can use your own function generator and observe the output with either the sound card spectrum analyzer or the USB 'scope. Connect the equipment in your preferred configuration and you're ready to go!

Components of Complex Signals

Starting with a 2000 Hz sine wave on the screen, set up the FFT on the USB 'scope

(running version 4 of the software) or the PC audio spectrum analyzer's frequency display to view a range of 0 to 20 kHz. You should see a single peak on both displays, representing the sine wave at 2000 Hz. (And you should hear it in your headphones, too!) You may also see some low-level peaks representing small amounts of distortion of the waveform.

Now change the waveform to a square wave — wow! You will now see the fundamental at 2000 Hz, plus four additional harmonics near 6000, 10000, 14000, 18000 and 22000 Hz. (The USB 'scope's FFT display will probably start to show other signals at the even harmonics due to its limited 8-bit resolution, compared to the 16 bits or higher of most PC sound cards. Figure 3 shows the spectrum of the square wave. Use the cursor functions of the spectrum display to verify the frequency of each harmonic. Listen carefully to the difference in sound of the sine and the square wave — can you hear the higher harmonics now present?

Notice that each successive harmonic has a lower amplitude. Use the cursor functions (if available) to compare the harmonic amplitudes. Determine the amplitude ratio of the harmonic to the fundamental component using ratio = [log^{-1} (dB)] / 20. Open the spreadsheet from the Hands-On Radio Web site and enter the ratio values into the cells at the top of the chart labeled AMPLITUDE. The fundamental amplitude should be 1.0, the third harmonic near 0.33, and so forth. Does this create a square wave?

Now change the function generator to output a triangle wave. The signal in your headphones will change and the relative amplitudes of the higher harmonics will decrease, although the odd harmonics are still the only ones present. Repeat your component ratio calculations and enter them into the spreadsheet. You'll probably see a rounded waveform on the spreadsheet graph not at all like a triangle wave. What happened? The phase of the harmonics is also important and the FFT or analyzer displays don't give you that information. Try entering 180 into the spreadsheet cell labeled PHASE for the third harmonic and observe the effect. Even though this harmonic has only about 1/10 the amplitude of the fundamental, its proper phasing makes a big difference in wavelength shape.

Continue to experiment with both the function generator waveforms and the spreadsheet. Listen to the characteristics of the signal as you change its shape. Try to duplicate the waveform on the spreadsheet, then change the amplitudes and phases of the components to see what happens.

Next Month

We'll explore the effects of amplifier and filter frequency response by combining the PC based function generator and the analyzer functions of your new tools.

8   Electronic Fundamentals

Experiment #65 — Spectrum Modification

In last month's experiment, waveforms were created by adding sinusoidal components together.[1] We learned that both amplitude and phase of the components is important in determining the shape of the resulting waveform. This month, we'll turn the process around by altering a set of components to make a new waveform. First we'll simulate the process, then build an actual circuit and compare the results.

Ramp Waveform Review

The negative-slope ramp waveform (or n-ramp) is made up of a fundamental component and all of its harmonics, each in phase with the fundamental. On the spreadsheet, cut and paste the phase and amplitude values for the n-ramp waveform into the PHASE and AMPLITUDE areas for the different components. Verify that the resulting waveform is an n-ramp.

Now see what happens when you change the amplitude of the fundamental or one of the harmonics. For example, set the amplitude of the second harmonic to 0.1 instead of 0.5. Observe the effect on the ramp waveform as shown in Figure 1. The smooth ramp begins to break up as the components no longer add up to create the original waveform. This is the general process we'll follow during this experiment. Now reset the second harmonic amplitude to 0.5.

Low-Pass Filter

What would happen to the n-ramp waveform if the harmonics were attenuated? Try it by reducing the amplitudes of the harmonics on the spreadsheet. When only the fundamental is left, the waveform becomes a sine wave. This is just what would happen if the

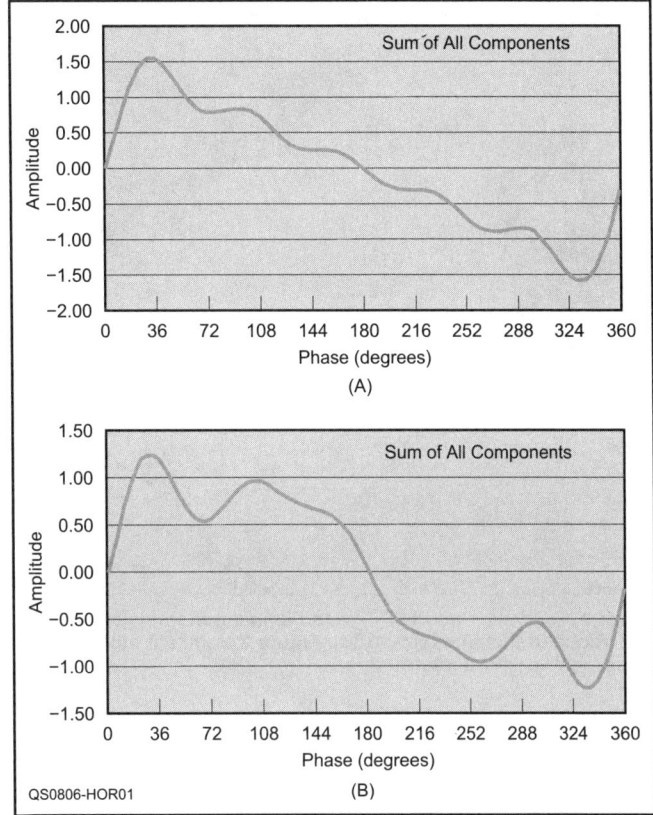

Figure 1 — A shows an n-ramp waveform with all components at their proper amplitude. B shows the distortion caused by reducing the amplitude of the second harmonic component to a tenth that of the fundamental.

n-ramp was passed through a low-pass filter, of course, so let's try it with a real circuit.

Before beginning — a short digression. We're about to perform a four step dance done by just about every product designer — design, simulate, build and compare. I touched on some of these issues in "Experiment #56 — Design Sensitivities."[2] There are many powerful design and simulation tools available to the experimenter. They are so powerful that it is easy to forget real circuits may not behave exactly like the models, just as an antenna doesn't always work just like the design software says it will.

Start with the design equations, model the circuit (or waveform) on a computer, build the circuit, and then be sure to compare the predicted and actual results. Look especially closely at high and low frequencies at which the limitations and parasitic effects of the components or your construction techniques affect circuit operation. The circuit's quiescent dc voltages and currents may also differ from those predicted because of component variations. By performing the all-important final comparison, you'll develop your intuition and experience about design and about modeling.

The design of the low-pass filter circuit is described in Figure 1 of Experiment #4 — Active Filters.[3] Use the design equations given for f_C = 2.8 kHz and a gain of –1. Build the filter and test it with a sine wave input from a function generator or sound card program. Use the sound card analyzer or 'scope as shown in Figure 2 to verify that the output voltage is 0.707 times the input voltage at f_C.

What would happen to a 1.4 kHz n-ramp waveform passed through such a filter? The harmonics will all be attenuated, more at higher frequencies, but by how much? To predict the amount of attenuation, calculate the filter's response versus frequency. Start by calculating the magnitude of the filter's frequency response:

$$|A_V| = 1 / \sqrt{(1 + (f / f_C)^2)} \quad [\text{Eq 1}]$$

Our fundamental, 1.4 kHz, equals 0.5 × f_C, the second harmonic is f_C, the third harmonic 1.5 × f_C, and so forth. Calculate the amplitude of all six simulated components by multiplying the magnitude of the response times the amplitude of the unfiltered component. (Table 1 shows the values for filter response, the response in dB, the amplitude of the unfiltered component, and the amplitude of the component after filtering.) Enter the filtered component amplitude values into the spreadsheet and see what happens to the waveform. Notice that the waveform

---

[1]All previous Hands-On Radio experiments are available online to ARRL members at **www.arrl.org/hands-on-radio**.

[2-5]See Note 1.

is smoother and has less high-frequency ripple. (If you are limited to p-ramp waveforms, the effect on the waveform will be similar.)

Now input a 1.4 kHz n-ramp waveform from a function generator, then compare the output waveform to the simulated waveform. Does the output look like what the spreadsheet predicts? You should see some distortion near the positive and negative peaks of the circuit's output not present on the simulated waveform. This difference is caused because Equation 1 ignored the filter's phase shift at different frequencies.

As signal frequency increases, the filter causes more phase shift from input to output, reaching 90° at high frequencies. The equation for the filter's phase shift is:

$$\theta = \tan^{-1}(f / f_C) \qquad [\text{Eq 2}]$$

Calculate the phase shift for each component. (Table 2 shows the values.) Now enter all of those phase values into the spreadsheet and see if the resulting waveform looks more like what you see on your scope or analyzer. The simulated waveform graph "Components" shows the components all slightly shifted and no longer lining up neatly at the zero-crossings of the fundamental.

Removing Individual Components

The low-pass filter attenuates the entire frequency spectrum. What about removing just one component? That's easy to do on the simulation spreadsheet — just set the amplitude of that component to zero! Restore the n-ramp component amplitude and phase values to the original, unfiltered values. Now set the amplitude of the second harmonic to zero as at the beginning of this experiment. The distortion of the resulting waveform (see Figrure 1) is obvious! Set each component's value to zero, one at a time, and sketch the resulting waveform.

To do the same thing with a real circuit

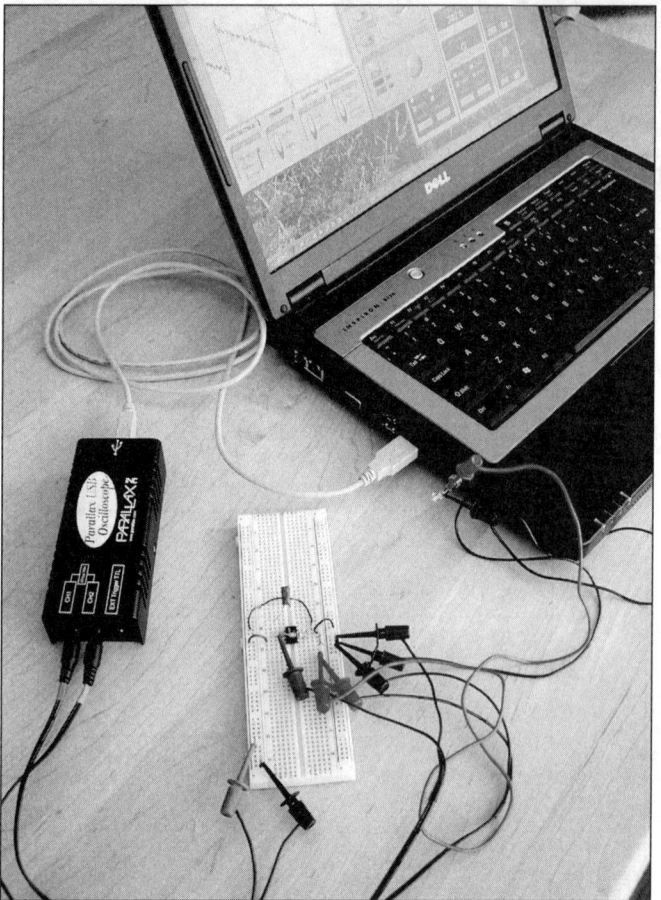

Figure 2 — Connecting the PC-based signal generator and USB scope to the filter circuit. See Figure 2 from last month's experiment for more information.

takes a notch filter, just as described in Experiment #42.[4] Build the active notch filter in Figure 2 of that experiment with the feedback resistor set so that Q = 2 (K = 0.875). The notch frequency of the filter, f_0, will be approximately 1 kHz and the bandwidth will be $f_0/Q = 500$ Hz. Input a sine wave and vary its frequency to find the exact value for f_0 at which the output is a minimum. Now input an n-ramp waveform.

We're not going to vary the filter's center frequency — we'll vary the waveform's frequency instead. Start with the waveform's frequency set to f_0. This corresponds to minimizing the value of the fundamental component. Does the notch filter's output waveform correspond approximately to what the simulation spreadsheet predicts?

Table 2
Low-Pass Filter Phase Response

f/f_C	Phase Shift (°)
0.50	26.57
1.00	45.00
1.50	56.31
2.00	63.43
2.50	68.20
3.00	71.57

(It won't be exact because of phase shift.)

Now lower the fundamental frequency to ½ of f_0 so that the notch becomes centered on the second harmonic. Note the shape of the output waveform and look for correspondences with the spreadsheet's simulation with the second harmonic set to zero. Repeat this process for the third, fourth, fifth and sixth harmonics, one at a time.

What all this mangling of a perfectly good waveform should demonstrate is how complex waveforms are constructed from individual sinusoidal components. You can experiment with square and triangle waves, watch the spectra change on the audio analyzer and compare with time domain scope waveforms. That Fourier guy was pretty bright, huh?

Parts List

Use the parts lists for experiments #4, #42 and #64.[5] For those readers who are just building up their stock of electronics parts, I'm pleased to announce that a parts kit for all of the Hands-On Radio experiments is now available through the ARRL (**www.arrl.org/catalog**) and Kanga US (**www.kangaus.com**). Kits are available with and without the solderless breadboards.

Recommended Reading

There is a good discussion of Fourier methods in *The ARRL Handbook* chapter on Digital Signal Processing. Wikipedia also has an extensive article on Fourier Analysis (**en.wikipedia.org/wiki/Fourier_analysis**). Follow the various hyperlinks in the article to find some unexpected applications of this powerful technique.

Next Month

Let's get back to building a circuit or two as we explore mixers. We will use the PC-based sound card audio analyzer to observe the mixing process and products.

Table 1
Low-Pass Filter Response

f/f_C	Response	Response (dB)	Unfiltered	Filtered
0.50	0.89	–0.97	1.00	0.89
1.00	0.71	–3.01	0.50	0.35
1.50	0.55	–5.12	0.33	0.18
2.00	0.45	–6.99	0.25	0.11
2.50	0.37	–8.60	0.20	0.07
3.00	0.32	–10.00	0.17	0.05

Experiment 117

Laying Down the Laws

Most amateurs tend to think of *wireless* as beginning with Marconi in the mid-1890s — he transmitted a message over a distance of a bit less than 2 miles in 1895. The historically minded ham might travel farther back through the experiments and papers of well known and not so well known names such as Nikola Tesla, Heinrich Hertz, Nathan Stubblefield and Mahlon Loomis to arrive at James Clerk Maxwell's electromagnetic theory, published in 1864. Yet their work required, as Isaac Newton had characterized it, "standing upon the shoulders of giants" who explored the *terra incognita* of electricity and magnetism from the early 1600s. Who were these giants and what did they discover?

My interest in prewireless was sparked, so to speak, by a recent article in the *IEEE Antennas and Propagation Magazine* giving a chronology of how wireless communications came to be.[1] (The article may be available through your local library or from an *IEEE* member.) A huge number of discoveries and explanations of basic concepts were required before Maxwell could synthesize them into his theory of electromagnetic waves.

For many of us, our electrical education began with Ohm's law, first stated by Georg Ohm in 1827.[2] We know it today as the familiar I = E / R, but R was a brand new idea in those days. In fact, Ohm's ideas were not well received at all! From Ohm's law, we progressed through the equation for power (P = E × I) and then into circuitry such as capacitance, inductance, reactance and impedance that quickly followed. But capacitance and inductance are treated as *givens* in equations we memorize for time constants, turns ratio, resonant frequency and so forth. Where do these *proto wireless* concepts come from?

In this column, we'll begin reviewing several of the most important advances listed in the

[1]Salazar-Palma, et al, "The Father of Radio: A Brief Chronology of the Origin and Development of Wireless Communications," *IEEE Antennas and Propagation Magazine*, Vol 53, No 6, Dec 2011, pp 83-114.
[2]en.wikipedia.org/wiki/Ohm's_law

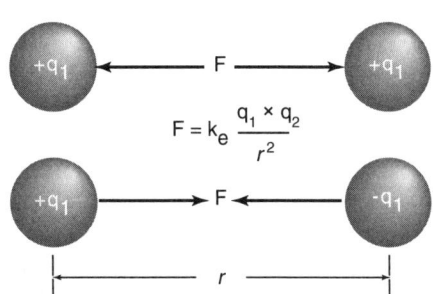

Figure 1 — Coulomb's law describes the force, F, between two electrically charged particles (q_1 and q_2). The force is proportionally weaker with the square of the distance between the particles, r. If the charges have the same polarity, the force is positive and they repel each other. If they are oppositely charged, the force is negative and attracts the particles together.

article then progress to some simple experiments you can do yourself. The goal is to more fully understand what is meant by the familiar symbols and units in the design equations and in schematics. It is one thing to memorize an equation or paragraph and quite another to experience it for yourself on the workbench!

The Beginnings

There is a long history of experimentation with static electricity and magnetism leading to the invention of the capacitor in the mid 1700s. Perhaps the best known example of an early capacitor is the Leyden jar.[3] Since static electricity was fairly easy to generate, the capacitor and its ability to store electrical energy were well known by the end of the 18th century. The relationship between electricity and magnetism, however, was quite unclear and that relationship lies at the root of electromagnetic phenomena — such as wireless.

The IEEE article begins its journey to wireless with Charles-Augustin Coulomb's determination in 1785 that electric forces varied proportionately to the inverse square of distance — now known as Coulomb's law (Equation 1) as illustrated in Figure 1.

$$F = k_e \frac{q_1 q_2}{r^2} \quad [\text{Eq 1}]$$

[3]en.wikipedia.org/wiki/Leyden_jar

where
F = the electric force between two particles with charges, q_1 and q_2,
r = the distance between them, and
k_e = a "constant of proportionality." It is this constant that turned out to have the most for reaching implications because it is determined solely by the properties of free space:

$$k_e = \frac{1}{4\pi\varepsilon_0} = \frac{c^2 \mu_0}{4\pi} \quad [\text{Eq 2}]$$

where
c = speed of light in vacuum
ε_0 = the permittivity of free space (roughly, the ability of free space to contain electrical energy) and
μ_0 = the permeability of free space (the ability of free space to contain magnetic energy).

As it turns out, the speed of light (electromagnetic energy) traveling in free space is also determined by these two quantities:

$$c = \sqrt{\frac{1}{\varepsilon_0 \mu_0}} \quad [\text{Eq 3}]$$

Not only does Coulomb's simple relationship contain the beginnings of wireless but it is also the first step in the studies of electromagnetic waves that led to relativity and its profound effects on our understanding of the universe. Coulomb did not know this at the time, of course. He only knew that he had discovered a relationship between electrical charge and electrical force.

Meanwhile (as the narrator often intones) other investigators were developing new ways of creating electricity. Up until this time, electrical experiments had to be performed with static electricity created by mechanical friction. In 1799, Alessandro Volta created an electrochemical *battery* based on chemical principles.[4] This was a major advance because experimenters then had a source not only of what Volta called the *electromotive force* (abbreviated *EMF*) but a source of current they could then control and study. Prior to that current was mostly available as pulses from electrical discharges — *sparks*.

[4]en.wikipedia.org/wiki/History_of_the_battery

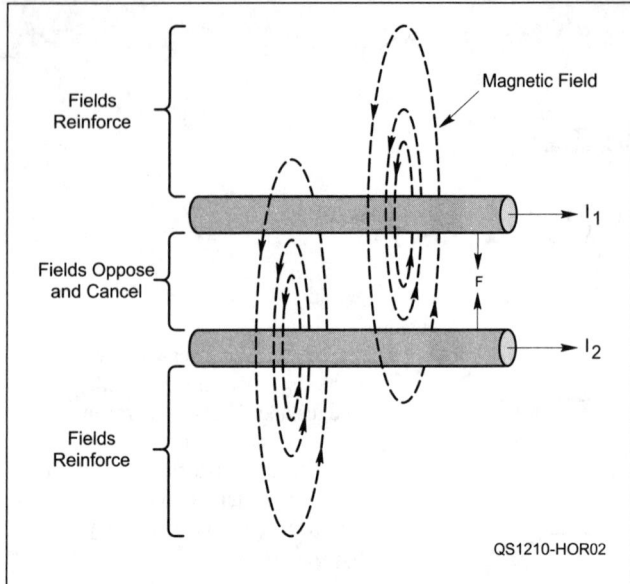

Figure 2 — Ampère's force law describes the force between two parallel, current-carrying wires. If the currents are flowing in the same direction, the fields are oriented in opposite directions between the wires, partially cancelling each other. Since the fields reinforce elsewhere, the result is a force pushing the wires together.

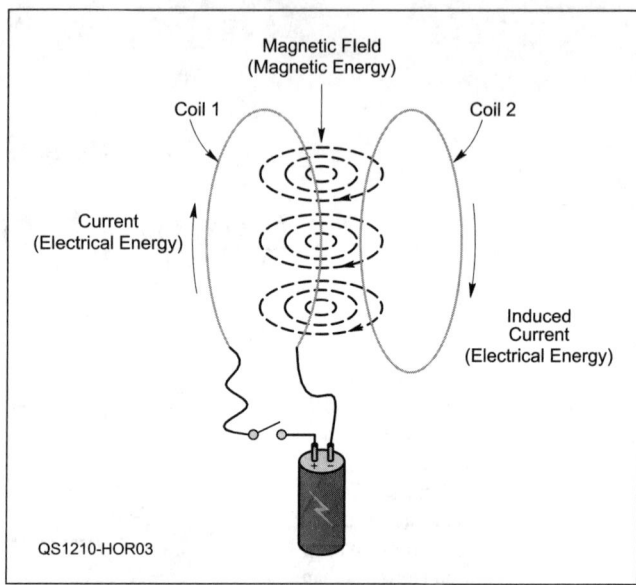

Figure 3 — Faraday demonstrated electromagnetic induction by showing how changing current in one coil induces a similar current in a second coil through a shared magnetic field. When the switch is closed, current in coil 1 will cause the current shown in coil 2. Lenz's law states that the current in coil 2 will be oriented to oppose the magnetic field from coil 1.

Magnetism was considered a separate phenomenon from electricity until 1820 when Hans Christian Ørsted discovered that current flowing through a wire caused a magnetic compass needle to deflect and created a circular magnetic field around the wire.[5] François Arago then demonstrated that not only did current flowing through a wire affect a magnet but that the current carrying wire itself became a magnet! Within days, André-Marie Ampère also demonstrated that parallel currents attract each other and opposing currents repel due to those magnetic fields:

$$F = 2k_a \frac{I_1 I_2}{r} \qquad [Eq\ 4]$$

where $k_a = k_e/4\pi$, where F is force per unit of conductor length

with similar definitions to Coulomb's law and illustrated by Figure 2. Note the similarity of Ampère's force law and Coulomb's law above.

The linkage of electricity and magnetism through the motion of electrical charge — current — led Ampère to create a theory of *electrodynamics* that is at the heart of wireless. After all, it is the continual acceleration and deceleration of electrons in our antennas that cause electromagnetic waves to be radiated. The movement of electrons in response to incoming waves allows us to hear those waves in our receivers. 1820 was a very good year!

In 1825 and 1826, Ampère published a collection of material on magnetism including what is now known as Ampère's law, the general relationship between currents and magnetic fields. This relationship was extended by Maxwell and forms one of Maxwell's equations that describe electromagnetic fields.

Getting Ready for Maxwell

Almost immediately, Ørsted's discovery and Arago's extension of it led to practical inventions. In 1821, American physicist Joseph Henry invented the *electromagnet* by winding the current carrying wire into a coil. While doing these experiments, he also discovered the need for insulation between the wires making up the coil. His experiments led to refinement of the electromagnet into the *electromagnetic telegraph* in 1831.

The really big news of that year, however, came from Michael Faraday, a self taught scientist who had been experimenting with electricity and chemistry since 1812. Faraday demonstrated *electromagnetic induction* by showing how changing currents in one circuit (later *ac current*) could induce similarly changing currents in another circuit without any direct connection between them.[6]

Figure 3 shows that in doing so, Faraday converted the electrical energy of current in the first circuit into magnetic energy in the surrounding field and back into electrical energy in the second circuit. This led Faraday to predict the existence of electromagnetic waves, as well.

Faraday refined his explanation of induction into the following formula known as Faraday's law:

$$\varepsilon = -\frac{\text{change in } \Phi_B}{\text{change in time}} \qquad [Eq\ 5]$$

where ε is the electromotive force (or *EMF*) and the fraction represents the change in magnetic flux (Φ_B) with time. The faster the magnetic flux changes or the larger the amount of change in one circuit, the larger the voltage that is *induced* in the other circuit.

The minus sign in the equation means that the current caused by the changing magnetic field flows in the direction that creates an *opposing* magnetic field. This is otherwise known as Lenz's law and it describes the *back EMF* we observe in motors and the *kickback voltage* in a relay coil when the relay is deenergized. If you look closely at Figure 3, you can see that the induced current flows in the opposite direction to the current caused by the battery.

In the next experiment, we will follow in the steps of Coulomb, Ampère, Ørsted and Faraday by performing some simple experiments that demonstrate the various effects they described. Is this purely a historical exercise? Not at all! These phenomena are at the heart of every radio — without them we would be wireless less.

[5]en.wikipedia.org/wiki/Hans_ Christian_%C3%98rsted

[6]en.wikipedia.org/wiki/Electromagnetic_ induction

Experiment 118
The Laws at Work

Last month's column ended in the year 1831 as Michael Faraday discovered *electromagnetic induction*. In fact, the day on which I finished this column was the 181st anniversary of that discovery — August 29th.

As a review, Faraday showed how changing currents in one circuit could induce similarly changing currents in another circuit without any direct connection between them. This occurs by first converting the first circuit's electrical energy into magnetic energy in the surrounding field and then back into electrical energy in the second circuit. Electromagnetic induction is described by Faraday's Law as explained in last month's column.[1] The faster the magnetic flux changes because of motion or changes in current, the larger the voltage that is induced in the other circuit.

The minus sign in Faraday's law gives rise to Lenz's law: The *electromotive force* (EMF) induced by the changing magnetic field causes current to flow in the direction that creates a magnetic field *opposing* the original change in magnetic flux. Lenz's law describes the *back EMF* we observe in motors and the *kickback voltage* in a relay coil when it is deenergized. Last month's Figure 3 shows the relative direction of the currents in both circuits.

There are several ways to cause the magnetic field linking the two coils to change. The most obvious way is to simply turn the current generating the field on and off with a switch. Another way is to move one of the coils so it encounters more or less of the magnetic field. This is the principal behind an electrical ac generator. An *armature* coil is rotated inside a current-carrying stationary coil (or *stator*) causing the armature voltage to vary as a sine wave. Similarly, changing the current in the stator causes the armature to turn, creating a motor.

In this month's column we will follow in the steps of pioneers André-Marie Ampère, Hans Christian Ørsted and Faraday by performing some simple experiments demonstrating the various effects they described. Is this purely a historical exercise? Not at all!

[1]All previous Hands-On Radio experiments are available to ARRL members at **www.arrl.org/**

These phenomena are at the heart of radio — without them we would be wireless-less.

Experiment #1 — Ørsted's Observation

During a lecture, with experimental apparatus scattered across a table, Ørsted noticed that a compass needle deflected away from north when he switched current on and off in a nearby circuit. This was the first observation linking electricity and magnetism and it was the proverbial *Big Deal* in 1820. So let's repeat it.

Head for the kitchen and "borrow" a strong refrigerator magnet capable of holding up a calendar. You'll also need a source of dc power such as a power supply — 12 V is fine — and a coil of several mH with a dc resistance of 10 to 30 Ω. You can either buy or scrounge a suitable inductor but the coil of a 12 V relay will do fine and you may have one in your junk box. (The relay used in experiments #107 through #110 is a good choice. Alternately, a RadioShack 275-001 will work — remove the plastic case to see the coil.)

Working on a nonferrous table (plastic or wood), suspend the magnet a few inches above the table using thread or dental floss. Mark one face with pencil or tape. Connect the inductor to the 12 V supply through a voltmeter configured to measure current — start with the scale for measuring several amps and use a more sensitive scale after confirming the current is low enough not to

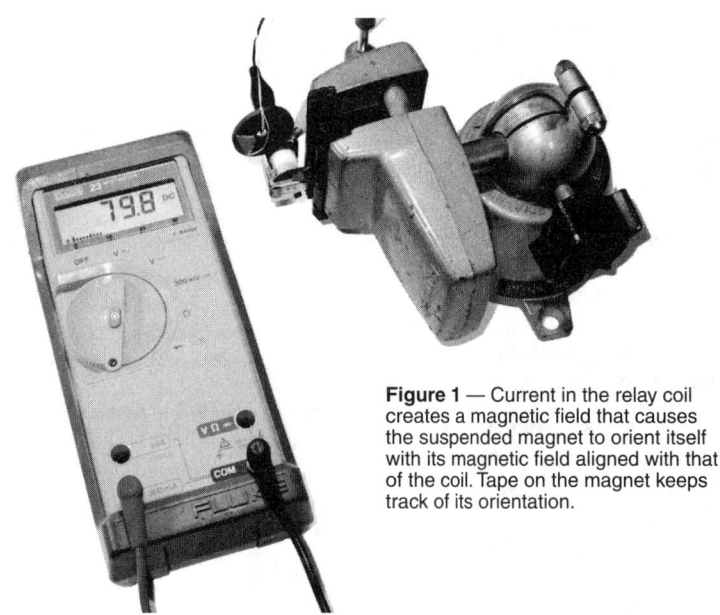

Figure 1 — Current in the relay coil creates a magnetic field that causes the suspended magnet to orient itself with its magnetic field aligned with that of the coil. Tape on the magnet keeps track of its orientation.

When Is E a V and V an E?

Beginners in electronics are often confused about the interchange of V and E in equations and formulas. The term "electromotive force" further muddies the water. When should each be used? Unfortunately, there is no standard definition or convention as described at **en.wikipedia.org/wiki/Electromotive_force**. Nevertheless, in ham radio E is usually used if referring to an electric field (such as the E field of an antenna) or if an electric field causes some effect or action (such as back EMF of an inductor). V is used to describe the difference in voltage between two points in a circuit, or the terminal voltage of a power supply or battery. V is also used as an abbreviation for volts, the unit of voltage.

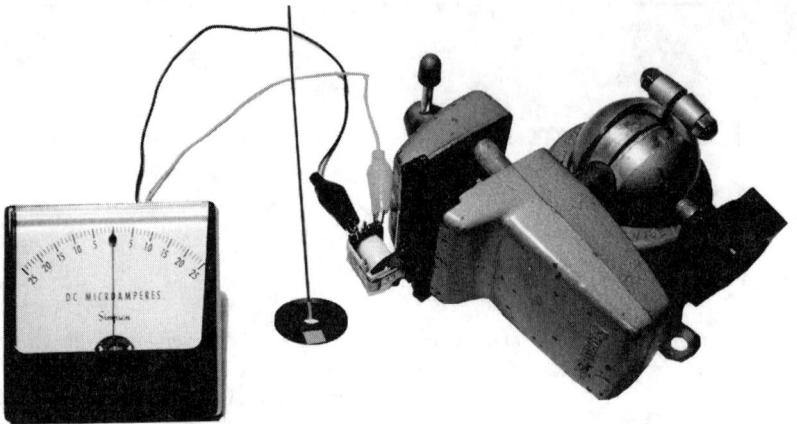

Figure 2 — Moving the magnet toward the coil changes the magnetic field in the coil and induces a current of one polarity in the circuit, causing the meter to deflect. Moving the magnet away from the coil causes an opposite change in the field and reverses the deflection of the meter.

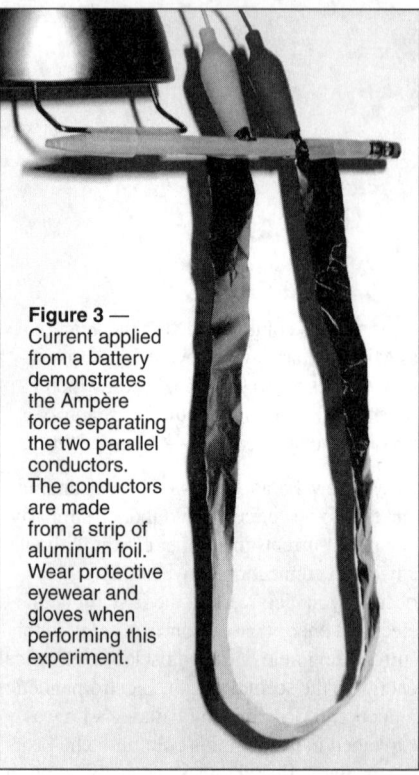

Figure 3 — Current applied from a battery demonstrates the Ampère force separating the two parallel conductors. The conductors are made from a strip of aluminum foil. Wear protective eyewear and gloves when performing this experiment.

overload the meter. Place the coil near the magnet but in a position where it cannot contact the coil or the connecting wires. Wait for the magnet to reach complete rest.

Switch on the current while watching the magnet — the magnet will pivot and move, eventually stabilizing in a fixed position as in Figure 1. (My relay coil is drawing about 80 mA.) When you switch off the current, the magnet will return to its original position. The magnet is moving so that its magnetic field is aligned with the magnetic field of the energized coil.

Cycle the current on and off several times. The magnet will always stabilize with one face of the magnet in the same position. Slowly rotate the coil and verify that the magnet rotates to follow the orientation of the coil. Reverse the power supply leads and verify that the magnet stabilizes with the marked face reversed as well.

Experiment #2 — Faraday's Law

Technically, we're not going to demonstrate Faraday's law exactly so as to keep the setup simple. However, you will be able to see the effect of the orientation of the field on the direction of the induced current in a circuit. You will need a sensitive microammeter or millivoltmeter (also known as a *galvanometer*) capable of showing current or voltage of both polarities. Digital meters work okay, but if you have or can borrow a sensitive analog meter, the effect is easier to see. I found a mint condition 25-0-25 µA meter for $5 at a hamfest so keep your eyes open.

Figure 2 shows the basic setup. Disconnect your coil from the power supply and connect it directly between the meter terminals. Polarity is not important. (If you have a digital meter, use the most sensitive voltage scale, usually 200 or 300 mV.) Hold the magnet in your fingers and move it toward the coil and then away from the coil while watching the meter. You should see the meter deflect or indicate in one direction as the magnet moves toward the coil and then in the other direction as the magnet moves away. Experiment with different orientations of the magnet as you move it past the coil.

There are a number of great YouTube videos that demonstrate Lenz's law in which the induced current creates an opposing magnetic field.[2] You may have seen someone drop a magnet into a copper pipe through which it then falls very slowly. The induced currents are called *eddy currents* and they set up a magnetic field that is almost, but not quite, strong enough to stop the magnet. If the magnet did stop, there would be no eddy current and opposing field so, as a result, the magnet moves just fast enough to overcome the resistive losses in the pipe.

Experiment #3 — Ampère's Force Law

The final experiment is at the heart of what makes an electromagnetic rail gun work — the magnetic force between two current-carrying conductors. Figure 2 in the previous experiment shows how the magnetic fields of two parallel conductors carrying current in the same direction align to force the conductors towards each other. In this experiment, we'll show opposite currents forcing the conductors apart.

Head back to the kitchen and find the aluminum foil. Use a measuring stick to tear or cut off a strip about 18 to 24 inches long and about ¾ inches wide. Back in the lab, suspend the aluminum foil so that it forms a long, narrow U above the table as in Figure 3. Attach heavy clip leads to the open ends of the foil strip.

Get four fresh D cell batteries and use a battery holder or plastic tube to connect them in series — *do not* use a larger battery as the resulting higher currents may be unsafe. Wear eye protection and work gloves for the following step although it is unlikely that any significant heating will occur. Attach one clip lead to the battery's positive terminal cap. Tape or secure both clip leads so that moving the free ends does not cause the foil to move. When the foil is completely still, brush the remaining clip lead's free end against the battery's negative terminal momentarily — you'll see the foil loop expand slightly as several amperes of current flow through it, creating a force separating the parallel conductors. Experiment with different orientations of the foil or create two parallel strips that carry current in the same direction and are forced closer together.

Applying the Law

Do these experiments have any practical application in ham radio today? Most certainly! The effect observed by Ørsted is replicated in every analog meter movement. Faraday's and Lenz's laws are the foundation of transformers and shielded wires. If you've ever seen the inside of a high-voltage power supply after a short circuit, you've also seen the effects of Ampère's force law at work.

[2]YouTube videos **www.youtube.com/watch?v=kU6NSh7hr7Q** and **www.youtube.com/watch?v=G7ysnXH53Wo** are particularly good.

Experiment 95 — Watt's In a Waveform?

As a beginner in ham radio and then again in my introductory electrical engineering courses (between operating stints at the college radio club, WØEEE) I remember struggling to understand the differences between all of the different ac waveform metrics and terminology. More than once I crisscrossed peak, RMS and average — and all the other possible combinations. In this experiment — it's always good to check up on the foundation!

Before we begin, note that this article will use degrees for angular values instead of the *radians* used in most engineering calculations. There are 2π radians in a circle so each degree equals $2\pi/360 = 0.0174$ radians and each radian equals $360/2\pi = 57.3°$. If you are using a calculator for the exercises here, be sure it is set to use the right units.

What's Your Sine?

It seems that ac waveforms are nearly always shown as sine waves. Why is the sine wave so ubiquitous? Why not a square wave or an irregular waveform? The key is rotation. If you imagine a point on a circular wheel rotating counterclockwise as in Figure 1, beginning at point 1 a point on the rim will rise and fall with a vertical height above or below the X axis equal to the sine of the angle through which it has rotated — a total of 360° in one rotation. At location 3, for example, it has rotated through 90° and reached maximum height. If we specify a radius of 1 for the wheel (the units of measurement don't matter), the height of the point is $\sin(90°) = 1$. At any other angle, θ, the height of the point = $\sin(\theta)$. As the point continues past 180° to 360°, which is the same as 0°, the point is below the X axis and $\sin(\theta)$ is negative.

Starting from $\sin(0°) = 0$ at location 1, plotting the height of the point against the angle of rotation (θ) on the X axis forms the familiar sine wave. When the point is rotating at a constant rate each degree of rotation always takes the same amount of time, so the sine wave is the same whether plotted with angle or time on the X axis. If each rotation takes T seconds, then the angle through which the point has rotated in t seconds, $\theta = 360 \times t/T$. Since 1/T is the frequency, f, of the wheel's rotation, $\theta = 360 \times f \times t$ and the value of the sine wave is $\sin(360 \times f \times t)$.

The cosine also makes an appearance as the point's horizontal distance from the Y axis. Positive is assigned to the right so that the cosine wave begins with a maximum value at $\theta = 0°$ of $\cos(0°) = 1$. The cosine wave looks just like the sine wave, but offset by 90°, starting at 1 and decreasing.

But why is the sine wave the "standard" ac waveform? Aside from the important fact of mathematical convenience, which is of primary concern to engineers and scientists, the sine wave also describes the output voltage from a rotating generator. As the generator's armature coil rotates between the poles of its field magnet, the voltage induced in the coil is a sine wave. That sine wave then appears at your ac wall outlet and everywhere else on the utility grid. Most of the metrics we use today for all ac waveforms were originally developed to describe generator output, the first application of ac power.

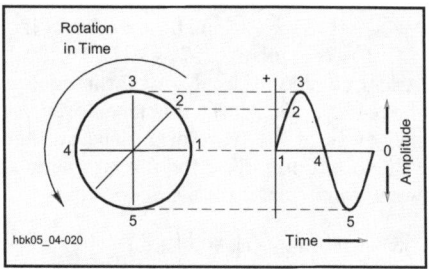

Figure 1 — The relationship between rotation and sine waves. The height of the rotating point is given by the sine of the angle of rotation.

A Peek at the Peaks

The first stop on the journey is Figure 2 which illustrates the primary points of interest using the voltage of an ac sine wave. At any single point in time, the value of the sine wave is called its *instantaneous value*, E_{INST}. The maximum value of the sine wave is its *peak value*, E_{PK}. It is also useful to know the difference between the maximum positive and negative values — this is called the *peak-to-peak value*, E_{PK-PK} or just E_{P-P}. The same subscripts apply to values of current or any other quantity varying this way.

Why E and not V? E is used because it represents *electromotive force* or *EMF*, the original term that was renamed *voltage* in honor of Alessandro Volta. EMF and voltage are the same. V is often used to represent a specific value of voltage in volts. For current, I is used as an abbreviation for intensity (or impetus, depending on your sources of information) because the letter C was already used to represent charge in the early days of electrical experimentation. The letter A is used to represent a specific value of current in amperes.

Here are the key equations relating these values for a sine wave, using voltage and

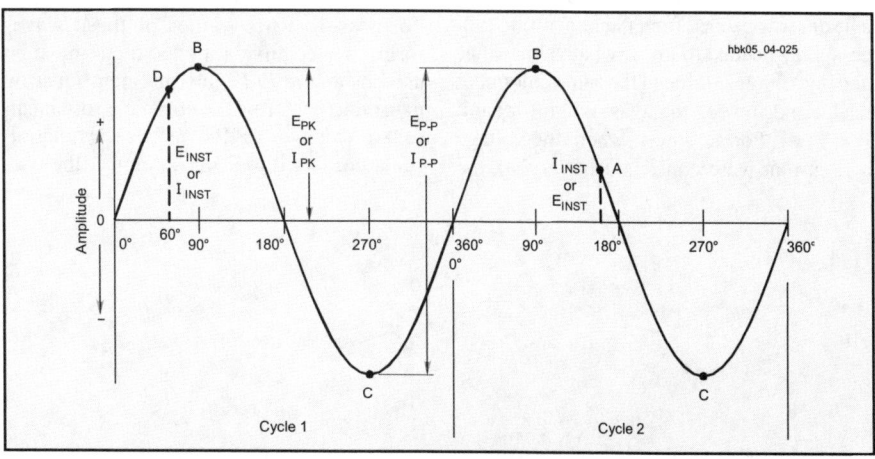

Figure 2 — Using a sine wave illustrates the difference between instantaneous, peak, and peak-to-peak values.

Electronic Fundamentals **15**

remembering that the same applies for current:

$E_{INST} = E_{PK} \times \sin(\theta)$
$= E_{PK} \times \sin(360 \times f \times t)$
$E_{P-P} = 2 \times E_{PK}$ and $E_{PK} = \frac{1}{2} \times E_{P-P}$
$\theta = 360 \times f \times t$ and $t = \theta/(360 \times f)$

Got your calculator ready? Here are a few problems to work out:

If $E_{PK} = 150$ V, what is E_{P-P}? (300 V)

If $E_{P-P} = 28$ V, what is E_{INST} when $\theta = 45°$? $(28/[2 \times \sin(45°)]$
$= 14/1.411 = 9.898$ V)

What is the period of a 60 Hz waveform? (T = 1/f = 16.67 ms)

What is E_{INST} of a 60 Hz sine wave with $E_{P-P} = 100$ V if t = 1 ms? $(100/2 \times \sin(360 \times 60 \times 0.001) = 50 \times \sin(21.6°) = 18.4$ V)

Power of a Waveform

Once the voltage and current of a waveform are known, the next step is to figure out how much power the waveform can deliver. After all, that's why the waveform was created in the first place — to do some useful thing and that requires power.

Just as with the values of voltage and current, power has an instantaneous value, $P_{INST} = E_{INST} \times I_{INST}$, and a peak value, $P_{PK} = E_{PK} \times I_{PK}$ (assuming voltage and current peak values occur at the same time as they do for a completely resistive circuit without reactance).

While it would seem natural to calculate a "peak to peak power", a more useful measurement is to compare the power supplied by the ac waveform to the power that would be supplied by a dc waveform with the same *effective value* of voltage or current. Back in the early days of electrical utilities some supplied dc and some supplied ac voltage. Engineers needed to be able to design equipment that would work with either type. For example, if a heater was needed for a manufacturing process, they needed to know what values of ac and dc voltage produced the same amount of heat. For dc power, that was straightforward: $P = V \times I$. For ac power, where the voltage and current were continuously varying, the answer wasn't so straightforward.

By applying some mathematics to add up the total power available at each instant of the waveform, it was determined that for any continuous, regularly varying waveform the effective value could be calculated by squaring the instantaneous values of voltage or current, finding their average or *mean* over one complete cycle, and taking the square root of the result. This was called the *root mean square* or *RMS* value because it represented the square root of the mean of the squared values.

While this could get complicated, the regular variations of a sine wave make the calculations easy if the peak value is known: $V_{RMS} = 0.707 \times V_{PK}$ and $I_{RMS} = 0.707 \times I_{PK}$. You may recognize 0.707 as $\frac{1}{2} \times \sqrt{2}$ or $1/\sqrt{2}$. A less commonly used waveform value is the *average value*, which is taken over one half cycle of the waveform. (The average value over a full cycle for any symmetrical waveform is zero.) For a sine wave:

$V_{AVG} = 0.637 \times V_{PK}$ and $I_{AVG} = 0.637 \times I_{PK}$

The value 0.637 is equal to $2/\pi$.

It is important to remember that these simplified calculations apply *only* to sine waves. The RMS and average values of a non-sinusoidal waveform, such as a square or triangle wave or speech, is quite different. To measure RMS values of these waveforms, a specially calibrated meter must be used or a *true RMS* measurement must be made that performs the complete root mean square calculation. If you have a function generator, set it to 1 kHz, connect the output to your DVM set to measure ac voltage. Switch the waveform between sine, square, triangle and pulse to see how the displayed value changes.

The power of any ac waveform can be calculated as the product of RMS voltage and current:

$P = V_{RMS} \times I_{RMS}$

The equivalence in available power between a waveform with a specific RMS voltage and the same numeric dc voltage is why RMS voltage is used to specify the ac voltage of today's utility grid, even though dc power is very rarely encountered outside of special applications. RMS is so widely used that it is the voltage displayed by voltmeters unless specially configured or a *peak* setting is used.

Hams need to be able to convert between peak and RMS voltages, especially when trying to determine the ratings of components such as capacitors connected to an ac voltage or a rectified ac voltage.

$V_{PK} = 1.414 \times V_{RMS}$

For example, what is the peak voltage at your wall outlet if the utility is delivering 117 V_{RMS}? $V_{PK} = 1.414 \times 117 = 165$ V

If you are trying to filter RF from the ac line with a capacitor rated at 150 V, be prepared to watch some fireworks when the capacitor fails from the overload! (Note: Always give yourself plenty of safety margin when working with ac line voltage because of transients and over voltage conditions.)

Here are a few more problems to exercise the information:

What is the RMS value of a 220 V_{PK-PK} sine wave? $(220/2 \times 0.707 = 77.8$ V)

What is the power of a sine wave with $E_{PK} = 24$ V and $I_{PK} = 2$ A? $(0.707 \times 24 \times 0.707 \times 2 = 24$ W)

What is the peak voltage of 240 V ac power? $(240 \times 1.414 = 339$ V)

Table 1 gives a number of useful conversions between peak, average, and RMS. You might want to keep those handy around your workbench or toolbox!

Table 1

Conversion Factors for AC Voltage or Current

From	To	Multiply By
Peak	Peak-to-Peak	2
Peak-to-Peak	Peak	0.5
Peak	RMS	$1/\sqrt{2}$ or 0.707
RMS	Peak	$\sqrt{2}$ or 1.414
Peak-to-Peak	RMS	$1/(2 \times \sqrt{2})$ or 0.35355
RMS	Peak-to-Peak	$2 \times \sqrt{2}$ or 2.828
Peak	Average	$2/\pi$ or 0.6366
Average	Peak	$\pi/2$ or 1.5708
RMS	Average	$(2 \times \sqrt{2})/\pi$ or 0.90
Average	RMS	$\pi/(2 \times \sqrt{2})$ or 1.11

Note: These conversion factors apply only to continuous pure sine waves.

Experiment #75 — Series to Parallel Conversion

Radio hardware designers, like magicians, have a repertoire of interesting and useful tricks. As with magic, the simple and familiar tricks are often the most useful because they can be used in many ways and many situations. So it is with circuit transformations.

Equivalent Circuits

Transformations are really all about *equivalent circuits*, the ability to replace one section of a circuit with a different set of components without changing any of the voltages and currents. Most equivalent circuits are *one-port* equivalents in that they are applied at a pair of terminals called a *port*.

Let's start with a very simple equivalent circuit. Suppose you have a circuit composed of a single 12 Ω resistor. It's sealed up in an opaque box and you connect to it through a pair of terminals. This pair of terminals is the port. What equivalent circuits can you think of that would appear to be a 12 Ω resistor from outside the box? I'm sure you can think up several: twelve 1 Ω resistors in series, two 24 Ω resistors in parallel, a 6 Ω resistor in series with the parallel combination of a 24 Ω and 8 Ω resistor, etc.

There are an infinite number of resistor combinations that produce a 12 Ω resistance between those terminals. These are all equivalent circuits for the single 12 Ω resistor. Using an ohmmeter, you could never know the difference between any of the combinations.

Why Equivalent Circuits?

Why are equivalent circuits so common in electronics? In a word, simplification. General class licensees learned to do simple equivalents to pass the exam — converting series and parallel combinations of resistance, capacitance and inductance into a single equivalent component that acted just the same as the combination of components they replaced. Why do calculations for multiple components, when a single equivalent component will represent the combination just fine?

Another simplification is that components, sections of circuits and sometimes even entire pieces of equipment can be replaced with a simple equivalent circuit with which it is easier to model or calculate behavior. Another example that comes quickly to mind is an RF signal generator. For all intents and purposes, to the external world it appears to be an ac voltage source in series with a 50 Ω resistor. That's oversimplifying the generator, but in most cases that simple model will do just fine. (The Thevenin and Norton equivalent circuits are good examples of this type of simplification and are discussed in Hands-On Experiment #32.[1])

Equivalent circuits can also change the form of a circuit so that it is easier to analyze or use in a design. An equivalent circuit might provide a better way to describe the behavior of the circuit. At any rate, having a set of mathematical tools to change one type of circuit into another is a valuable skill.

[1]Previous Hands-On Radio columns are available to ARRL members at **www.arrl.org/hands-on-radio**.

AC Equivalents

What about ac circuits with capacitors and inductors — do they have equivalents, too? Yes, but because the behavior of those circuits changes with frequency, the equivalent circuit is generally only an exact replacement at one frequency.

A simple component (R, L or C) is more complex than you might think. For example, an inductor has loss associated with the resistance of the wire. A capacitor's dielectric dissipates some of the stored energy as heat. A resistor's leads act like small inductors. All of these *parasitic* effects can be significant in certain applications. Consequently, test instruments that measure component value also measure the parasitic values and can provide the measurement as one of several different equivalent circuits.

The most common equivalents for measurements of capacitors and inductors are parallel R and series R, respectively, as shown in Figure 1. For capacitors, a parallel resistor most closely represents the effects of loss in the dielectric. For inductors, losses in the wire are best represented by a series resistance. Resistors are measured in the same way, except that R is the primary component and series L or parallel C are the parasitic effects.

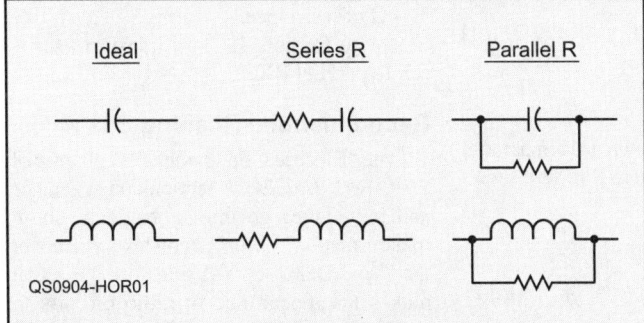

Figure 1 — Losses in non-ideal capacitors and inductors are modeled as series and parallel resistances. Parasitic capacitance and inductance in resistors are modeled as parallel and series components.

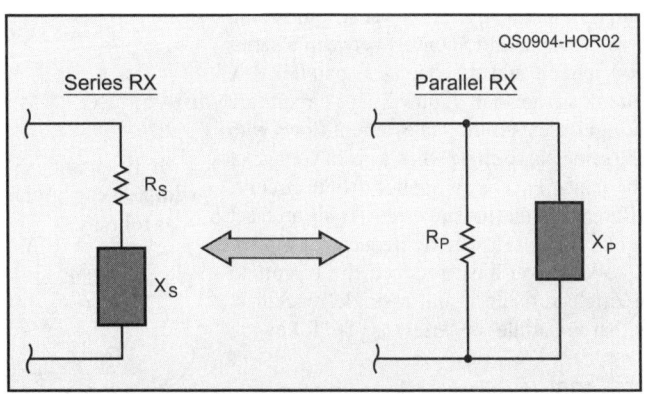

Figure 2 — Series circuits containing resistance and reactance can be transformed into parallel circuits (and vice versa) to simplify circuits and aid in design tasks.

Electronic Fundamentals 17

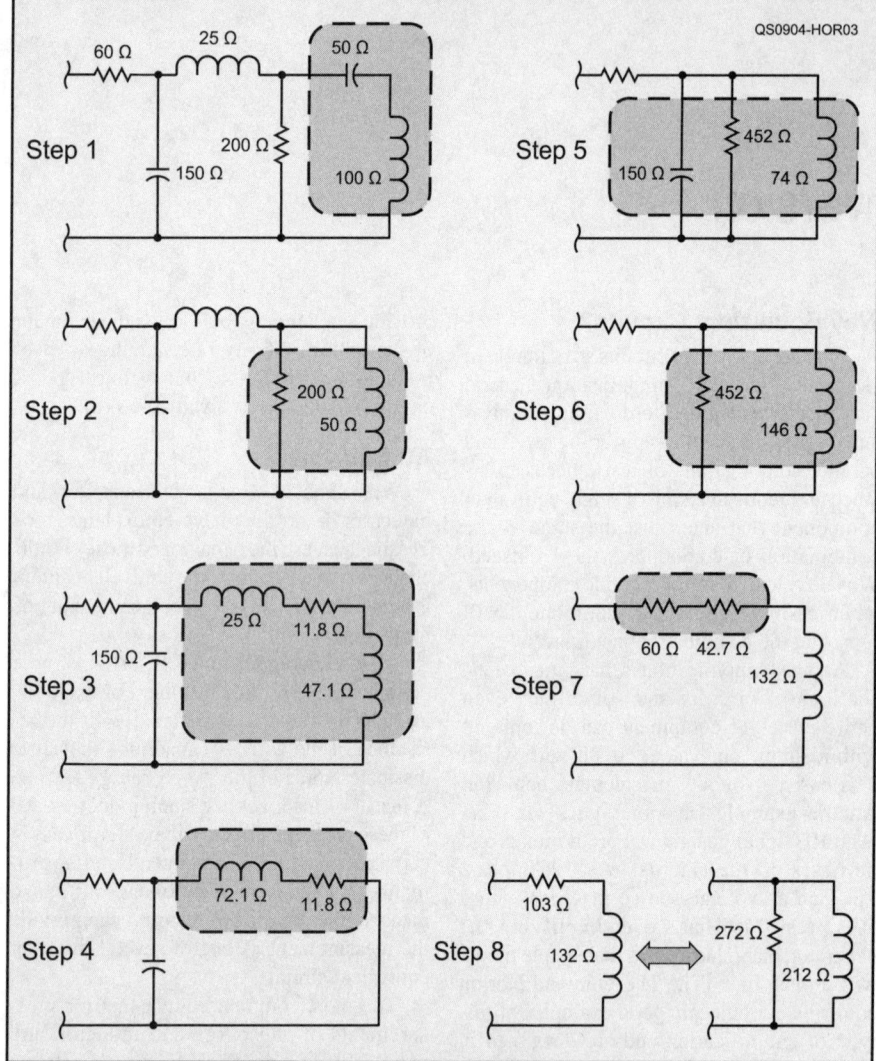

Figure 3 — A series of step-by-step transformations simplifies the complex circuit at the upper left into a simple series-RX or parallel-RX equivalent.

Series to Parallel Transformation

You already know how to work with combinations of series and parallel resistances to turn them into their simplified equivalents. Parallel and series reactances can be combined, as well. Yet at some point, it is often useful to convert between a series R-X circuit and its equivalent parallel R-X circuit as shown in Figure 2. Two circuits are ac equivalents if the same current flows with the same phase angle when a given voltage of the same frequency is applied to both circuits. (Remember that the equivalent is only an exact replacement at a specific frequency.)

When you have reduced the circuit to a single resistance and a single reactance, either in parallel or series, use the following formulas to convert between series and parallel circuits. If the circuit can be drawn as R in series with X, it can be converted to R in parallel with X as follows:

$$R_P = \frac{R_S^2 + X_S^2}{R_S}$$

$$X_P = \frac{R_S^2 + X_S^2}{X_S}$$
[Eq 1]

If the circuit can be drawn as R in parallel with X, it can be converted to R in series with X as follows:

$$R_S = \frac{R_P X_P^2}{R_P^2 + X_P^2}$$

$$X_S = \frac{R_P^2 X_P}{R_P^2 + X_P^2}$$
[Eq 2]

These formulas assume that the absolute value of reactance is used, so that X is always a positive number. The reactance will be of the same type (capacitive or inductive) before and after the series-parallel transformation.

Let's start with two simple examples. In the circuit of Figure 2A (series RX), let $X_S = -50\ \Omega$ (capacitive) and $R_S = 50\ \Omega$. Use formula set 1 to convert to the parallel combination of $X_P = -100\ \Omega$ (still capacitive) and $R_P = 100\ \Omega$. In the circuit of Figure 2B (parallel RX), work the problem backwards, using formula set 2.

Step by Step

Figure 3 shows a more challenging problem — reducing the circuit shown at Step 1 to its series or parallel equivalent. This takes eight steps of simplification, the component values to be transformed or combined circled by a dashed line. The sequence shows the circuits getting progressively simpler until the only circuits left are the series RX and parallel RX equivalents. (Values shown are the results of the required operation.)

Wherever there are reactances in series, combine them by subtracting X_C from X_L. One more formula is required. Combine parallel reactances by using the formula

$$X_{total} = \frac{-X_L X_C}{X_L - X_C}$$
[Eq 3]

Again, use only the magnitudes of the reactances. If the result is negative, X_{total} is capacitive and if positive, X_{total} is inductive.

Start at the far right of the circuit, combining the two series reactances, leaving 50 Ω of inductive reactance.

■ Transform parallel RX into series RX with formula set 2.
■ Combine the series reactances, leaving 72.1 Ω of inductive reactance.
■ Use formula set 1 to transform series RX into parallel RX.
■ Use formula 3 to combine the parallel reactances.
■ Transform parallel RX to series RX using formula set 2.
■ Combine the series resistances, leaving 132 Ω of inductive reactance in series with 103 Ω of resistance.
■ Use formula set 1 to transform series RX to parallel RX.

Recommended Reading

The "Electrical Fundamentals" chapter of *The ARRL Handbook* sections on reactance and impedance go into more detail about transformations.[2] Most circuit texts will cover the "Wye-Delta" or Y-Δ transformation that makes an appearance in radio circuits to change T networks into π networks.

Next Month

The lowly diode, simplest of all semiconductor devices? Not quite so simple, as we'll see next month as you compare several types and find out how they're used.

---

[2]*The ARRL Handbook for Radio Communications*, 2009 Edition. Available from your ARRL dealer or the ARRL Bookstore, ARRL order no. 0261 (Hardcover 0292). Telephone 860-594-0355, or toll-free in the US 888-277-5289; **www.arrl.org/shop/**; **pubsales@arrl.org**.

Experiment #74 — Resonant Circuits

Resonant circuits are the oldest of all radio circuits. Hertz's transmitter, with which he discovered radio waves, was little more than a spark-excited resonant circuit and the receiver a similar circuit coupled to a simple detector. In this experiment we'll review the definition of resonance, explore how it comes about, and observe it in a real circuit. As a bonus, we'll also discuss the mysterious Q.

Definition of Resonance

In a circuit with both types of reactances connected in any combination of series and parallel branches, at some frequency the amount of inductive and capacitive reactance in the circuit will be equal. This is the condition called *resonance* and the frequency at which it occurs is called the *resonant frequency*, often designated f_0. A circuit that includes both types of reactance is a resonant circuit, even though it may not be operated at its resonant frequency.

Series and Parallel Resonant Circuits

Figure 1 shows the two basic types of resonant circuits: series (A) and parallel (B). In the series circuit, the same current, I, flows in both L and C. The applied voltage, E, appears partly across the capacitor, C, as E_C and partly across the inductor, L, as E_L. These two voltages add together to equal the source voltage, $E = E_C + E_L$. This is much like the familiar resistive voltage divider.

In the parallel circuit, instead of both components carrying the same current, the same voltage, E, appears across both L and C. It is the current that divides in this circuit, with the capacitor current shown as I_C and the inductor current as I_L and the source current $I = I_C + I_L$, similar to a resistive current divider.

Unlike the resistive voltage and current dividers, in a reactive circuit we must account for the phase of the voltages and currents when adding them together. Recall that for an inductor, the ac current through the inductor "lags" the applied voltage by 90°. In a capacitor, ac current "leads" the applied voltage by 90°.

As an example, let's examine what happens in the parallel resonant circuit shown in Figure 1B by looking at the current and voltage relationships in Figure 2. In Figure 2A, the waveforms show what happens if the frequency of the applied voltage, E (shown in black), is below the circuit's resonant frequency. That is, when the circuit's inductive reactance, X_L, is lower than the capacitive reactance, X_C. If $X_L < X_C$ and both L and C have the same applied voltage, E, then the inductor current, I_L, will be larger than the capacitor current, I_C. That makes sense — if the reactances were replaced by resistors with the same numeric value of ohms, more current would flow through the branch of the circuit with the lower resistance. No surprises yet!

Remember, though, that we have to take the phase of ac currents into account when adding them together. I_L will lag the applied voltage, E, by 90° and I_C will lead E by 90°.

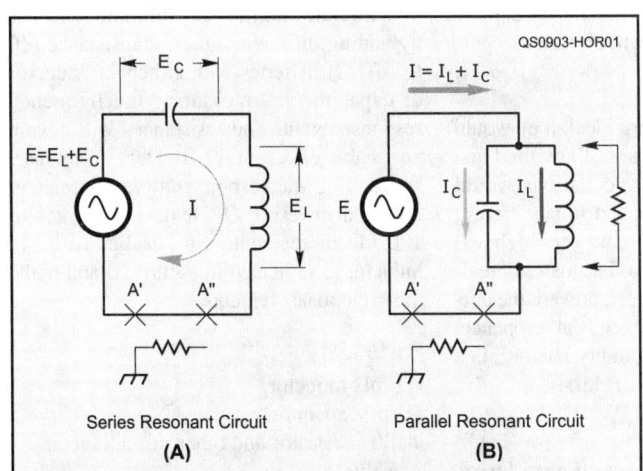

Figure 1 — Series and parallel resonant circuits have different behaviors above and below their resonant frequency, f_0.

Figure 2 — A parallel resonant circuit appears inductive below its resonant frequency and capacitive above because of partial current cancellation. At resonance, the currents cancel completely and the circuit presents an open circuit. For clarity, only portions of the steady state wave form are shown.

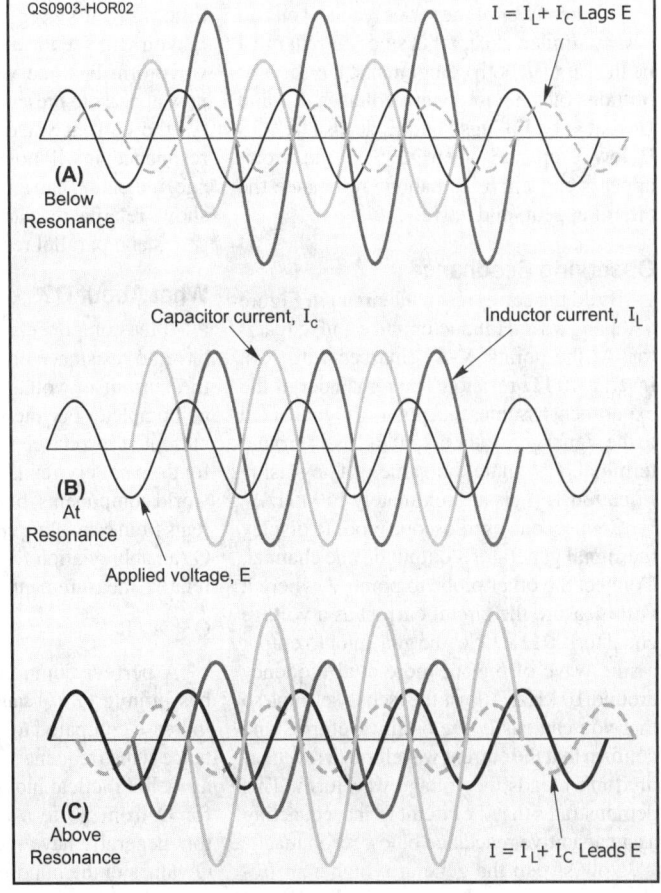

Electronic Fundamentals 19

To help visualize this, the I_L waveform is shown in red and I_C in blue. Because the two currents are offset from E by 90° in different directions, they have a 180° phase difference, causing them to partially cancel. Because I_L is larger than I_C, the dashed line showing the sum of the currents represented by the green dashed line, I lags E by 90°, and the voltage source sees the circuit as an inductive reactance.

Above resonance, the opposite case is shown in Figure 2C. The capacitive reactance, X_C, is now smaller (capacitive reactance decreases with increasing frequency) than X_L and $I_C > I_L$. The currents still have the 180° phase difference, partially cancelling, but the net result as shown by the dashed green line is that I leads E by 90° and the circuit presents a capacitive reactance.

Precisely at resonance, the frequency at which $X_L = X_C$, something very interesting happens as shown in Figure 2B. Both I_L and I_C are now equal, so they cancel completely! No current flows from the voltage source at all ($I = I_L + I_C = 0$), giving the appearance of infinite resistance, with resistance and no reactance at all. It is only if the frequency of E is increased or decreased that current begins to flow once again. You can determine the resonant frequency, f_0, by solving for the value of f at which $2\pi fL = 1/(2\pi fC)$:

$$f_0 = 1 / \left[2\pi \sqrt{(LC)} \right] \quad [\text{Eq 1}]$$

The behavior of the series resonant circuit is very similar, except that since current must be the same in both components, it is the amplitudes of the component voltages, E_C and E_L that vary. E_C lags I and E_L leads I by 90°. Below resonance, where $X_C > X_L$, the circuit appears capacitive and above resonance the circuit appears inductive.

Observing Resonance

Build the series resonant circuit in Figure 1A using a 1 mH inductor and a 1 nF capacitor. At the points A'-A" (marked with Xs), insert a 100 Ω resistor. The remainder of the experiment assumes that point A' is connected to the signal generator output's chassis ground terminal. Calculate the value of f_0 by using Equation 1. (f_0 is approximately 159 kHz.)

Connect one oscilloscope probe to display the signal generator's output on one channel. Connect the other probe to point A" where it will measure the circuit current as a voltage equal to 100 Ω × I. Set the generator to output a sine wave of 5 V or more at a frequency around 10 kHz. Adjust the 'scope settings so that you can easily see both waveforms and confirm that the current waveform (a few hundred mV) leads the voltage waveform. This demonstrates that the circuit is, indeed, acting as a capacitive reactance below resonance.

Now sweep the generator higher in frequency while watching the amplitude and phase relationship of the applied voltage and circuit current. As you approach f_0, the amplitude of the current waveform will increase dramatically and you will have to adjust the 'scope to keep the signal entirely on-screen. The phase difference between the voltage and current waveforms will get smaller, until precisely at resonance, they will be in phase.

Reconfigure the circuit to the parallel circuit shown in Figure 1B and repeat your measurements. Confirm that below resonance the circuit is inductive, with the current waveform lagging the applied voltage. This time, as you sweep the frequency through f_0, you will see the amplitude of the current waveform become very small. The circuit now appears to present an infinite impedance. As the current waveform reappears above resonance, it will now lead the applied voltage, indicating capacitive reactance. Figure 3 shows relative reactance changes for generic series and parallel resonant circuits.

What About Q?

If the components were ideal, they would have no resistance or losses of any kind and the current or voltage cancellations would be complete. For the series resonant circuit, current at resonance would be very high and for the parallel circuit, zero. The losses of real-world components, however, prevent the currents from cancelling completely at resonance. Q (an abbreviation for "Quality Factor") is a relative measurement of that loss.

$$Q = X / R \quad [\text{Eq 2}]$$

A perfect component with zero losses has infinite Q. (Q has no units.) As losses increase compared to the component's reactance at the frequency of interest, Q decreases as well. Practical inductors will have values for Q from 20 to a few hundred. Capacitors generally have much lower losses with Q values in the hundreds to thousands.

Q also affects resonant circuits. The component losses that prevent currents or voltages from cancelling completely can be observed as a more gradual transition of the circuit's impedance through the resonant frequency.

To observe the effect of Q, plot the frequency response of the parallel resonant circuit from 50 to 500 kHz using the frequency response spreadsheet provided on the Hands-On Radio Web site FAQ entry for experiment #18.[1] At each frequency, the spreadsheet will calculate 20 dB (E_{OUT}/E), where E_{OUT} is the voltage at point A". Far from resonance the response will be close to 0 dB, but at resonance, depending on the type of component you are using, the circuit may show a drop in output of up to 30 dB or more.

Take note of the points at which the response is 3 dB below the 50 and 500 kHz values. This is the circuit's bandwidth (BW). Q of a resonant circuit is f_0/BW. The narrower the bandwidth, the higher the circuit's Q.

Simulate additional component losses by adding different values of resistance (10 to 470 Ω) in series with either the inductor or capacitor and measuring the frequency response again. The resistances will have a noticeable effect on the frequency response, broadening the response curve, increasing BW and lowering Q. Next, add 4.7 kΩ to 100 kΩ of resistance in parallel with the inductor as indicated in Figure 1B and make the same measurements.

Parts List

- 1 mH inductor.
- 1 nF capacitor.
- 100 Ω resistor and other miscellaneous values.

Next Month

As long as the words "series" and "parallel" are fresh in our minds, next month's column will show you the trick of circuit shape-shifting: the series-parallel conversion.

[1]Previous Hands-On Radio columns are available to ARRL members at **www.arrl.org/hands-on-radio**.

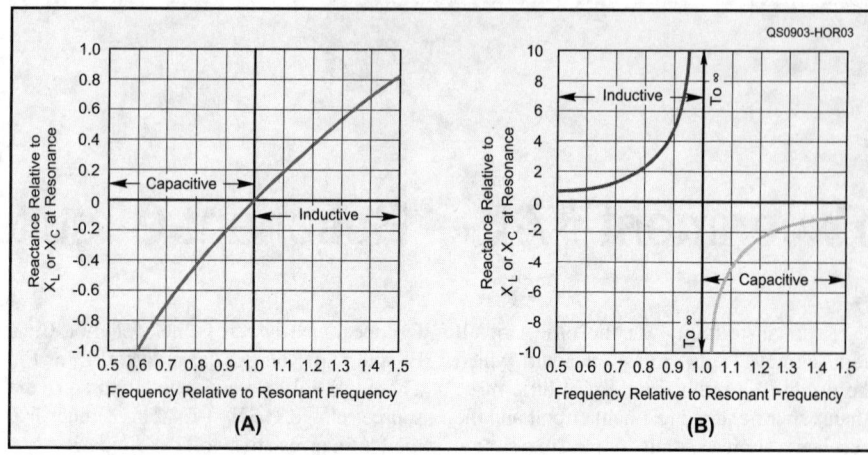

Figure 3 — At (A), a series resonant circuit's impedance relative to the reactance of the components at resonance. (B) shows circuit impedance for a parallel resonant circuit.

Experiment #77 — Load Lines

Once past the very basic levels of transistor amplifier circuits, you'll encounter the *load line*, a graphical method of circuit design. This experiment shows you how the load line is determined and applied to circuit behavior.

Diode Load Line

A diode is the simplest semiconductor device for which a load line can be drawn. Figure 1(A) shows a diode in series with a resistor load, R_L. For any given combination of V_S and R_L, if we know the diode's forward voltage, V_F, we can solve for the diode current, $I_F = (V_S - V_F) / R_L$. V_F, however, depends on I_F, so we must solve the exponential equation for I_F as a function of V_F, shown as the diode's *characteristic I-V* curve in Figure 1(B).[1]

Figure 1(B) also shows the less precise, but easier to use, graphical method of load lines. The load line describes what happens to voltage and current in R_L. It is drawn between the maximum and minimum possible values of current and voltage across R_L. For example, if $I_F = 0$, there is no voltage drop across the diode and the voltage across R_L is V_S — that's point A on the load line. Similarly, if $V_F = 0$, then $I_F = V_S/R_L$ and that's point B on the load line.

[1]The Fundamental Diode Equation is presented in the "Analog Basics" chapter of any recent *ARRL Handbook*.

The only point at which the load line intersects the curve is point C — the *operating point* for the circuit. The intersection is the solution of the diode's characteristic curve equation with the known value of R_L and V_S. If either V_S or R_L change, the slope or placement of the load line will change along with its intersection with the diode's characteristic curve. Let's try it!

Operating Point Control

Build the circuit in Figure 1(A) using a 1N4001 silicon diode rectifier, $V_S = 3$ V, and $R_L = 1$ kΩ. Prepare a graph with the I_F axis showing 0 to 50 mA and the V_F axis showing 0 to 10 V. Draw the load line between point A ($I_F = 0$ mA, $V_F = V_S = 3$ V) and point B ($I_F = V_S / R_L = 3$ mA, $V_F = 0$ V).

Measure the diode's forward voltage, V_F, and use Ohm's Law to calculate I_F from the voltage across R_L or measure it directly with a meter. The values should be somewhere around 0.6 V and 2.4 mA. When that point is plotted, it should be very close to or on the load line.

Vary V_S from 1 to 10 V in steps of 1 V, calculating point A and B and drawing a new load line at each step. Measure the diode voltage and circuit current as before, plotting the combination on the graph and confirming that each point is on a load line. You will start to see the diode's characteristic curve appear as the sequence of plotted points!

Return V_S to 3 V and change R_L to each of the following values, drawing a new load line at each step: 100, 220, 470, 1000, 2.2 k, and 4.7 kΩ. Measure and plot V_F and I_F at each step. This will fill in even more points, each very close to the load line for that value of R_L. As you can see, if you had enough values of R_L and sufficient power supply range, you could determine the diode's characteristic curve exactly!

You'll also have noticed that while you were only varying V_S, the load lines were parallel, but when R_L was varied, the load line slopes changed. That's because the slope of the load line is $-1/R_L$. Lower load resistance results in a steeper load line.

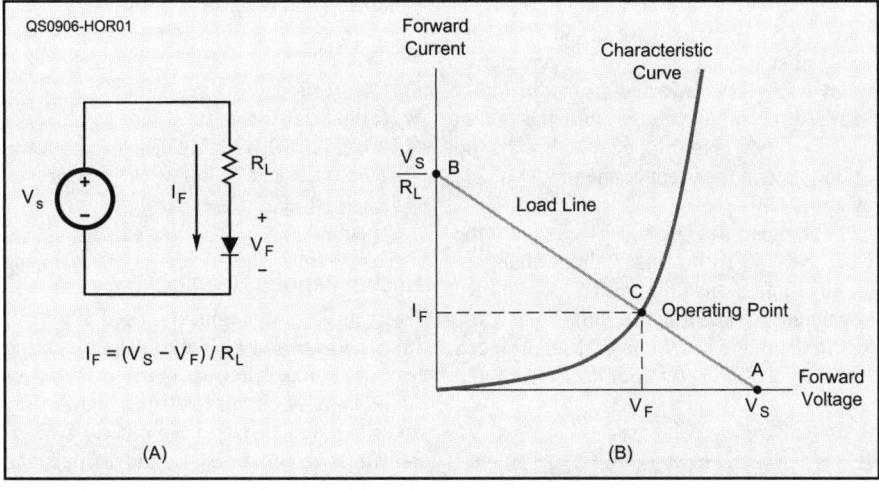

Figure 1 — The simple circuit at (A) can be used to determine the diode's characteristic curve at (B). The intersection of the load line and the characteristic curve is the circuit's operating point.

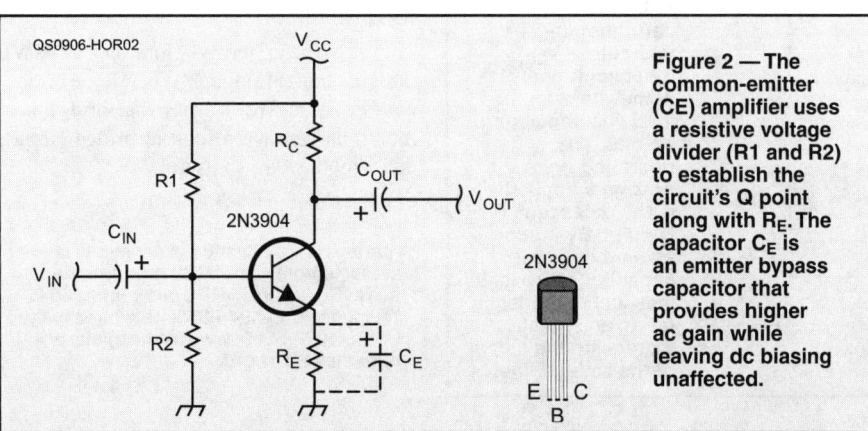

Figure 2 — The common-emitter (CE) amplifier uses a resistive voltage divider (R1 and R2) to establish the circuit's Q point along with R_E. The capacitor C_E is an emitter bypass capacitor that provides higher ac gain while leaving dc biasing unaffected.

Transistor Amplifier Load Line

The load line is much more useful in designing transistor circuits, since current and voltage can take wide ranges of values. The common emitter (CE) amplifier in Figure 2, from the first Hands-On Radio experiment is one you'll use frequently, so we'll use it as an example.[2] This circuit uses self bias and emitter degeneration to establish a stable *Q point* (the operating point with no input signal).

The characteristic curves for a typical 2N3904 NPN transistor in the CE configuration are shown in Figure 3. Instead of just having a single characteristic curve as did the diode, a transistor's I_C-V_{CE} characteristic curve can change. As base current varies, the height of the curve changes on the graph. The set of curves show "snapshots" of the transistor's characteristic curve, each at a different value of base current.

Because the load for the circuit is resistive (consisting of R_C+R_E), the operating point falls along the dc load line drawn on the characteristic curves. We'll get to the ac load line later. As with the diode circuit, the intersection of the load line with the characteristic curve corresponding to the value of base current is the circuit's operating point. If you imagine one of the constant base current lines moving up and down as an input signal varies the base current, you can see its intersection with the load line moving, too. When no signal is applied, the base current is fixed at the level of bias current chosen by the designer and that operating point is the circuit's Q point. In the case of our CE amplifier, the values of R_1, R_2 and R_E determine the location of the Q point by controlling the value of the base bias current.

V_{CC} and the values of R_C and R_E determine the orientation of the load line. The two end points of the load line correspond to transistor saturation [$I_{Csat} = V_{CC} / (R_C+R_E)$ on the I_C axis] and cutoff (V_{CC} on the V_{CE} axis). The slope of the load line is $-1/(R_C+R_E)$, because the output current of the transistor flows through both the collector and emitter resistors.

In order to experiment with the load line, here are a set of components that will result in a Q point of $I_{CQ} = 4$ mA, $V_{CEQ} = 5$ V and a voltage gain of -5 with $V_{CC} = +12$ V: $R_E = 270$ Ω, R1 = 39 kΩ, R2 = 6.8 kΩ, and $R_C = 1.5$ kΩ. (10 µF capacitors will be fine for C_{IN} and C_{OUT}.) Download and print the sample 2N3904 characteristic curves from the Hands-On Radio Web site and draw the load line between cutoff and saturation in this circuit. (The Q point should be on the load line.)

Build the circuit and verify that the values of I_{CQ} and V_{CEQ} are about right. Apply a 1 kHz, 0.5 V_{P-P} sine wave at the input and verify that the output signal is about five times larger and inverted from the input. Increase the input voltage until the output waveform becomes clipped at either the top or bottom and then reduce the input voltage by about half.

Now move the Q point by changing the value of I_{BQ}. To do this without changing the load line, adjust the ratio of R1 and R2 to change V_B, keeping the sum of the resistors in the range of 20 k to 50 kΩ. (You can substitute a 50 kΩ potentiometer for R1 and R2, with the wiper connected to the transistor base.) Measure the new values of I_{CQ} and V_{CEQ}, locate the new Q point on the load line, and observe the effect on the output waveform. For example, doubling the value of R2 will raise the value of I_{BQ} dramatically and probably cause the output waveform to be clipped at the bottom. This is because the higher bias current has moved the Q point farther along the load line toward saturation (left), making it easier for an input signal to drive V_{CE} lower into the saturation region.

AC Load Lines

Figure 2 shows an *emitter bypass* capacitor, C_E, next to R_E. When C_E is connected across R_E, the circuit has a different ac voltage gain $A_V = -R_C/r_e$ (r_e is the internal emitter resistance of a few ohms) than dc gain $A_V = -R_C/R_E$. For an ac signal, the circuit operates on a separate ac load line as shown in Figure 3, because R_E has been effectively short circuited for ac signals. Without R_E, the slope of the ac load line is $-1/R_C$, steeper than for the dc load line. The ac and dc load lines intersect at the circuit's Q point because the circuit's ac and dc operation is the same if the ac input signal is zero.

Parts List[3]

- 1N4001 diode
- 2N3904 transistor
- 100, 220, 270, 470, 1000, 1.5 k, 2.2 k, 4.7 k, 6.8 k and 39 kΩ, ¼ W resistors
- 3 each 10 µF, 25 V electrolytic capacitors

Recommended Reading

Even for non-engineers, used copies of first- and second-year circuit engineering textbooks make fine workbench references for all sorts of circuit questions. Two of my favorites are Hayt and Kemmerly's *Engineering Circuit Analysis* and Millen and Grabel's *Microelectronics*, both published by McGraw-Hill. The former is good for basic R-L-C circuit mechanics and the latter for semiconductor circuits.

Next Month

You've heard terms before such as "SWR bridge," "noise bridge," "Wheatstone bridge" and so forth. We'll cross that bridge next month as we take a look at bridge circuits and why they are so useful.

---

[2]Previous Hands-On Radio columns and a complete parts list for all experiments are available to ARRL members at www.arrl.org/hands-on-radio and in Experiment #76 (see next note).

[3]A parts kit for the first 61 experiments is available from your ARRL dealer or the ARRL Bookstore, ARRL order no. 1255K. Telephone 860-594-0355, or toll-free in the US 888-277-5289; www.arrl.org/shop/; pubsales@arrl.org.

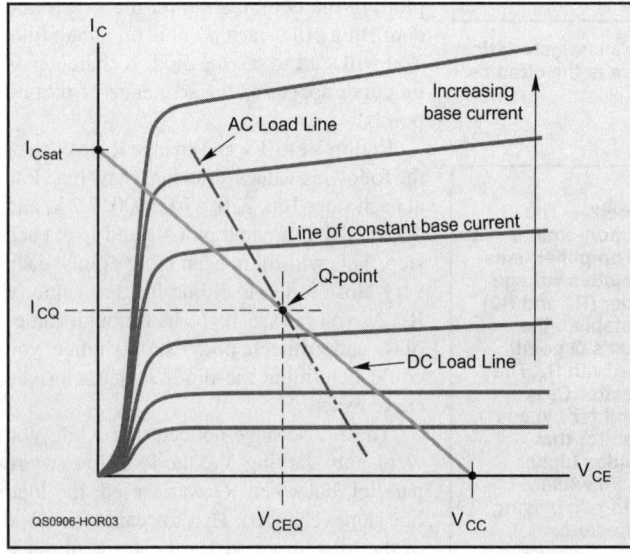

Figure 3 — The dc load line's end points correspond to transistor saturation (I_{Csat}) and cutoff (V_{CC}). The circuit operates somewhere on the load line depending on bias. The operating point with no signal applied is the quiescent, or Q point. By adding an emitter bypass capacitor, there is less load impedance for an ac signal, increasing the slope of the ac load line.

Experiment 105
Gain-Bandwidth Product

"I feel the need for speed!" went the line in the movie *Top Gun* and that might apply to electronics designers, too, in their quest for circuitry that operates at higher and higher frequencies. The *speed* of a device also has an effect at frequencies far below the cutting edge of high-speed electronics and that is the subject of this month's column.

Gain-Bandwidth Product

Often abbreviated GBW or GBP, the *gain-bandwidth product* of a device such as an op-amp is used to specify how much gain the device can muster at different frequencies. "Faster" devices have higher GBW specifications. (Because gain is unitless, GBW has units of frequency — usually MHz.) The relationship between gain and bandwidth is a constant for any particular device because its ability to amplify a signal at a particular frequency is determined by the internal structures of the components that make up the device.

GBW is defined primarily for devices that have a simple, *single-pole* frequency response such as is illustrated in Figure 1 for a low-pass RC circuit. (For more information about frequency response and poles, review Hands-On Experiment #18 and *The ARRL Handbook's* chapter on Analog Fundamentals.[1,2]) Above the cutoff frequency, output amplitude rolls off at 6 dB for every doubling of frequency (an *octave*), equivalent to 20 dB/decade.

Op-amps behave similarly to the RC circuit because their manufacturers build in a *dominant pole* — even though it reduces gain at high frequencies. An op-amp (or other complex IC) is composed of very many components, each with its own frequency response and interacting with other components and the device's structure in different ways. By creating a dominant pole in the device's circuitry, the manufacturer and designer can be confident that all ICs of a specific model will behave in approximately the same way.

Individual transistors have a GBW, too, although it's usually specified as the *transition frequency*, F_T. For bipolar transistors, current gain is used, and for FETs, voltage gain is used. Transistors are generally used at frequencies well below their F_T for consistent performance. The basic concept is the same — how much gain at how high a frequency.

GBW of an Op-Amp

If we measure the *open-loop gain*, A_{VOL}, of an op-amp (without any feedback from an external circuit to reduce circuit gain), we would see that it has a very high value at very low frequencies. For example, the common TL082 op-amp has a typical A_{VOL} of 100 V/mV or 100,000 (100 dB). It also has a GBW of 4 MHz. Thus, we can predict that the TL082 can only produce that high gain up to a frequency of 4 MHz / 100,000 = 40 Hz.

The GBW can also be imagined as the frequency at which the gain of the device falls to 1, or its *unity-gain frequency*. If we measure the open-loop gain of a TL082 over a range of frequencies, the curve will look a lot like that in Figure 1 with f_C being approximately 40 Hz. The open-loop gain of the op-amp will fall to unity at approximately 4 MHz, rolling off at about 20 dB per decade beginning near 40 Hz. (4 MHz / 40 Hz is five decades of frequency and 5 × 20 dB = 100 dB.)

GBW in a Feedback Circuit

The relationship between gain and bandwidth of a device in a circuit that uses feedback to control frequency response, such as a band-pass amplifier, is a simple one:

Bandwidth = GBW / closed-loop gain

If you design the circuit for higher gain, the resulting bandwidth of the circuit will go down and vice versa. For example, if a device with a 10 MHz GBW is used in a circuit with a gain of 10 the bandwidth will be approximately 10 MHz / 10 = 1 MHz.

Unlike the high open-loop gain of a device such as an op-amp, which is difficult to measure, closed-loop gain and frequency response are both easy to observe. You can perform the experiment with actual components or by using a circuit simulator. I'll provide the circuit and then show the simulation results — it would be *really* great if you do both so as to "close the loop" of learning by designing, simulating, building, and then comparing!

Measuring GBW by Simulation

We're going to use the *LTspiceIV* circuit simulator described in Experiments #83 through #86. If you haven't yet downloaded and tried this straightforward simulator, now would be a good time. The price is right (free) and it does not occupy large amounts of disc space or require high speed processors to run. I'll assume from here that you're running the latest version of the program. (*LTspiceIV* automatically checks for the latest version each time it runs.)

The circuit we're going to simulate is shown in Figure 2. It's a simple amplifier circuit with gain $A_V = -R2 / R1$. (See Ex-

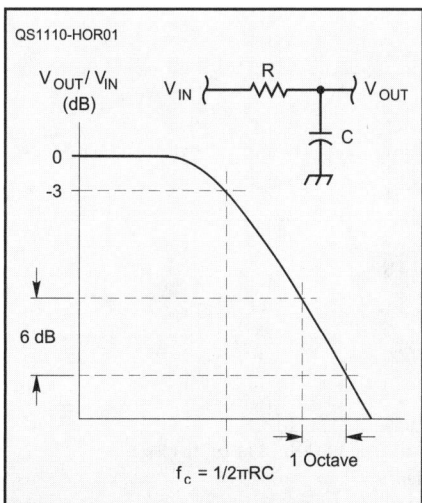

Figure 1 — The single-pole frequency response of a low-pass RC circuit showing the 6 dB/octave = 20 dB/decade rolloff above the cutoff frequency.

[1] All previous Hands-On Radio experiments are available to ARRL members at **www.arrl.org/hands-on-radio**.

[2] *The ARRL Handbook for Radio Communications*, 2011 Edition. Available from your ARRL dealer or the ARRL Bookstore, ARRL order no. 0953 (Hardcover 0960). Telephone 860-594-0355, or toll-free in the US 888-277-5289; **www.arrl.org/shop**; **pubsales@arrl.org**.

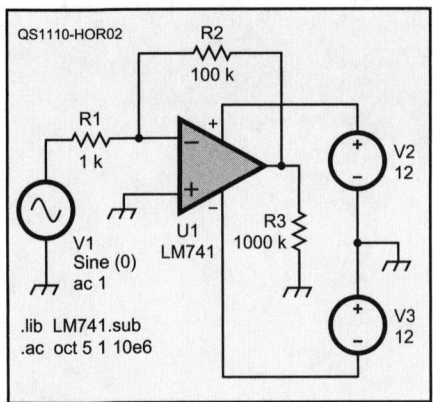

Figure 2 — An inverting op-amp amplifier circuit entered in the *LTspiceIV* schematic window. *SPICE* directives that define the LM741 op-amp model and perform an ac small-signal sweep are shown at the lower left.

periment #3 to learn how the circuit works.) Connect the circuit in the simulator, using the "opamp2" device to start with. Each of the dc voltage sources should be set to 12 V — note how the sources are connected to supply ±12 V. The input source should be configured as a sine source with a dc offset value of 0 V and the small signal ac amplitude set to 1.

To observe the effects of GBW on circuit performance, we're going to use two different types of op-amps; the LM741 and the LM318. Neither of these models is supplied with the simulator, but they are available on-line and are easy to add to your library. First, browse to the National Semiconductor Parametric Catalog for General Purpose Op-Amps at **www.national.com/cat/index.cgi?i=i//10**. The GBW column at the left of the window will be highlighted. Scroll to the LM741 and click the part number to open a window with all the information for that part. Download the PDF datasheet (the link is at the upper right) and then click the TOOLS tab at left. That opens another window for DESIGN TOOLS, including a *SPICE* model for the LM741. Download that file and save it to your *LTspiceIV* directory (usually C:\Program Files\LTC\LTspiceIV\lib\sub) as "LM741.sub."

Now go back to your simulation schematic and right click on the OP-AMP symbol. In the window that opens, change the VALUE to "LM741." Click the SPICE DIRECTIVE button on the far right toolbar symbol labeled .OP and enter ".lib LM741.sub." That text should appear on your schematic. From the SIMULATE menu, select EDIT SIMULATION COMMAND and select the AC ANALYSIS tab. The sweep type should be set to OCTAVE with five points per octave. The start frequency should be "1" and the end frequency "10e6" (1 Hz to 10 MHz). The schematic should now look just like Figure 2.

From the SIMULATE menu, click RUN and a blank oscilloscope window will appear above the schematic. Move the cursor over the schematic to the ungrounded end of R3 — a voltage probe symbol will appear. Position the probe on the wire connecting R3 to the output of the op-amp and click. Two traces will appear in the upper frequency response window — a solid one for gain and a dashed line for phase. With these initial values for R1 and R2, the circuit should have a maximum gain of 40 dB (gain of 100) and a cutoff frequency of around 10 kHz. The unity-gain frequency will be just below 1 MHz.

Move the cursor over the left-hand vertical axis. It will change into a ruler icon. Click, then set the maximum axis value to 40 dB and the minimum to 0 dB with tick marks every 5 dB. I also turned off the phase display for clearer viewing by moving the ruler cursor over the right-hand vertical axis, clicking, and then selecting DON'T SHOW PHASE. Print or otherwise save the frequency response — turn on the gridlines for easier reading of the frequencies and gains.

Now go back to the schematic and change the value of R2 to 10 kΩ (a voltage gain of 10 = 20 dB). Simulate and save the frequency response again. Change R2 to 2 kΩ (a voltage gain of 2 = 6 dB) and repeat. Compare the three graphs, as in Figure 3, and you will see that for all three circuits, the slopes of the responses very nearly line up and cross the horizontal axis quite close to each other. You can clearly see the op-amp's GBW limiting the gain at higher frequencies. (The responses don't line up exactly because of various circuit limitations internal to the op-amp.)

Finally, change the op-amp to an LM318 by downloading its *SPICE* model as described above and storing it in your library. Change the symbol value and the .LIB command as well. (I had to edit the *SPICE* model file's first non comment line to remove "/NS" from the device name — an edited version is available on the Hands-On Radio website.) Compare the circuit's frequency response using this high GBW op-amp (15 MHz) to that when using an LM741. You can experiment with many different op-amp types and circuits to see just how important GBW is to the final circuit performance.

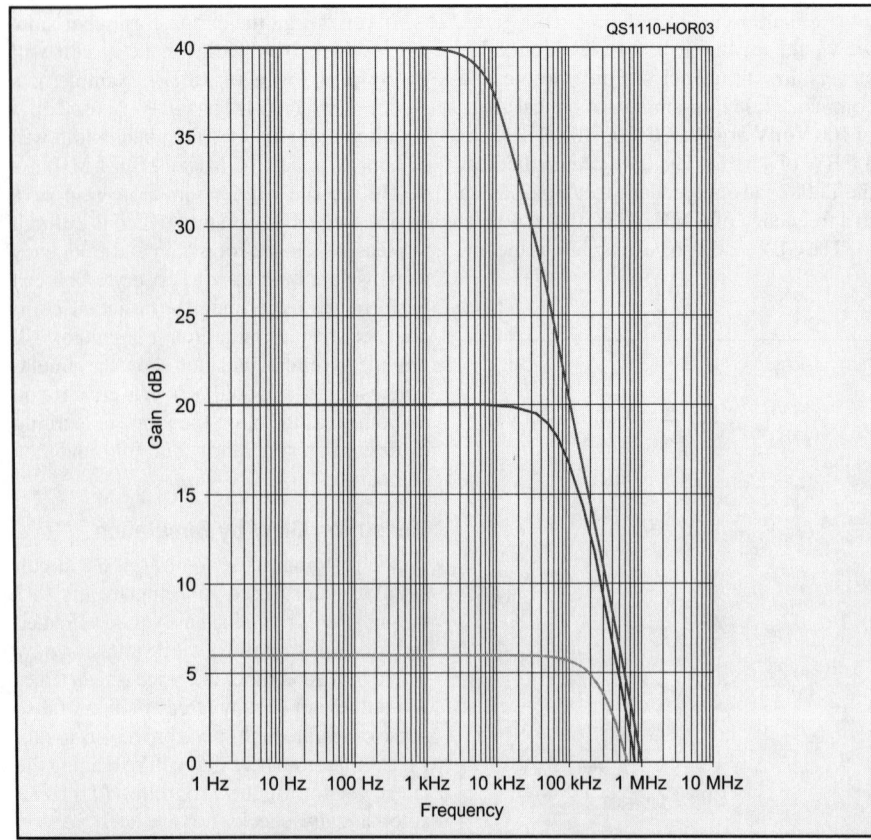

Figure 3 — Frequency response of the circuit in Figure 2 for three different gains of 40, 20 and 6 dB. The op-amp's GBW limits the circuit gain at high frequencies.

Experiment 106
Effects of Gain-Bandwidth Product

Last month, we discussed gain-bandwidth product (GBW or GBP) and how it affects the ability of an op-amp to amplify signals of different frequencies. That's important, because op-amps are used as the active element in signal processing and filter circuits. What effect does GBW have in that kind of application? We'll use *LTspice* to illustrate the effects of GBW in a band-pass filter circuit as an example of the issues the circuit designer has to consider.

Gain and Q

In the experiment portion of the previous experiment, you built a simple amplifier circuit and substituted op-amps with different GBW to see the effect. Clearly, as GBW increased, so did the gain of the circuit at higher frequencies. What about circuit performance at much lower frequencies? Does GBW affect performance there, as well? Yes!

The effects are most easily seen in band-pass filters because requirements for steep filter "skirts" and narrow bandwidths require a lot of gain. Why do they require a lot of gain? Let's take a look at the multiple-feedback band-pass filter in Figure 1.[1] (This design was created by Jim Tonne, W4ENE, using the professional-level version of his *ELSIE* filter design software.[2]) It shows a two-pole band-pass filter with a center frequency, f_0, of 10 kHz and a bandwidth, BW, of 1 kHz. Thus, the filter's Q is

$$Q = f_0 / BW = 10 \text{ kHz} / 1 \text{ kHz} = 10$$

In this example, the software requires values for f_0 and BW, the capacitor values (using the equal C-method), and the order and type of filter response (second order Chebyshev in this case). Figure 1A is the filter design if an *ideal* op-amp is used. That means an op-amp with an infinite GBW and infinite dc gain. Each filter section has the same gain ($A_V = 5.6$ dB) and Q (18.24). The section's center frequencies are slightly dif-

[1]Multiple-feedback band-pass filters are discussed in Hands-On Radio experiment #4. All previous experiments are available to ARRL members at **www.arrl.org/hands-on-radio**.
[2]Tonne Software, **www.tonnesoftware.com**.

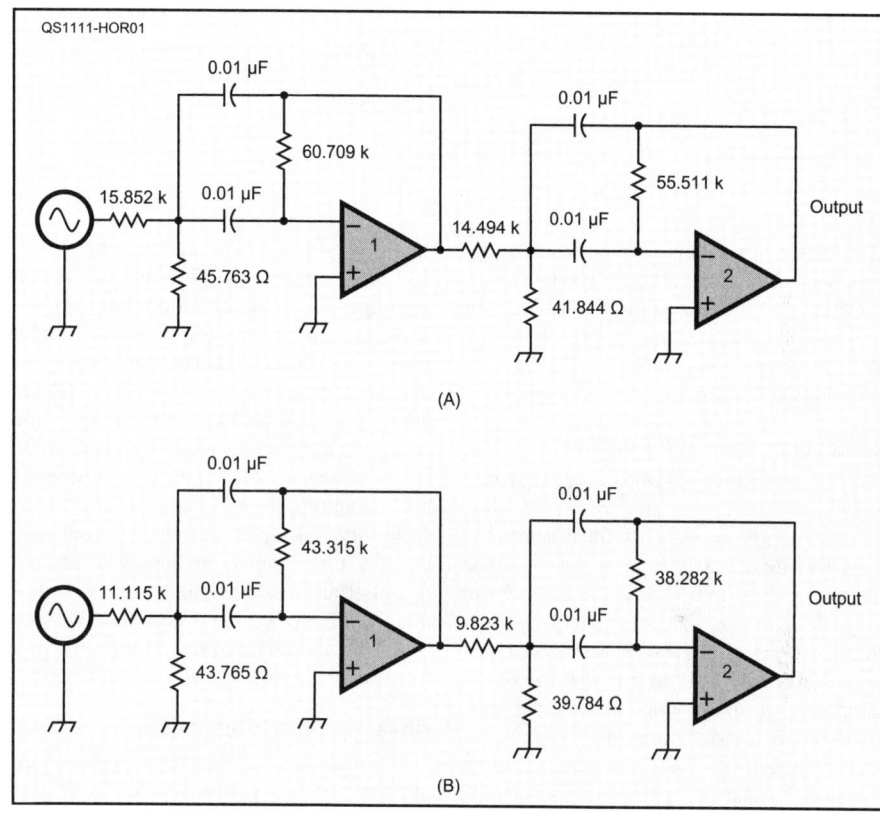

Figure 1 — Schematic of a two pole multiple feedback band-pass filter with a center frequency of 10 kHz and bandwidth of 1 kHz for a Q of 10. (A) shows the design for an ideal op-amp while (B) provides adjustments needed for practical op-amp performance (see text).

ferent: $f_{0-1} = 9.56$ kHz and $f_{0-2} = 10.46$ kHz. Each section then acts as a narrow filter (Q = 18.24) tuned to a single f_0.

If the two filter sections are *cascaded* as shown, the result is the band-pass frequency response as shown in Figure 2. The *pole* for each section is shown by the small, red lines on the frequency axis to either side of 10 kHz. The extra gain is required because the individual filter sections work against each other away from their respective center frequencies. To create the passband of the filter the total response has to add up to 0 dB at the filter's overall center frequency of 10 kHz, which is between the two individual f_0 values. The result is that each filter has to have a gain of greater than 0 dB at its individual f_0.

All well and good, but it's kind of hard to buy an ideal op-amp. They are always out of them when I go to the store! Jim's software, though, allows you to specify the performance of the op-amp and compensates

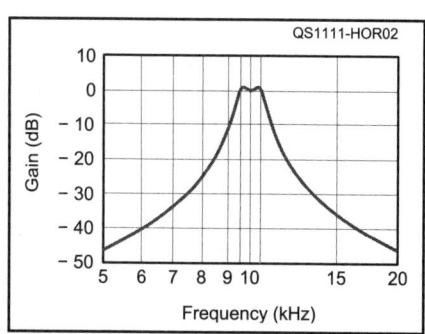

Figure 2 — Frequency response of the filter in Figure 1.

Electronic Fundamentals 25

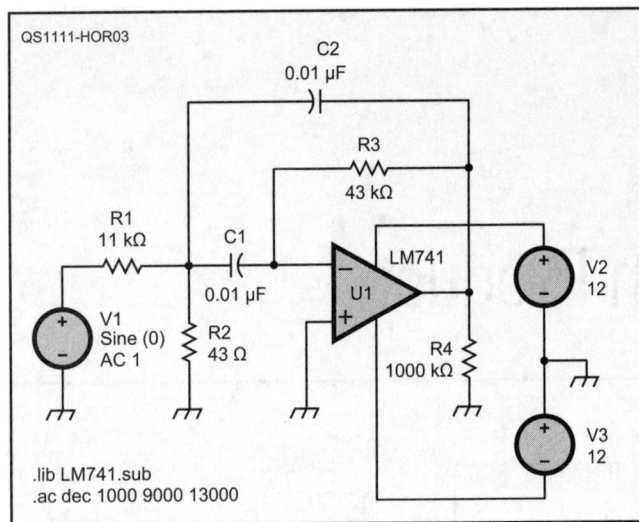

Figure 3 — *LTspice* schematic for Section 1 of the multiple-feedback band-pass filter.

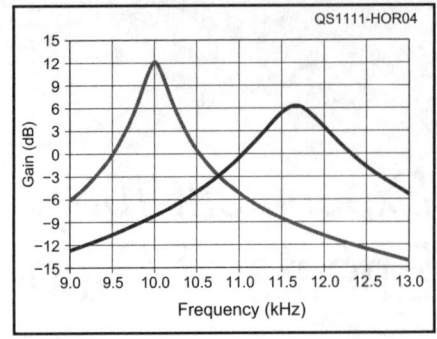

Figure 4 — Frequency response of the filter with an LM741 op-amp (red) and an LM318 op-amp (blue). The higher GBW of the LM318 results in performance that is closer to that of an ideal op-amp.

for its behavior. In this case, Jim used an op-amp with a dc gain of 100 dB (100,000 V/V), a GBW = 1 MHz and an input impedance of 1 MΩ. Figure 1B shows the result — the resistor values are a little smaller and the overall frequency response is the same.

Necessary Gain and Peaking

The non-ideal op-amp selected would seem to have plenty of gain at 10 kHz: 1 MHz / 10 kHz = 100. Each filter section has a gain of about 5.6 dB = 1.9 so we should be in good shape, right? Well, not really. From page 5.70 of the Analog Devices *Op-amp Applications* online book referenced last month: "A rule of thumb is that the open-loop gain of the op-amp should be at least 20 dB (×10) above the amplitude response at the resonant (or cutoff) frequency, including the peaking caused by the Q of the filter... $A_0 = H Q$, where H is the gain of the circuit."[3] (For a discussion of filter response peaking, see experiment #41.)

If each stage has a gain of 5.6 dB = 1.9 at f_0 and a Q of 18.24, then the op-amp must have a gain of $10 \times (1.9 \times 18.24) \approx 348$ at f_0. We're short of gain by a factor of about 3.5 to be able to ignore the effects of the op-amp's 1 MHz GBW. That's why the circuit values have to change a little bit.

Why does GBW make a difference at such a low frequency? What happens if the op-amp's GBW is too low? Quoting from page 5.106 of the *Op-amp Applications* book: "Without sufficient...gain, the op-amp virtual ground is no longer at ground. In other words, the op-amp is no longer behaving

as an op-amp. Because of this, the [filter] no longer behaves like [a filter]." A virtual ground exists at the op-amp's inverting (−) input *only* if the op-amp's output signal causes all of the currents flowing into and out of those connections to balance. That allows the voltage at the inverting input to be the same as at the non-inverting (+) input, which is connected to ground. If the op-amp doesn't have enough "oomph" (gain and output drive capability) to keep those currents in balance, the inverting input is no longer at ground potential and that invalidates the assumptions on which the filter design equations are based. The circuit may provide some filtering function but it won't perform as designed.

Observing the Effects of GBW

You can simulate Section 1 of the circuit of Figure 1A to see the effects of GBW. Use the closest standard 5% series resistor values, such as 11 kΩ, 43 kΩ, 43 Ω, 10 kΩ, 39 kΩ and 39 Ω. This will shift the center frequency to nearly 12 kHz from the software's precision design. Retrieve the amplifier circuit you simulated for last month's experiment and add the necessary resistor and capacitors to make the multiple-feedback circuit as shown in Figure 3.

To change the values of the components, move the cursor over the symbol until it takes the shape of a hand, right click, then edit the value. (Use "u" for micro.) Start with the LM741 op-amp. You can change the op-amp library model by moving the cursor over the ".lib" library model identification line so that it becomes a text cursor, then right clicking and editing. Don't forget to change the op-amp part number as well, using the same process.

Because we want to see the frequency response of the circuit close to 10 kHz and not spread out from 1 kHz to 1 MHz, edit the simulation command line by right-clicking

over the ".ac" line. I found that a span of 9 kHz to 13 kHz made it easy to see the effects of changing the op-amp. Figure 4 shows the result in red. (Click on the horizontal axis cursor to change the plot to linear and use 500 Hz tick marks. Click on the vertical axis cursor to turn off phase plotting.) Now change to the LM318 op-amp used for comparison last month and rerun the simulation. You'll get a response shown in blue in Figure 4 — quite a change!

First, the center frequency shifts from 10.1 kHz and a bandwidth of 300 Hz for the LM741 to 11.7 kHz and 800 Hz with the LM318. Gain also changes from 12 dB with the LM741 to 6 dB with the LM318. Because we're using standard values for the resistors, the design center frequency is now approximately 12 kHz, but the Q and gain values for the LM318 circuit are much closer to what is expected for an ideal op-amp.

You can see the effect even more clearly if you use one of the low cost high GBW op-amps available today, such as the LM7171 with a GBW of 200 MHz. (Download and use the model file as explained last month.) Another way to see big changes in performance is to increase the filter's center frequency. To change f_0 to 100 kHz, reduce the two capacitors by a factor of 100 kHz / 10 kHz = 10 for a value of 0.001 µF. The higher-speed op-amp is required to get anything close to expected performance.

The moral of this story is that sensitive circuits such as moderate- to high-Q filters can be very dependent on the performance of the components used to implement them. Although our junk boxes are full of op-amps with 1, 4 or 10 MHz GBW, they will probably give confusing results in circuits for which they are not suited, or if the tools we use to design the circuits make too many assumptions about their capabilities!

---

[3]www.analog.com/library/analogDialogue/archives/39-05/op_amp_applications_handbook.html

Experiment #72 — Return Loss and S-Parameters

If you look up the performance specifications for RF equipment, such as filters, in the literature from a commercial manufacturer, you may be surprised not to find SWR in the data tables. In its place, you'll find a parameter called *return loss*. What's up with that and how does it relate to SWR? That's the topic of this month's column and we'll also touch on *S-parameters*, not commonly used by hams but well-known in industry.

The Basics

Let's start with basic transmission line operation. If the load attached to a transmission line has the same impedance as the line's *characteristic impedance*, Z_0, all of the power flowing as electromagnetic waves toward the load in the line will be transferred to the load. If the load has some other impedance, higher or lower than Z_0, some of the power will be reflected back toward the power source (aka — the generator).

The forward wave (generator-to-load) and the reflected wave (load-to-generator) interfere with each other to create stationary patterns of voltage and current in the line called *standing waves*. The greater the difference in impedances between the line and the load, the larger the amplitude of the interference pattern will be.

If the pattern of voltages in the line is measured, the ratio of peak to minimum voltage is the *voltage standing wave ratio* or VSWR. The pattern's amplitude can also be measured as current to give ISWR, but voltage is more easily measured. VSWR and ISWR have the same value and are usually given the more general term, SWR.

Voltage and current in the transmission line are rarely measured directly outside the laboratory. Amateur SWR meters use directional coupler circuits to create voltages proportional to the power flowing in each direction and use a special meter scale to convert the ratio of the voltages to SWR. (Hands-On Radio Experiment #52 discusses how garden-variety SWR meters work.[1]) Wouldn't it be easier to just display forward and reflected power? Yes, but SWR is the mental model that amateurs use to describe the relationship between Z_L and Z_0, so our equipment displays SWR.

Outside Amateur Radio the situation is different. What RF engineers usually want to know is how much of the power in a transmission line will be delivered to a device or antenna and how much will be reflected. These engineers think in terms of power, not SWR. Their mental model is different because they are more concerned with system efficiency and other similar calculations. They also tend to use more sophisticated instruments that measure power directly.

Thus the term *return loss*, measured in dB, is used. Return loss and SWR measure the same thing — how much power in the transmission line is sent to the load and how much is reflected by it — but state the result differently.

Return loss (RL) =
$-10 \log_{10} (P_{REFL}/P_{FWD})$ dB [1]

Because P_{REFL} is never greater than P_{FWD}, RL is always positive. The greater the value of RL, the less the amount of power reflected from the load compared to forward power. If all the power is transferred to the load because $Z_L = Z_0$, RL = ∞ dB. If none of the power is transferred to the load, such as at an open or short circuit, RL = 0 dB. For practice, calculate RL for the following values of P_{FWD} and P_{REFL}:

(A) P_{FWD} = 100 W, P_{REFL} = 25 W
(B) P_{FWD} = 100 W, P_{REFL} = 1 W
(C) P_{FWD} = 1 kW, P_{REFL} = 50 W
(D) P_{FWD} = 5 W, P_{REFL} = 0.1 W

(Answers are provided at the end of the article.) Table 1 contains a series of values for P_{FWD} and P_{REFL} and the corresponding value of RL. Note that RL only depends on the ratio of power values, not the absolute values of the powers involved. RL is the same wherever P_{REFL} / P_{FWD} has the same value.

RL can also be calculated directly from power ratios, such as dBm (decibels with respect to 1 mW) or dBW (decibels with respect to 1 W). In this case, RL = P_{FWD} − P_{REFL} because the logarithm has already been taken in the conversion to dBm or dBW. (Ratios in dB are computed by subtraction, not division.) For example, if P_{FWD} = 10 dBm and P_{REFL} = 0.5 dBm, RL = 10 − 0.5 = 9.5 dB. Here are some more practice exercises:

(E) P_{FWD} = 25 dBm, P_{REFL} = 4 dBm
(F) P_{FWD} = 12 dBm, P_{REFL} = 6 dBm
(G) P_{FWD} = 10 dBW, P_{REFL} = 1 dBW
(H) P_{FWD} = 1 dBW, P_{REFL} = −20 dBW

Both power measurements must have the same units (dBm, dBW, etc) for the subtraction to yield the correct results. For example, one can't subtract dBW from dBm directly. Bonus exercise — what if P_{FWD} is 10 dBW and P_{REFL} is 20 dBm? (1 W = 1000 × 1 mW, so to convert dBW to dBm, add $\log_{10}(1000)$ = 30 dB. The answer to the bonus exercise is then 40 dBm − 20 dBm = 20 dB.)

As you can see, the more positive the value of RL, the smaller the fraction of forward power that is reflected towards the source. Higher values of RL are "better" in the same sense that lower values of SWR are "better."

SWR to Return Loss Conversion

If SWR and RL measure the same thing — reflected power as a fraction of forward power — can one be converted to the other? Of course. There are tables of those conversions, but how about an equation instead?

Start by converting RL back to a power ratio as follows:

Table 1 — Return Loss Versus Power

P_{FWD} (W)	P_{REFL} (W)	P_{REFL}/P_{FWD}	RL (dB)
1	0.1	0.1	10
1	0.2	0.2	7
1	0.5	0.5	3
1	1	1	0
10	0.1	0.01	20
10	0.2	0.02	17
10	0.5	0.05	13
10	1	0.1	10
100	0.1	0.001	30
100	0.2	0.002	27
100	0.5	0.005	23
100	1	0.01	20
100	2	0.02	17
100	5	0.05	13
100	10	0.1	10
100	20	0.2	7
100	50	0.5	3
100	100	1	0

---

[1]Previous Hands-On Radio columns are available to ARRL members at **www.arrl.org/hands-on-radio**.

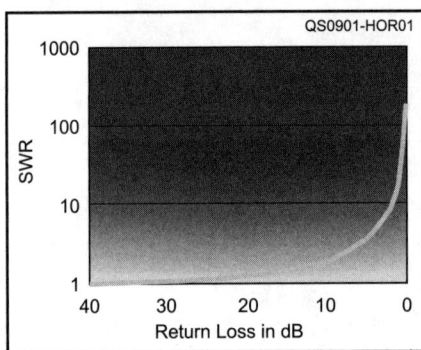

Figure 1 — A graph of SWR versus return loss (RL) shows that RL, measured in dB, is more useful at low values of SWR.

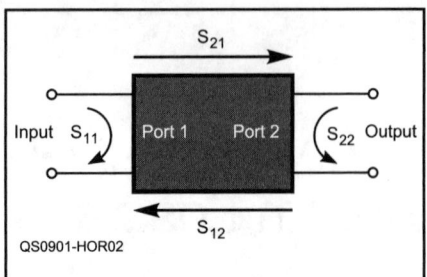

Figure 2 — A two-port device with its associated s-parameters. The s-parameters describe what happens at a port when a signal is applied to the same or the other port.

$$P_{REFL}/P_{FWD} = \log^{-1}(0.1 \times RL) \quad [2]$$

Now use the equation for computing SWR from forward and reflected power:

$$SWR = \left[1 + \sqrt{P_{REFL}/P_{FWD}}\right] / \left[1 - \sqrt{P_{REFL}/P_{FWD}}\right] \quad [3]$$

Table 2 contains a series of values that show the relationship between P_{REFL}/P_{FWD}, RL, and SWR that is also graphed in Figure 1.

What about going the other way, from SWR to RL? Start with the equation for power ratio in terms of SWR:

$$P_{REFL}/P_{FWD} = [(SWR - 1)/(SWR + 1)]^2 \quad [4]$$

Then convert that to RL using equation [1]. Table 3 shows a series of values for P_{REFL}/P_{FWD} and RL based on the value of SWR. You can use the equations above to make your own table or automatic RL to SWR converter in a spreadsheet!

Two-Port Devices

Why do RF engineers prefer to work in terms of RL and dB instead of the more familiar SWR? Most instrumentation used professionally is calibrated in dB (and related units, such as dBm) for amplitude such as on spectrum analyzers, a common instrument in the professional world. A measurement in dB "fits better" than a measurement such as SWR that is calculated linearly, without logarithms. Figure 1 gives another good reason — for large values of RL, SWR becomes very close to 1.0 and changes in value become smaller and smaller. It's much easier to work with the larger values of RL in dB, just as hams do for gain and attenuation.

Another reason is that RF engineers tend to think of their systems as a network of "black boxes" called *two-port devices* as shown in Figure 2. (Power supply connections are ignored in this model.) Each two-port device has an input (Port 1) and an output (Port 2). The behavior of the device is then described mathematically by the relationships between signals at the various ports.

There are a number of techniques to describe the relationships and each technique relies on a set of mathematical constructions called *parameters*. There are Z-parameters, H-parameters, T-parameters, and so forth, all designed to describe the device in a way that is of most use to the designer for a specific type of product or system. For example, T-parameters describe the device in terms that relate to the transmission of signals. Z-parameters describe the device in terms of impedances. Each type of parameter is a mathematical tool to be applied in the appropriate environment.

S-Parameters

In the RF design world, the most common set of parameters is that of the *s-parameters*, or "scattering parameters." The word scattering is used because S-parameters describe what happens when a signal is applied to one port and "scatters" to the other port, or even from the port to which the signal is applied.

There are four S-parameters: S_{11}, S_{12}, S_{21} and S_{22} shown in Figure 2. The numbers indicate the direction of the scattering. The first number is the port at which the scattered signal is measured and the second number the port at which the signal was applied. So S_{21}, for example, describes the signal at port 2 that results from a signal being applied at port 1. Hmmm...does that ring a bell? If the device is a circuit, then S_{21} describes its gain! If the device is a transmission line, S_{21} describes its loss.

Well, why didn't they just call it *gain* or *loss*, I hear you asking. It's because the math involved with these four parameters often operates on all four at once (as a matrix) and it's easier to keep everything straight if a consistent symbol naming convention is used as opposed to common language names.

In an amplifier, S_{12} describes the isolation between the output and input ports. S_{11} describes what happens when a signal is applied to port 1 and then the resulting signal that comes back from port 1 is measured. Ah, hah! That's our definition of return loss, isn't it? (Similarly, there is an equivalent return loss at the output port, S_{22}.) So the set of four S-parameters — gain, isolation and two return losses — describes an amplifier (or filter or transmission line or...) pretty well. And now you know what you're looking at in those data sheets!

Exercise Answers

(A) 6 dB, (B) 20 dB, (C) 13 dB,
(D) 17 dB, (E) 21 dB, (F) 6 dB,
(G) 9 dB, (H) 21 dB

Recommended Reading

There are detailed discussions of return loss and scattering parameters on Wikipedia. Browse to **www.en.wikipedia.com** and enter either of those terms. Unfamiliar terms are often hyperlinked from Wikipedia pages for even more information.

Next Month

If you browse the tables of op-amp data in the Component Data and References chapter of *The ARRL Handbook* or in an electronic distributor's catalog, you'll discover a cornucopia of devices types. Next month, we'll talk about op-amp parameters and how to decide which common op-amps are right for your project.

Table 2
SWR Versus Return Loss

P_{REFL}/P_{FWD}	RL (dB)	SWR
0.0001	40	1.02
0.0010	30	1.07
0.0100	20	1.22
0.0316	15	1.43
0.1000	10	1.92
0.3162	5	3.57
0.6310	2	8.72
0.7943	1	17.39
0.9772	0.1	173.72

Table 3
P_{REFL}/P_{FWD} **Versus SWR**

SWR	P_{REFL}/P_{FWD}	RL (dB)
1.01	0.00002	−46.0
1.1	0.0227	−26.4
1.2	0.0826	−20.8
1.5	0.04000	−14.0
2	0.11111	−9.5
3	0.25000	−6.0
5	0.4444	−3.5
10	0.66942	−1.7
100	0.6079	−0.2

Experiment 104
Words to Watch For

At the end of nearly every "Hands-On Radio" column are one or more references or suggestions for follow-up reading that expand on the topic at hand. Behind most of those references lies a seasoned author, providing clear explanations that encourage the reader to go further and dig deeper. The recent and completely unexpected passing of two of these authors, both highly respected by your columnist and many others, prompted me to list of a few favorites. Perhaps some of my readers will find them as valued as I do.

The first of the giants to fall was Jim Williams, staff scientist for Linear Technology and author of dozens of technical articles over a span of more than 35 years.[1] Most hams can instantly identify with Jim's practical down to earth approach to "the art, science, and personalities" of circuit design, the subtitle of *Analog Circuit Design*, his collection of essays from fellow analog travelers — especially since he was a self-taught individual for the most part.[2] I know that I am not alone in having saved every article that I noticed was written by Jim on any topic, regardless of whether or not it pertained to my technical specialty.[3] They were simply fun to read because I reveled in his enjoyment and craft of the work. Read his classic essay "Max Wien, Mr. Hewlett, and a Rainy Sunday Afternoon" from the collection to experience the sheer savor of what we all search for on our own rainy Sunday afternoons.

Still contemplating a post-Jim Williams world, I was even more shocked to learn of the death of Bob Pease in a car accident on the way home from Jim's memorial service![4] This simply could not be — but it was. Anyone who has ever used a National Semiconductor op amp or voltage reference (among many other products) has been touched by Bob's work, which spanned the era of vacuum tubes to nanoengineering. His "What's All This…" and "Pease Porridge"

columns in *Electronic Design* attracted legions of engineers.[5] In common with Jim, he also contributed a classic book on electronics, *Troubleshooting Analog Circuits*, that speaks with the authority of someone who has done battle with legions of bugs, large and small.[6] His autograph reads, "…may all your Troubles be Middle Sized: so you can find 'em!" Jim and Bob filled a library between them with articles, columns, applications notes, data books and other contributions.

Enough with the remembrances! The point of this column is not so much to remember Williams and Pease — many words do that job elsewhere and more personally. The point of this column is to also identify some of my favorite authors in the worlds closer to Amateur Radio and its many practitioners. So many of us rely on the guidance of others and loan their treasured books, growing a little more worn with each project and bearing the brown badges of honor from an errant soldering iron. Here are some of my guides — I owe them a lot.

Bill Orr, W6SAI

No list of influential ham radio authors could possibly leave out Bill Orr, in whose honor the ARRL's Bill Orr, W6SAI, Technical Writing Award is dedicated. The many

---

[1]en.wikipedia.org/wiki/Jim_ Williams_%28analog_designer%29
[2]J. Williams, *Analog Circuit Design*, Butterworth-Heinemann, 1991.
[3]Jim Williams archive at *EDN*, www.edn.com/ article/472111-Jim_Williams.php.
[4]en.wikipedia.org/wiki/Bob_Pease
[5]www.national.com/rap
[6]R. Pease, *Troubleshooting Analog Circuits*, Butterworth-Heinemann, 1991.

editions of his *Radio Handbook* provided friendly competition to *The ARRL Handbook* with complementary coverage and presentation of mutual subjects from a different perspective. Other major Orr titles on my bookshelf include: *The Beam Antenna Handbook*, *The Quad Antenna Handbook*, *The VHF-UHF Manual*, *Wire Antennas* and *The W6SAI HF Antenna Handbook*. Amplifier builders continue to refer to his many technical notes about Eimac tubes and his byline is seen in archives of *QST* articles and *Ham Radio* and *CQ* magazine columns.

Don Lancaster

If you are an active builder or want to be, you'll no doubt have at least a couple of Don's "Cookbooks" on your shelf. And if you are like me, they will be well used. His *Active Filter Cookbook*, *TTL Cookbook* and *CMOS Cookbook* are three must-haves for the electronic hobbyist. Don's style is direct and hands-on with few digressions — and a very clear way of visually grouping and presenting material so that it is easily grasped.

Walter Jung

Walter Jung extends the cookbook style another layer deeper in his widely read *Op Amp Cookbook*, first published in 1974 and now in its third edition. Far more than a collection of op amp circuits, the book opens up the op amp itself, explaining the subtleties of how this circuit behaves (and misbehaves). Supported by this detailed description of the op amp's inner working, the book continues with a smorgasbord of op amp circuitry: filters, signal conditioning, audio processing, regulators, oscillators and much more.

A member of the Analog Devices team, Jung also wrote the *Op Amp Applications Handbook*, free for downloading in PDF format from the Analog Devices website.[7] While you're shopping, keep an eye out for his *IC Timer Cookbook*, an endless source of ideas and how-to information on the venerable 555 IC timer and its variants.

Shelf Stackers

Another series of prolific electronics authors churned out volume after volume of basic electronics how-to books that you will find in libraries everywhere. Three of my favorite such authors are Joe Carr, K4IPV (SK), Howard Berlin and Forrest Mims III.

Joe Carr's many books dealt mainly with radio related topics: antennas, receivers, measurements, power supplies and other topics. The two still in print (*RF Components and*

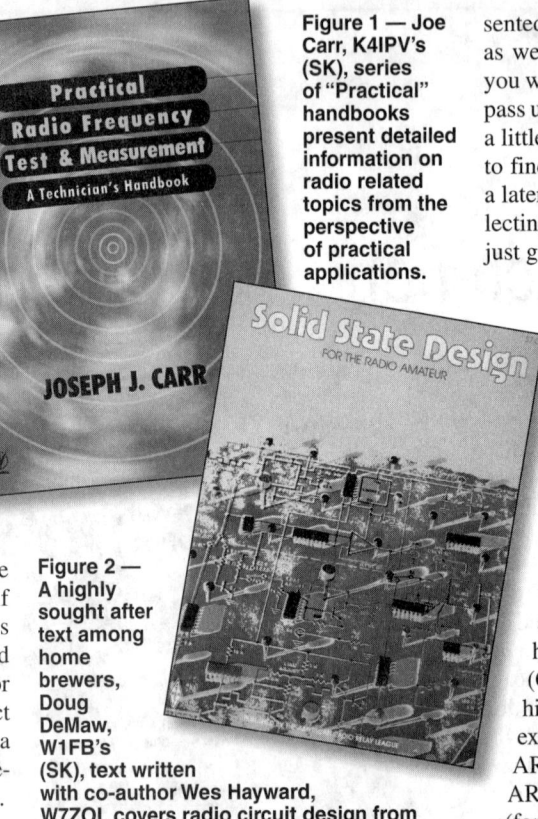

Figure 1 — Joe Carr, K4IPV's (SK), series of "Practical" handbooks present detailed information on radio related topics from the perspective of practical applications.

Figure 2 — A highly sought after text among home brewers, Doug DeMaw, W1FB's (SK), text written with co-author Wes Hayward, W7ZOI, covers radio circuit design from the ground up. The successor publication is *Experimental Methods in RF Design*, ARRL order no. 9239.

Circuits and *Practical Radio Frequency Test & Measurement*) are more detailed than the bench top guidebooks with which his writing career began.[8]

Howard Berlin has a long list of publications dealing with electronics, some of which include: *555 Timer Applications Sourcebook*, *Design of Phase-Locked Loops*, *Guide to Complementary Metal Oxide Semiconductors*, *Digital Electronics*, *Design of Active Filters*, *Instrumentation and Measurements* (partial titles). Nearly all of them emphasize "with experiments" to let you know that the information in the book can be replicated on your workbench — no excuses!

There are volumes of Forrest Mims III (www.forrestmims.org) books on any electronics topic you care to name. I first encountered his books when I became interested in digital design and quickly discovered his tutorials were useful and interesting to read. His compilations of circuits in the *Notebooks* series and *Circuits and Projects* series are huge time savers when you need a starting point for a project — each one verified to work as designed.

All three of these authors are well represented at every used bookstore in the land as well as on the hamfest table. Whenever you wander the aisles of a flea market, don't pass up those piles and boxes of books — do a little prospecting for the gems you're sure to find. Even if you already have a copy or a later edition, spend a couple of bucks collecting them to give away to someone else just getting started.

Doug DeMaw, W1FB

My goodness, where would ham radio be without Doug DeMaw? Most regrettably a Silent Key, Doug basically raised a generation of us with article after article in *QST*, supporting *The ARRL Handbook* and creating timeless classics such as *Solid State Design for the Radio Amateur*, co-authored with another titan, Wes Hayward, W7ZOI. Copies of that book are worth their weight in gold amongst homebrewers such as the low power (QRP) community. He also went out of his way to provide in-person support. An experimenter would do well to browse the ARRL's online *QST* archives (available to ARRL members) for articles with "Part" (for multipart tutorials) or "Basics" in the title and DeMaw as the author. (A search on "DeMaw" alone turns up 640 entries!)

George Grammer, W1DF

Unknown to most hams today, George Grammer's call sign was a fixture in *QST* for many years with feature articles and Technical Correspondence items. His "Simplified Design of Impedance Matching Networks, Parts I-III" beginning in the March 1957 issue are as useful in explaining basic concepts today as they were a half century ago. Another prolific writer, with 438 entries in the archive, his instructional articles are easier to find if you start with the older issues and work your way toward the present. You can find long-delayed echoes of W1DF in the "Electrical Fundamentals" chapter of every *ARRL Handbook*.

To the Giants

I know that I have omitted many excellent authors, including those who are busy writing today. They deserve your support and thanks. In closing, I relay the opening quote from Jim Williams' second volume of essays, *The Art and Science of Analog Circuit Design*:

"MIT building 20 at 3:00 A.M.

Tek[tronix] 547 [oscilloscope], pizza, breadboard.

That's Education."

Yes, Mr Williams truly "got it" and I hope you do, too.[9]

---

[7]www.analog.com/library/analogdialogue/archives/39 05/op_amp_applications_handbook.html

[8]J. Carr, *RF Components and Circuits*, ARRL order no. 8759, and *Practical Radio Frequency Test & Measurement*, ARRL order no. 7954. Telephone 860-594-0355, or toll-free in the US 888-277-5289; www.arrl.org/shop; pubsales@arrl.org.

[9]J. Williams, *The Art and Science of Analog Circuit Design*, Butterworth-Heinemann, 1995.

Experiment 100
A Hands-On Hundred

"Why don't we try it for a few months and see how it works out?" reads the fall of 2002 e-mail from then ARRL Chief Operating Officer Mark Wilson, K1RO (now my partner in editing various ARRL publications). Armed with verified bits and pieces of my lab teaching experiences for Seattle University's Electrical and Computer Engineering Department, the first two page column featuring the common emitter amplifier was created and launched. I guess it's worked out okay since we have now reached Experiment 100, 8⅓ years along the Hands-On Trail. The response has been wonderful and I've enjoyed the challenge of coming up with a new topic each month — the to-do list seems far from empty, so I suppose I'll keep at it!

The Hundred

Naturally, this being the 100th column in the series, it should feature something to do with the century mark. After some consideration of what to write about, I came up with the idea of featuring the number 100 in a variety of different experiments gleaned from the history of Hands-On Radio. (All Hands-On Radio experiments are available to ARRL members from the Hands-On Radio website — **www.arrl.org/hands-on-radio**.) How many different ideas can be worked into a single column? Well, probably not 100.

One Hundred

Experiment 56 on Design Sensitivities explains the difference between the various types of values encountered while working with circuits — nominal, actual, typical and so forth. The experiment also covers tolerance — the allowed variation of a component's actual value around some nominal value. An awareness of tolerance in component value should also extend to tolerance in instrument readings. To demonstrate both, here's an experiment that illustrates both and is designed for a club or team.

Start by purchasing a batch of 100 resistors, all 100 Ω, ¼ W, 5% tolerance. They can be any type, but carbon film resistors are a good choice since they are inexpensive and they have leads that are easy to handle. Recruit 10 members to each bring a digital multimeter and 10 or more 100 Ω resistors of any type from their stock at home. Divide the batch of 100 resistors into 10 smaller batches of 10. Download the spreadsheet named "Tolerance Worksheet" from the Hands-On Radio website for this experiment. (It's an *Open Office* document that can be read by many different spreadsheet programs.)

Give each batch to a person with a multimeter and have them measure the value of each resistor and enter it into the spreadsheet's red box for individual resistor values (cells A10 through A20 — delete the sample values before beginning). You'll see the top chart, "Distribution of Values," create a *histogram* (also known as a *frequency distribution* chart) showing how the resistor values are distributed. An average value will be calculated in cell A8. With all 10 values entered, the average value should be close to 100 Ω and none of the resistors should have a value below 95 Ω (100 Ω × 95%) or above 105 Ω (100 Ω × 105%). Take a minute to look at the distribution of the resistor values — they should be centered on and closely packed around 100 Ω. Record the average value in the AVERAGE VALUES box (cells N10 through N20).

Now transfer the individual and average values into one of the BATCH boxes below the charts. You'll reuse those values later. Repeat the process of entering individual resistor values from each remaining batch into cells A10 to A20 in turn, entering the batch's average value in the AVERAGE VALUES box. Once all 10 batches are done, transfer all individual batch values back into column A, cells A10 through A110. Again, the top chart will show the distribution of all resistors but with a much higher set of samples. You'll see a classic bell curve centered near 100 Ω (see Figure 1).

What about the instrument tolerance? It's instructive to observe how much variation exists between meters. Save the spreadsheet under a different name for later reference and delete the values you entered. Now take one single batch of 10 resistors and measure it with all 10 meters, recording the numbers and averages as before. Look for variations from meter to meter. What was the maximum and minimum average value? Did the distributions look the same for each meter?

Your first set of measurements were taken on one batch of resistors that were probably made by the same machine within a few seconds of each other. In addition, they had been stored under identical conditions until you took them out of their package — that's the farthest they'd been from each other in their whole lives! As a result, they are closely matched with a minimum amount of manufacturing variation. Let's try a different scenario.

Save the spreadsheet under a different name for later reference and delete the values again. Now take the set of resistors each member brought and repeat the first part of

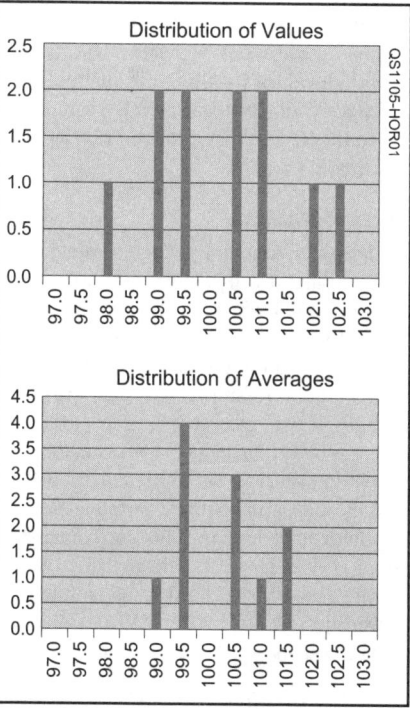

Figure 1 — A typical distribution of 5% resistor values from a batch of 10. By measuring more resistors, a better picture of component variation is obtained. The average value from different batches also varies in roughly the same way. If different measuring instruments are used, the values will also include variations in instrument accuracy.

Electronic Fundamentals 31

this experiment, comparing the resulting distributions to those of the siblings in the one big batch — you'll see a lot more variation!

Now that you've had a chance to observe variation between resistors and between measuring instruments, consider what the circuit designer must contend with. All of the components vary about some nominal value, yet the circuit has to meet its performance specifications every time. Even the instruments used to determine performance vary! The individual homebrewer has it easy — only the one copy of the circuit has to work!

Two Hundred

The venerable 555 timer IC is now in use in a fifth decade, having been invented in 1970 and sold by Signetics beginning in 1971. Happy 40th birthday! There are now hundreds of variants and it still occupies a useful niche in many areas of electronics. The 555 can be configured to operate in monostable (one shot or single pulse) and astable (free running oscillator) modes with pulse widths and frequencies controlled by the ratio of a capacitor and one or two resistors.

Review Experiment 5 and design circuits that use the same capacitor value — the first being a monostable circuit that outputs a 100 µs pulse and the second an astable circuit with a frequency of 100 Hz. Use the astable circuit to trigger the monostable circuit for a double 100! (Hint — in the original experiment, replace T with the appropriate values for time in Equation 1 and solve for C in terms of R. Pick a value of C, solve for R, and that's the sum of $R_1 + R_2$ in Equation 3.)

Three Hundred

The three terminal voltage regulator is so versatile that it can even be used to regulate current as explained in Experiment 19 on current sources. The regulator will do its best to maintain a constant voltage between its output and reference pin. You, the clever circuit designer, can use Ohm's law to change that constant voltage into a constant current as shown in Figure 2. Scramble around in your junk box and find a three terminal positive regulator and design a current source that supplies 100 mA. Be sure that the current setting resistor, R, is adequately rated for power dissipation, which is V_{REG}^2/R. Use a 100 Ω load and verify that the current source

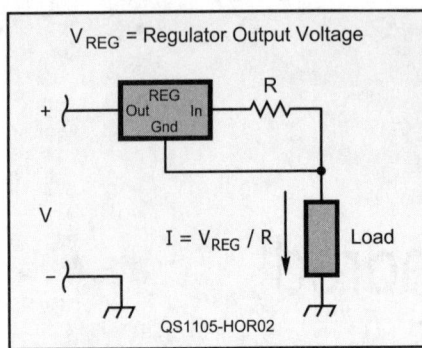

Figure 2 — The three-terminal regulator can be configured to regulate current by placing a current-set resistor between its output and reference terminals. The resulting current through the load is then the regulator's design voltage, V_{REG}, divided by the current set resistor, R.

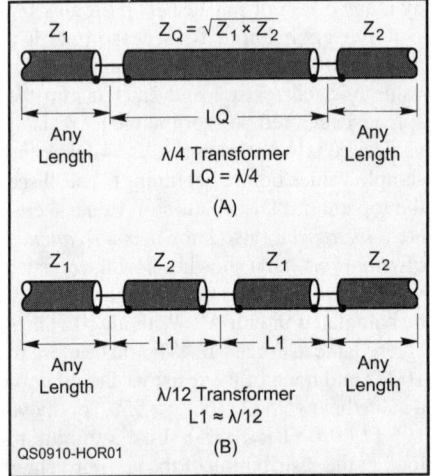

Figure 3 — The ¼ wave and ¹⁄₁₂ wave synchronous transformers. A series of carefully phased reflections create an impedance match. The ¼ wave version or Q section is often used to match the impedance of a loop antenna to 50 Ω coaxial cable.

works by measuring the current directly with your meter or by measuring voltage and using Ohm's law.

Four Hundred

Experiment 33 featured the transformer — a simple but amazing component with many useful properties. Let's exercise our transformer know-how:

1) What turns ratio is required for an impedance transformation of 100? (10:1)

2) If the turns ratio is 10:1 (primary to secondary) and a 100 Ω load is connected to the secondary, what is the impedance seen at the primary winding terminals? (10,000 Ω = 100 Ω × 10²)

3) If the primary winding has 100 V across it and 100 mA through it and the secondary winding has 10 V across it, if delivering the power, what is the current in the secondary winding? (100 V × 100 mA = 10 V × I_S; I_S = 100 × 0.1 / 10 = 1 A) What is the turns ratio? (100 V / 10 V = 10:1)

Five Hundred

What else is there with a value of 100? Hmm…sounds like a transmission line's characteristic impedance and didn't we talk about that in Experiment 81 on Synchronous Transformers? A quarter wavelength of transmission line — or a *Q section* — transforms a load impedance at one end, Z_2, to an impedance Z_1 at the other by *inverting* Z_2 about the Q section's characteristic impedance, Z_Q. That is,

$$\frac{Z_2}{Z_Q} = \frac{Z_Q}{Z_1} \text{ or } Z_Q = \sqrt{Z_1 Z_2}$$

Putting that equation to work, if I have a loop antenna with a feed-point impedance of 100 Ω, what should the characteristic impedance of my Q section be to create a match to 50 Ω feed line?

$$Z_Q = \sqrt{Z_1 Z_2} = \sqrt{100 \times 50} = 70.7 \text{ Ω}$$

Similarly, if I had a Q section with Z_Q = 100 Ω and connected a 150 Ω load to one end of it, what impedance would I see at the input?

$$Z_1 = \frac{Z_Q^2}{Z_2} = \frac{100^2}{150} = 66.7 \text{ Ω}$$

See Figure 3.

The Next Hundred

Next month launches Column 101. While I can't say for sure that there will be a complete second century, we made it this far, didn't we? See you next month — keep your workbenches busy and your reading list full!

Simulation

Experiment #83 — Circuit Simulation, *Part 1*

This month, we're going to begin a multi-part article to get you started with circuit simulation. The 2010 edition of *The ARRL Handbook* has a brand new chapter on computer aided design (CAD) by Dave Newkirk, W9VES. Many readers will want to try out some of the suggestions and techniques in that chapter, so this seems like a good time to get started with a professional quality circuit simulator. This column will lead you through the process of downloading and installing your very own simulation software, Linear Technology's *LTspiceIV*, which comes at a ham-friendly price — free![1]

Simulation tools are ubiquitous in the engineering world these days. In fact, it's quite unusual for a product design to be created purely on paper or amid swirls of rosin smoke on the workbench. Circuit designers stay at the computer until they are confident in their creation before picking up a soldering iron. Hobbyists also use simulation tools, but before we start, some general cautions are in order.

"Here be Dragons!"

That legend prominently indicated *terra incognita* on early maps. As with exploring a new world, simulation is often full of dragons for the beginning circuit designer for whom everything is unfamiliar. Nevertheless, this is no reason to avoid taking the simulation plunge! Amateurs use antenna modeling programs — electromagnetic simulators — to great effect. They have learned to recognize the trickery of a "trouble dragon," unrealistic gain; excessive bandwidth; extreme sensitivity to small changes in frequency, orientation, or size, bizarre impedances and so forth. *Caveat simulator!* By knowing and respecting the limits of the tool, excellent and useful results can be obtained.

Trouble dragons live at the numerical limits of the models and mathematics on which the simulation is based. Circuit simulators create equations for voltage and current at each point in the circuit. All of the equations are then solved simultaneously for one instant in time, called a *time step*. This data is stored and used as input to solving them all again one time step later. By repeatedly solving, storing and solving again, a numerical picture of the circuit's behavior is built up.

Computers have their limits. The equation variables have limited precision so that every value is rounded by a tiny amount. Time steps, while very short, are still finite, leading to problems if their duration becomes significant with respect to the frequency at which the simulation is performed. The circuit models used to describe the components themselves approximate how a real world component actually behaves. These small cracks are how the trouble dragons get in.

If your design involves very high or very low power, very low noise or very high frequency signals, then you should be using a simulator designed specifically for that type of application. For example, the *Handbook's* CAD chapter uses examples generated by the free SV2 student-version of *Ansoft Designer*.[2] That software package is designed for use at RF and has a number of special tools for use in RF systems.

The Simulation Cycle

There is no smell of burning resistor or overheated transistor in a simulation. The placement of components on the screen has no effect on the behavior of the circuit, so a high-gain stage whose input is too close to its output will never break into oscillation. The dc power sources are free of ripple and noise. These effects and many more can only be experienced (and remedies learned) by building real circuits.

Figure 1 shows the process by which you really, really learn circuit design — from concept to finished project. The first step is to select a type of circuit and describe what it is supposed to do — the *performance requirements*. For example, an amplifier will need to achieve some level of gain over some frequency range. You may need a certain input impedance and output impedance. Armed with that information, choose a circuit and come up with a preliminary set of component values by using pencil and paper or a computer design tool. This is your *design*.

Next, *simulate* the circuit's performance. If the result satisfies your performance requirements, you can move to the next step. If not, change the circuit in some way (or change your requirements) until you are satisfied.

Now *build* your design as a real world collection of components and verify that the circuit works. This is where the real fun begins as the effects of construction and actual component variation take effect. Are you done? Not yet!

To soak up every bit of design experience and know-how, go back and *compare* your actual measured performance to what the simulator predicted, particularly near the limits of the circuit's function. Look for design sensitivities by substituting different parts or values. If the circuit's behavior diverges from the simulator's predictions, now is the time to take a closer look. You may not be able to say exactly why differences are present, but you'll be aware they exist and that will map a bit more of the unknown coastline.

---

[1]Linear Technology, **www.linear.com**. *LTspiceIV* is the simulator portion of the switchmode power supply design package, *SwitcherCAD III*. It has been made available for use subject to the restrictions described in the *LTspice User's Guide*.

[2]Ansoft, **www.ansoft.com**. *Ansoft Designer SV2* is a student version of the full-featured *Ansoft Designer* simulator package for signal analysis and communications applications.

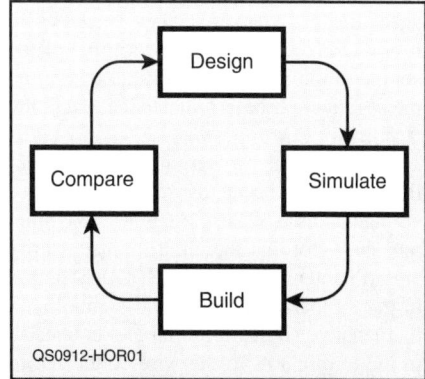

Figure 1 — Getting the most out of circuit simulation requires that you compare what the simulator predicts with how the actual circuit behaves.

Setting Up Your Simulator

Okay, enough philosophy! Start by browsing to Linear Technology's Web site, **www.linear.com/software**. Download the following three items: *LTspiceIV*, the *LTspice User's Guide*, and *LTspice Getting Started*. Register with Linear Technology in order to receive notices of new versions of the software and other related information.

To run *LTspiceIV* effectively, your PC will need to have at least 128 kB of RAM and at least 200 MB of free hard drive space available as simulations can generate a large amount of data. If the program runs out of space, you'll get an OUT OF MEMORY message. See the FAQ section of the *User's Guide* for information on system requirements. (*Linux* users can run *LTspiceIV* in the emulator software *WINE*.) The file containing *LTspiceIV* is a self-extracting .exe file. Double click on the file to begin the installation process. The process takes only a few moments and installs a shortcut to *LTspiceIV* on your desktop.

You immediately clicked on the shortcut and launched the program, didn't you? Admit it! Okay, so did I! What you need to do now is to open the PDF document *LTspice GettingStarted*. For general-purpose use, review pages 14 to 26 that show the simulator's basic operating tools.

Entering a Circuit

This month, we will just enter a very basic circuit to get used to the controls. Start a new schematic as shown on page 14. (The section "Schematic Capture" in the *LTspice Userguide* will provide additional information.) You should see the *LTspice* toolbar and status bar at the top of the screen, and window tabs at the top left. Under the VIEW menu, left click to turn on a field of guiding dots. (From here on, "click" means "left-click" unless stated otherwise.)

Now you'll create the two resistor voltage divider shown in Figure 2. Click the RESISTOR button in the toolbar and move the cursor into the schematic area. You'll see a black resistor symbol with two blank boxes next to it. Click once to create R1, then move the cursor and click again to create R2. (Creating R2 below R1 will make wiring the circuit a little easier.) Press the ESC (escape) key to turn off the resistor tool.

Move the cursor over R1 and the hand symbol appears. Right-click to open the PROPERTIES window for R1, enter 1000 in the RESISTANCE (Ω) window and click OK. R1 is now shown on the screen as having a value of 1000 (Ω assumed). Assign R2 a value of 2200 Ω in the same way.

Click the WIRE button in the toolbar and a set of crosshairs will appear. Move the center point of the crosshairs over one of the R1 terminals and click. Move the crosshairs to one of the R2 terminals and click again, then press the ESC key. A blue "wire" now connects R1 and R2.

Click the 'Component' button in the toolbar, scroll right to find the word VOLTAGE and click once to highlight it. The voltage source symbol will appear above the list of components. Click OK, then place the voltage source on the schematic by clicking once at the desired location and press the ESC key. Open the voltage source's PROPERTIES window and assign a value of 10 V for voltage and 0 for series resistance. Use the WIRE tool to connect the positive terminal of the source to the remaining terminal of R1 and the negative source terminal to the remaining terminal of R2.

Circuit simulators require that you identify a specific point to use as a reference voltage. This is what the GROUND symbol means — not that the circuit is necessarily at Earth potential. Click the GROUND button in the toolbar, place a ground symbol near the negative source terminal and connect it to the terminal or to the wire connected to the terminal, which forms a connection dot. You should now have a circuit that looks something like Figure 2.

Running a Simulation

Each of the connections between com-

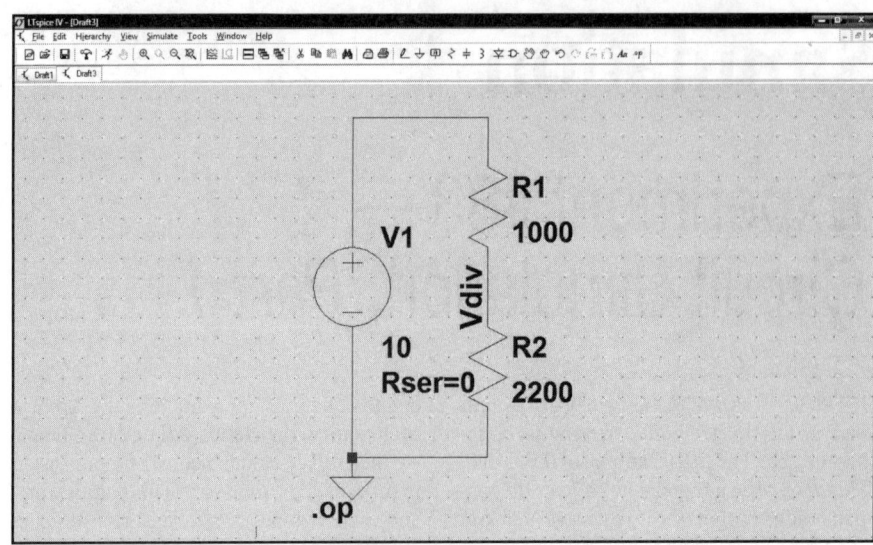

Figure 2 — A simple two resistor voltage divider. Each component is labeled with a designator and value. The ground symbol establishes a reference voltage point for the simulation. The simulator command to compute the circuit's dc operating point is .op.

Table 1
DC Operating Point

V(n001):	10	Voltage
V(Vdiv):	5.52	Voltage
I(R2):	0.003125	Device_Current
I(R1):	0.003125	Device_Current
I(V1):	–0.003125	Device_Current

ponents is called a *node*. The simulator calculates current through every circuit branch and voltage at every circuit node, assigning each node a label. To make it easier to read, the computer's output labels the nodes. In the divider circuit, label the connection between R1 and R2 as *Vdiv* by clicking on the LABEL NET toolbar button, entering Vdiv into the text window, click OK, placing the attachment point (the small box) on the "wire" between R1 and R2, then click once, followed by pressing the ESC key.

Under the SIMULATION menu, click RUN. In the EDIT SIMULATION window, select the DC .op PNT tab. Click OK and two things happen: the label .op appears on the schematic and a window appears with the results of the simulation in Table 1.

Here's what Table 1 is telling you, line by line: V(n001) is the source's 10 V output voltage. V(Vdiv): 5.52 is the voltage at the Vdiv label you placed and is equal to 10 × 2200 / (1000 + 2200) = 5.52 V. I(R2) is the 3.125 mA current through R2, which in a simple series circuit like this is also equal in magnitude to the current through the source, I(V1), and R1, I(R1). The current in the source is negative because it flows *out* of the positive terminal. This is the circuit's dc operating point — you've just run your first simulation! Under the FILE menu, click SAVE AS and save the schematic. You can now experiment to your heart's content!

Further Reading

This would be a good time to join the *LTspice* User's Group at **groups.yahoo.com/group/LTspice** where you'll find tutorials and other useful information. The CAD chapter of *The 2010 ARRL Handbook* will also start making sense.

Experiment #84 — Circuit Simulation, Build and Test

Last month, we downloaded and installed the *LTspice* circuit simulator program.[1] Now, let's move on to bigger and better things — a working transistor circuit.

Building a Common Emitter Amplifier

It seems fitting that the first real circuit to be simulated also be the subject of the very first Hands-On Radio column, "Experiment #1 — Common Emitter Amplifier." Download that article, take a good look at its Figure 1, then build the circuit in *LTspice*. Use the same techniques that we used last month to build the voltage divider. (See the "Further Reading" section below for additional tutorials and texts on using *LTspice* or its cousin, *PSPICE*.)

To create the transistor, click the toolbar control to the right of the diode symbol that looks like an AND gate — this is the component tool. A selection window will appear — click NPN and OK. Place the transistor symbol on the schematic by clicking once and then press ESC. Move the cursor over the transistor symbol so that it changes to a pointing hand

[1] All previous Hands-On Radio experiments are available to ARRL members as downloadable PDF files at www.arrl.org/hands-on-radio/.

and right click. The information window for the transistor will appear. Click PICK NEW TRANSISTOR and in the list of transistor types that appears, highlight the entry for a 2N2222 (at the top of the list) and click OK.

When you are done, the circuit should look something like Figure 1 of this article. Add the labels for IN, OUT, Vcc, Vc, Ve and Vb. Make sure the labels are attached to a symbol or wire. After a symbol is connected to other symbols, using the DRAG control allows you to move the symbol, but maintain all connections. MOVE also works on wires, allowing you to make the final product look like a finished schematic.

Have the simulator calculate the circuit's dc operating point (the ".op" text in the schematic) in the same way that you did for the voltage divider. Check it against Table 1. You should see voltages corresponding to the labels attached to the different circuit nodes. For example, V(vcc), V(vc), and so forth. If you see labels such as "V(n001)" then your label didn't get attached to a circuit node. Find the unattached labels by activating the MOVE tool and clicking on the label. The smaller of the boxes should be on a component connection point or on a wire. If not, move the label so that it is. Rerun the dc operating point until

all of the voltages are properly labeled.

Compare!

Remember that comparing results to requirements is a key part of using a simulator. Let's take a look at what we expected the circuit to do. From the original experiment, our design was supposed to have a quiescent or "Q" point at which I_{CQ} = 4 mA and V_{CEQ} = 5 V. How did we do? According to Table 1, Ic(Q1) = 3.65 mA, V(vc) = 6.52 V and V(ve) = 1.68 V. Thus, V_{CEQ} = V(vc) − V(ve) = 4.84 V. That's pretty close! Note that Ie(Q1) = Ic(Q1) + Ib(Q1), but it's negative. That's because Ie flows *out* of Q1 (an NPN transistor). By convention, currents that flow out of a component are negative. (Quick — what's the Q point power dissipation of Q1? $V_{CEQ} \cdot I_{CQ}$ = 17.7 mW)

Checking against the calculations by which the resistor values were determined, we expected an I_B of 27 μA and got the lower value of 17 μA. Why might this be so? Part of the reason is different resistor values (we used standard values instead of the exact values) and the 2N2222's current gain (beta or BF in the *SPICE* model string at the right of the selection menu) is 200 instead of 150. Still, these are good results: The transistor

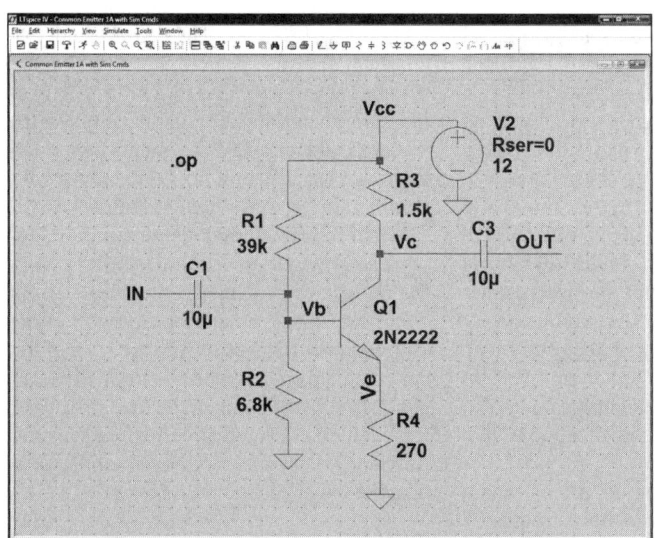

Figure 1 — The common-emitter circuit of Experiment #1, constructed in the simulator's schematic capture window.

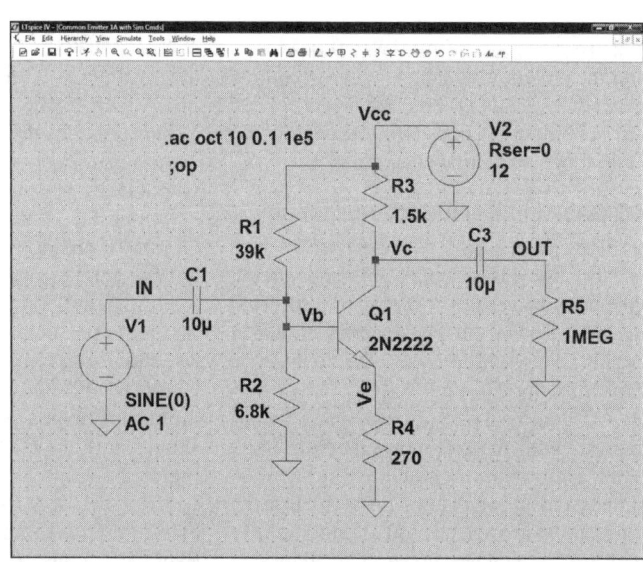

Figure 2 — Two simulations (.op and .ac) are added along with an input voltage source and an output load resistor.

Table 1

CE Amplifier Operating Point

Parameter	Value[2]	Data Type	Parameter	Value[2]	Data Type
V(vc):	6.52089	voltage	I(C2):	9.90998e–018	device_current
V(vb):	1.67962	voltage	I(C1):	1.6796e–017	device_current
V(ve):	0.990998	voltage	I(R5):	6.52089e–017	device_current
V(vcc):	12	voltage	I(R4):	0.00367036	device_current
V(in):	1.6796e–005	voltage	I(R3):	0.00365274	device_current
V(out):	6.52089e–014	voltage	I(R2):	0.000247002	device_current
Ic(Q1):	0.00365274	device_current	I(R1):	0.000264625	device_current
Ib(Q1):	1.76227e–005	device_current	I(V2):	–0.00391737	device_current
Ie(Q1):	–0.00367036	device_current			
I(C3):	–6.52089e–017	device_current			

[2]The simulator program uses double-precision variables when making calculations. These allow the simulation to handle up to 16 digit values with enormous exponents (10±307). This minimizes errors due to rounding or truncating. But we shouldn't really pay much attention to any of the digits beyond those that are considered to be significant digits. See **en.wikipedia.org/wiki/Significant_figures**.

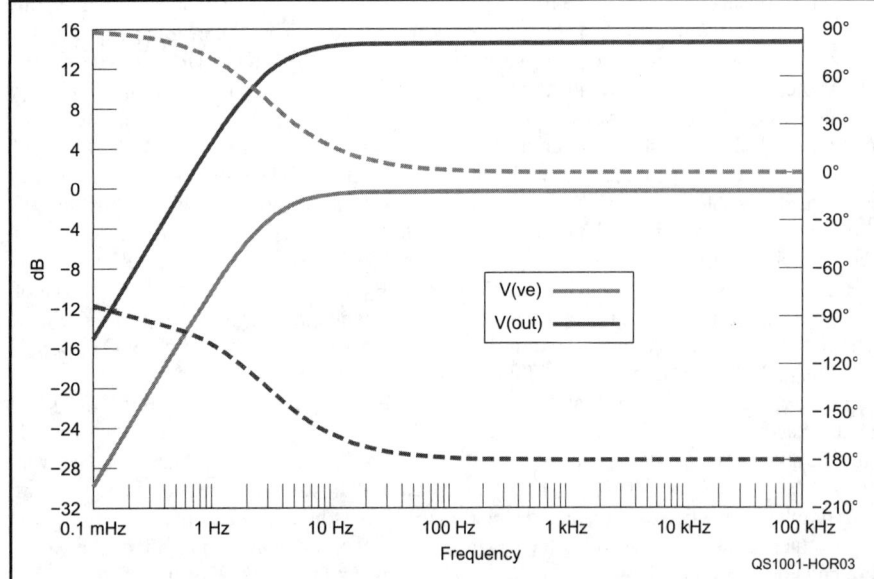

Figure 3 — The frequency response of the amplifier at the output node (red lines) and at the transistor emitter (blue lines). Magnitude response is shown by the solid line and phase response by the dashed line.

is active (V_{BEQ} = V(vb) – V(ve) = 0.69 V) and the collector voltage is almost centered between V_{CC} and ground. Always check the dc operating point of a circuit before moving on to more sophisticated simulations.

AC Gain and Frequency Response

The circuit's gain was expected to be 5 and, because this is a common emitter circuit, the output waveform is shifted by 180° from the input. To find out what gain our simulated circuit has, we need to use the ac analysis capability of *LTspice*.

AC analysis performs a *small signal* analysis with the circuit operating at its Q point. That means the simulation is based on signals that are small enough not to cause the transistor to approach cutoff or saturation. To perform this analysis, however, we'll need to give the circuit an input signal. Create another voltage source and connect its positive terminal to the input capacitor and the negative terminal to ground. Right click on the source, click ADVANCED, and then select SINE, set the dc offset to 0, and under SMALL SIGNAL AC (.AC) ANALYSIS set the AC AMPLITUDE to 1. This provides the ac analysis tool with an input signal. I also added a high value, 1 MΩ, output load resistor that we'll use later.

Set up the simulation by selecting EDIT SIMULATION CMD from the pull down SIMULATE menu. Select the AC ANALYSIS tab and select OCTAVE as the type of sweep, enter 10 points per octave, a start frequency of 0.1, and a stop frequency of 1e5 (100 kHz). Click OK and place the simulation text on the schematic at a convenient location. You should now have a circuit that looks like Figure 2.

Click the RUN tool on the toolbar and a blank graph will appear with a frequency span of 100 mHz (yes, that's millihertz) to 100 kHz and no vertical axis. Without clicking or any further keystrokes, move the cursor across the schematic and you'll notice that it changes to a red voltage probe when on a circuit node (wire) or a black current sensing probe when on a component. Carefully move the cursor to the wire labeled OUT and click once. *Voila!* Two response curves will appear on the graph, the left-hand vertical axis (gain) will be labeled –32 dB to 16 dB, and the right-hand axis (phase) –210° to +90°!

Move the cursor to the Ve node so it becomes a probe again and click. Two more response curves will appear and the top of your simulator screen will look like the graph in Figure 3. One pair of curves will be labeled V(out) — the solid line is the frequency response and the dotted line is the phase response — and the other pair as V(ve). (This assumes you haven't changed any of the default graphing settings. I used the TOOLS > COLOR PREFERENCES controls to change the plot background and graph controls.)

The *midband gain* of the circuit at V(out) flattens out above 10 Hz to about 14.5 dB, so A_{V-} = log^{-1} (14.5 / 20) = 5.31. Very close to the requirement and the phase of the output signal is shifted 180° as expected! Because of the coupling capacitors in series with the input and output circuits, the amplifier has a *high-pass response*. The cutoff frequency at which gain has been reduced by 3 dB is about 3 Hz and the phase shift is 45° less than in the midband, too. The gain and phase for V(ve) should remind you of another common transistor amplifier circuit — the unity-gain common-collector or voltage follower!

With the circuit and simulation running, you can now start to experiment after saving the basic circuit. Change the value of the coupling capacitors and observe the effect on frequency response. Alter the ratio of the collector and emitter resistors to get different gains. Lower the value of the load resistor to less than 10 kΩ and that will affect gain and frequency response, too. Next month, we'll take the experiment a little further!

Further Reading

Tutorials come in mighty handy for beginners. Here's one:

denethor.wlu.ca/LTspice is a simple overview of commands and controls.

Experiment #85 — Circuit Simulation, Complex Parts

After our first two simulation sessions, you should feel comfortable creating circuits made up of simple components. You also know how to find their dc operating points and determine their frequency responses. That covers quite a bit of territory. For readers who had never tried circuit simulation before, I hope this has helped you get up and running. There is a lot more, of course, and this month we'll learn how to use complex circuits as a single component.

First, you'll encounter the word *spriting* or *sprited* in the manual. Used in reference to the graphic selection of a component, spriting a symbol means to treat it and its connection points into a single graphic element that can then be moved around the schematic, rotated, mirror imaged, copied and so forth. The use of sprite in this manner began with early computer graphics chips.

Using Hierarchy

As you build larger and larger circuits, you'll find yourself using similar constructions over and over: a common-emitter amplifier, a full wave diode rectifier or a low-pass filter. Creating them anew with every use would be a waste of time. It's also better in many cases to treat the circuit as a component, in the manner of an IC. This is accomplished by *hierarchy*. You'll encounter hierarchy frequently in the *LTSpice* literature on circuit simulation and schematic capture. Block diagrams of equipment, such as in your license exams, are an example of the use of hierarchy to hide unnecessary complexity.

You also know about hierarchy if you've ever drawn an op-amp circuit. The op-amp is a complex circuit built from dozens of passive and active components. Obviously, we don't draw all those internal components every time we use an op-amp. The familiar triangle symbol with two inputs, an output and two power connections suffices for most op-amps. A schematic with an op-amp symbol has a two level hierarchy; the top level contains the op-amp symbol and the rest of the discrete components that make up the circuit. The bottom level contains the discrete components inside the op-amp.

LTSpice symbols use hierarchy to represent a discrete component, a sub-circuit or even another page of the schematic. In this way, it's easy to build up fairly extensive circuits without having all of the components crammed onto a single unwieldy schematic. This helps your schematic remain clear and uncluttered — valuable attributes when you or someone else tries to read them later!

We'll use the common emitter circuit from last month's session to show how hierarchy works and how complex circuits are turned into symbols.[1] The *LTSpice* manual shows an example of a very similar circuit being used as a preamp in a top-level schematic. (See page 35 of the manual in the section "Rules Of Hierarchy.") Let's play along.

Creating An Amplifier Symbol

Start by opening the schematic of the common emitter amplifier and immediately save it as a new schematic with the filename CE_AMP. This will become the circuit inside the symbol we are about to create with the same name. (Symbols and their associated schematics may not contain spaces or other characters not allowed in filenames.) Use Figure 1 as a guide for the following instructions.

Use the CUT tool (scissors symbol) to remove both voltage sources, the simulation command text, the output load resistors and all ground symbols. Reconnect the grounded ends of the bottom bias resistor and the emitter resistor. Reconnect the ends of the top bias resistor and the collector resistor. Attach a short wire to each of these reconnections and to the unconnected end of the input and output capacitors for the next step.

Use the LABEL NET tool (box symbol with ABC inside) to create ports to the circuit. Ports are the connecting points for the symbol of the amplifier. With the LABEL NET tool active, click on each unconnected wire end to create a port as follows:

- Positive power supply, labeled V+, bidirectional port
- Ground, labeled V–, bidirectional port

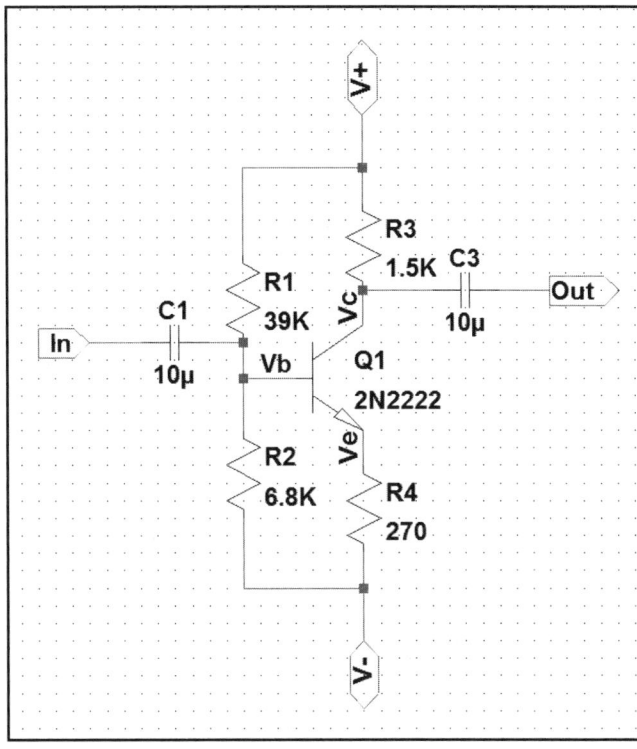

Figure 1 — The common emitter amplifier circuit from Experiment #83 has been stripped down to its basic circuit as CE_Amp. The ports IN, OUT, V+, and V– connect to the pins of the circuit's symbol and from there to other circuitry.

[1]All previous Hands-On Radio experiments are available to ARRL members as downloadable PDF files at **www.arrl.org/hands-on-radio/**.

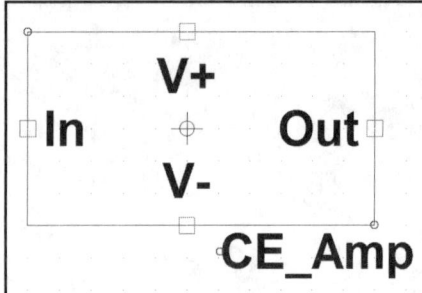

Figure 2 — The new common emitter amplifier symbol, CE_AMP. Because it has the same name as the schematic, the simulator will connect the ports of the schematic with the pins on the symbol.

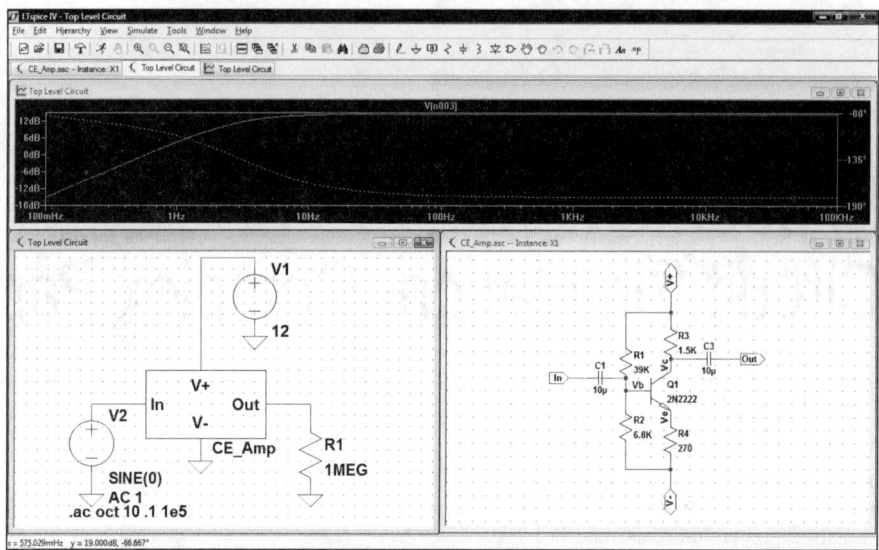

Figure 3 — The frequency response simulation (top) can be run on either the top level schematic that shows the amplifier as a single symbol (lower left) or in the actual schematic (lower right).

- Input, labeled IN, input port
- Output, labeled OUT, output port

(Creating ports is described on page 20 of the User's Manual in the section "Label a node name.") Enter the port names exactly as they are spelled here because the symbol and port names must match exactly, or the simulator will not be able to "make the connection." Your circuit should now look like Figure 1. Save the schematic file and close the schematic window.

Now we'll create a symbol that can be used by other schematics to use our common-emitter amplifier. Start by using the FILE > NEW SYMBOL command to open a blank window. Save it as a symbol with the name CE_AMP in the same directory as the schematic CE_AMP. Remember that the names of the symbol and of its internal schematic must match exactly.

Use the DRAWING menu to create a rectangle as shown in Figure 2 and add the text label CE_AMP. Use the EDIT > ADD PIN/PORT tool to create four symbol pins labeled IN, OUT, V+, and V− as shown in the figure. Remember to spell the labels correctly. Save the symbol. You have now defined a new graphic symbol and the circuitry it contains. Close the SYMBOL window.

To use your symbol, create a new schematic called TOP LEVEL SCHEMATIC shown at the lower left in Figure 3. Click the COMPONENT tool (symbol shaped like an AND gate) and then click once where you want to place the new symbol. You'll probably see a list of components in the default *LTSpice* library, so at the top of the window, use the TOP LEVEL WINDOW pull down menu to navigate to the directory in which you stored the new symbol and its associated schematic. Select CE_AMP and the symbol will appear on the schematic.

Create the input and power supply voltage sources and connect them to the symbol. Define the parameters for the sources as in the previous sessions. Create and attach ground symbols and add a 1 MΩ resistor to the symbol's output pin. Set up the ac sweep by selecting SIMULATE > EDIT SIMULATION COMMAND, then selecting the AC SWEEP tab. Enter the same values as in the previous experiment. Save the schematic.

Right-click on the CE_AMP symbol and click OPEN SCHEMATIC. A new window will open with the CE_AMP schematic of Figure 1. Click the TILE WINDOWS tool so that you can see both circuits. Right-click inside the TOP LEVEL SCHEMATIC window so that it is the active window and select RUN. Move the cursor over the wire connecting the CE Amp output pin to the 1 MEG resistor: It will change to a red voltage probe. Click once and the results window will open with a frequency response that should be identical to the one for the circuit obtained in the previous session. Move and resize to see all three windows arranged something like Figure 3.

Close the results window, right click once inside the CE_AMP schematic window, select RUN, place the voltage probe on the wire connected to the OUT port and click. An identical set of results will appear demonstrating that the symbol and the circuit it contains are the same.

You can now create symbols containing any circuit you know how to build and define. For example, create a symbol for the voltage divider schematic drawn in the introduction to simulation. Add it to the top level schematic and connect it between the input voltage source and the amplifier IN pin as you would a variable resistor, creating an adjustable gain amplifier. You could also replace the CE Amp schematic's pair of bias resistors with a voltage divider symbol. A symbol can be re-used at any level of the hierarchy, but avoid using the symbol inside of the same symbol. After all, the familiar resistors and transistors are just symbols themselves.

At this point, you are ready to begin using op-amps, timers, voltage regulators and all manner of ICs in your schematics. You can create sub-circuits using ICs, as well. For example, the simple low-pass active filter of Experiment #4 can be made into a low-pass filter symbol and used in an audio circuit's block diagram. You can begin to see the power of using hierarchy in circuit simulation.

Since the focus of this column is not exclusively simulation, we'll return to circuit topics but include simulation as one of the instruction tools. For example, in an experiment that involves pulses, we'll use the waveform viewing facilities of *LTSpice* to create an oscilloscope like view of circuit behavior.

Further Reading

You should continue to explore symbol creation and hierarchy on your own. Read the *LTSpice User's Manual*, pages 24 to 36 for more information.

Next Month

Next month I'll introduce a different kind of circuit design software that specializes in filters. Written by Jim Tonne, W4END, *ELSIE* is a powerful filter design package that's part of the CD ROM software included with every copy of *The ARRL Handbook*.[2]

[2]*The ARRL Handbook for Radio Communications,* 2010 Edition. Available from your ARRL dealer or the ARRL Bookstore, ARRL order no. 1448 (Hardcover 1462). Telephone 860-594-0355, or toll-free in the US 888-277-5289; **www.arrl.org/shop**; **pubsales@arrl.org**.

Experiment #86 — Viewing Waveforms in *LTspice*

In rereading the initial *LTspice* columns, I realized that the ability to view waveforms was too important not to cover in a column of its own. As a result, the introduction to W4ENE's filter design software, *ELSIE*, will be delayed for a month.

Time Domain and Scope View

So far in this series of columns, we've performed two of the simulations available in the *LTspice* software: calculating the dc operating point (the .*op* simulation) and determining the frequency response (the .*ac* simulation). The dc operating point is a static calculation, by definition, and shouldn't change over time. The frequency response, also called an *ac sweep*, determines how the circuit behaves over a specific frequency range.

Both simulations are quite useful, but neither shows you how the circuit behaves over time — plotting a signal's amplitude against a calibrated time scale. That type of graphical output is popularly known as a *scope view*, as if an oscilloscope were connected to the circuit. More correctly, it's called the *time domain* or *waveform* view.

The convention for *SPICE* based circuit simulations is to refer to time domain simulations as *transient analysis*. The word *transient* refers to the reaction of the circuit to a temporary (or transient) change, not necessarily a sharp pulse also called a "transient." For example, the *step waveform* is often used as the changing input in which a signal instantly changes from one voltage to another voltage. The resulting *step response*, when graphed in the time domain, tells a lot about the circuit's ac characteristics. Another common transient response input signal is the *impulse function* and it looks a lot more like a transient, consisting of an infinitely narrow pulse. The circuit's output is then called the *impulse response*. *LTspice* can simulate all of these responses and more.

Most of the time, we're interested in seeing an oscilloscope-like view of the voltage or current at a particular spot in the circuit. To become familiar with using the transient analysis simulation, we'll revisit a familiar circuit — the 555 timer.

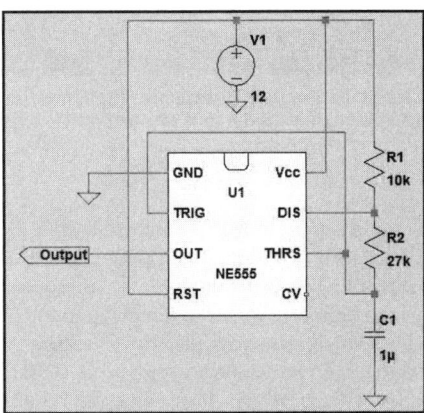

Figure 1 — The astable 555 timer circuit. Because this is a simulation, no bypass capacitor is required on the CONTROL VOLTAGE (CV) pin. In the real world, a bypass capacitor would be good practice.

Timer Circuit Review

Take a few minutes to download Experiment #5 and read about the 555 timer.[1] We're going to simulate the continuously oscillating *astable* version of the circuit. From the original experiment, we can follow the basic sequence of events. The timing capacitor, C1, charges toward V_{CC} through the combination of R1 and R2. As the capacitor voltage reaches ⅔ V_{CC}, the threshold comparator output turns on the discharge transistor (connected to the DIS pin) and the capacitor starts to discharge through R2. When the capacitor has been discharged to ⅓ V_{CC}, the trigger comparator turns off the discharge transistor and the cycle begins again. This creates a train of pulses at the output while C1 charges and discharges between ⅓ and ⅔ V_{CC}.

The total time it takes for one complete cycle is the charge time, T_C, plus the discharge time, T_D:

$T = T_C + T_D = [0.693 \times (R1 + R2) \times C1] + [0.693 \times R2 \times C1] = 0.693 \times (R1 + 2 \times R2) \times C1$

[1]All previous Hands-On Radio experiments are available to ARRL members as downloadable PDF files at **www.arrl.org/hands-on-radio/**.

and the output frequency is:

$f = 1/T = 1.443 / [(R1 + 2 \times R2) \times C1]$

The duty cycle of the output waveform is the ratio of time the output is ON (at V_{CC}) to the total period of the waveform, T:

duty cycle = T_C / T = (R1 + R2) / (R1 + 2R2)

Note that duty cycle is always greater than 50% because R1 is always greater than zero.

Setting Up the Timer Circuit

Build the circuit shown in Figure 1. It's a basic astable 555 timer circuit. You can find the 555 timer symbol by selecting MISC and then NE555 after clicking on the COMPONENT tool. The 555 symbol's pins are arranged differently than you might be accustomed to, so the circuit will look different when compared to Experiment #5. If you want an alternate arrangement for future use, you can edit the symbol. Assign values to all of the components and the voltage source as shown and create the OUTPUT port using the labeling tool as explained in the previous experiment.

You may have noticed a small change from a "real" circuit. The CONTROL VOLTAGE pin (CV) is left open circuited in our simulation. In an actual circuit, this would be bypassed to ground with a small value capacitor so that noise won't affect the switching thresholds. In our simulation world, though, there are no stray or unwanted signals and CV can be safely left unconnected.

Calculate the charge and discharge times (25.6 msec and 18.7 msec, respectively), the total period and frequency of the output waveform (44.3 msec and 22.6 Hz), and the duty cycle (57.8%).

Time Domain "Scope" Simulation

Now we are ready to turn on our simulated oscilloscope. Use SIMULATE > RUN and then select the TRANSIENT tab. The only required value to be entered is STOP TIME. A value of 0.5 seconds will allow the simulation to run for several waveform cycles. We will let *LTspice* use its default value for the time

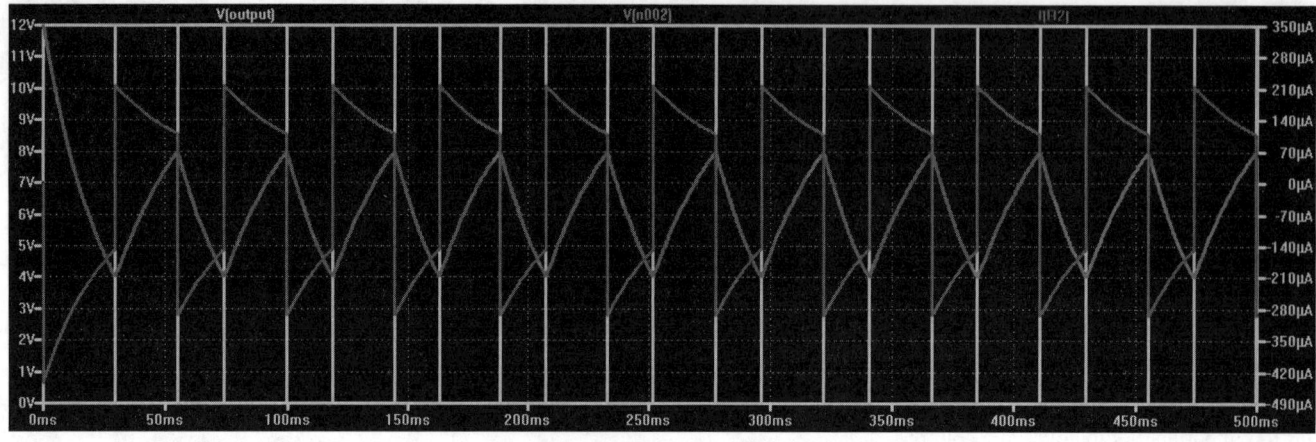

Figure 2 — Three waveforms in the 555 timer circuit. Green is the output voltage, magenta is the capacitor voltage, and red is the current through R2. Voltage is calibrated on the left Y axis and current on the right Y axis.

step. Click the OK button and the label .TRAN .5 will appear on the schematic. (In the parlance of *SPICE*, this is referred to as a *dot tran* command.) A blank waveform or *trace* window will appear above the schematic waiting for you to identify which voltage or current to plot.

Move the cursor to the OUTPUT port symbol so that it changes to a red voltage probe then click. The trace window should fill up with a pulse train swinging between 0 and 12 V. (If not, check your schematic wiring.) The label V(OUTPUT) will appear at the top of the window in the same color as the trace. Now move the cursor to the junction of R2 and C1 and click again. The charge-discharge waveform of the capacitor voltage will appear as a second trace labeled V(N002). Let's get fancy and move the cursor over R2, where it will change to a current probe, and click again. A third trace labeled I(R2) will appear along with a numeric scale at the right calibrated in µA. Using the capacitor voltage trace as a clue, which way is the current flowing in R2 when *LTspice* assigns it a positive value? (Top to bottom, causing C1 to charge.) You should now see a window with three traces as in Figure 2.

Customizing Your Time Domain View

If you right click anywhere in the trace window that is not on a trace or axis, a menu will appear in which you can click GRID to turn the X-Y gridlines on and off. My Figure 2 will look a little different than your trace window because you are probably using the default colors and trace widths. To change the color of a trace, move the cursor to the trace's label [such as V(N002)] and right click. The EXPRESSION EDITOR window will appear with a default color selection menu that you can change. To increase the trace thicknesses so that they are easier to see, select TOOLS > CONTROL PANEL. Then click the WAVEFORMS tab and click the box for PLOT DATA WITH THICK LINES.

Just as in a real 'scope, you can also change the scale of the axes. Move the cursor to just below the X axis so that it changes into a ruler. Click and a window will appear showing axis values for the minimum (LEFT) and maximum (RIGHT) times and the divisions (TICK). You can change the values to shrink or expand the time scale. The same controls are available for both Y axes on the left and right. Feel free to experiment with the scale settings.

Voltage and Current Measurement

If you've used a modern, digital scope, you know that one of its most handy features is the ability to use on-screen cursors to measure voltage and time. No more squinting and moving traces to use the graticule for imprecise measurements. The same feature exists in *LTspice*.

Move the cursor to the label for capacitor voltage, V(N002), and right click. In the EXPRESSION EDITOR window, from the ATTACHED CURSOR menu select 1ST. A pair of lines, one vertical and one horizontal, will appear in the trace window along with a small window showing the value of the trace where the lines cross. Move the cursor to the intersection of the lines and a yellow 1 will appear. Click and hold the left mouse button then move the mouse left and right. The intersecting lines will follow the capacitor voltage trace and the value window will show voltage and time. Verify that the capacitor charges and discharges between $\frac{1}{3}$ and $\frac{2}{3}$ V_{CC} and that the charging and total periods are as predicted above. Can you explain why the first output cycle of the 555 is longer than the rest? (The capacitor has to be discharged starting from 12 V, not 2/3 V_{CC}.)

Further Reading

Continue to experiment with the .*tran* command and read about it and other *dot* commands in the *LTspice User's Manual*. See if you can figure out how to plot the sum and difference of voltage traces as described in the section "Waveform Arithmetic." Keep an eye on the Hands-On Radio Web page for additional *LTspice* hints and tips.

Next Month

Okay, okay — I promise that next month we'll try W4END's filter design software, *ELSIE*. Remember that it is included in the CD that comes with *The ARRL Handbook*.[2] If you haven't yet tried this software, you'll have a great opportunity next time.

[2]The ARRL Handbook for Radio Communications, 2010 Edition. Available from your ARRL dealer or the ARRL Bookstore, ARRL order no. 1448 (Hardcover 1462). Telephone 860-594-0355, or toll-free in the US 888-277-5289; **www.arrl.org/shop**; **pubsales@arrl.org**.

Experiment #87 — *ELSIE* Filter Design — *Part 1*

After four consecutive columns about the *LTspice* circuit simulator software, it's time to move on — but not totally away from the world of virtual circuits. Your experience with *LTspice* has given you the opportunity to rapidly test circuit performance, but a simulator will not design the circuit for you. You are still responsible for selecting a circuit and determining a set of values for every component. That's where another class of software called *design tools* comes in.

A design tool helps you select the type of circuit and develop a set of values for the individual components. That circuit can then be simulated with a program such as *LTspice* and the result turned into a working prototype. If the design software is properly used, the prototype's behavior on the workbench will correspond closely to its behavior on the computer screen. That's where this month's column begins. As promised, we'll explore one of the terrific design tools available to hams — *ELSIE*, by Jim Tonne, W4ENE. *ELSIE* is a design tool for passive LC filters, thus the name.

The student version of *ELSIE* is included in recent editions of *The ARRL Handbook*.[1] The 2010 edition also includes seven other W4ENE design tools for diplexers, Pi-L networks, Class E amplifiers, meter faces and more. Assuming you have your *Handbook* CD ready, let's get started!

Installing and Running *ELSIE*

Load the *Handbook* CD ROM on your computer. (I assume the reader is using a *Windows* based computer.) Your computer may be configured to *Autorun* the installation software to install the complete *Handbook* document set onto your computer. If you want to install the complete set of documents, proceed as directed by the installer software. If the installation process did not begin automatically, run *Setup* in the top level or root directory of the CD ROM.

To install the *ELSIE* software, open MY COMPUTER or run *Windows Explorer*. Open the README file in the top level CD ROM folder and make sure your computer is suitable for installing the software. Open the folder PROGRAM FILES then ARRL 2010 HANDBOOK then COMPANION SOFTWARE and finally TONNE SOFTWARE (or the appropriate folders on the CD ROM for your edition).

Run *LCinstall* — the program name concludes with a number representing the program version; version 2.32 is shipped with the 2010 edition and so the program name is *LCinstall232*. Follow the program's installation directions: I recommend using the default folder names.

To run *ELSIE*, use the START button, select ALL PROGRAMS and navigate to the folder ELSIE. In that folder double click the ELSIE icon and off you go! The opening screen will display the version and note that you have the student version of the program. Take a look at the schematics on the screen and note all the different configurations of inductors, capacitors, transmission lines and even numeric values of reactance. All of these can be used to design filters in the *ELSIE* software.

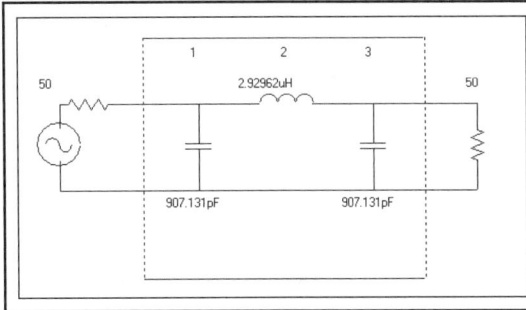

Figure 1 — This is filter design Demo1 from the design examples provided with the *ELSIE* filter design program. The program is included at no extra charge with recent editions of *The ARRL Handbook*.

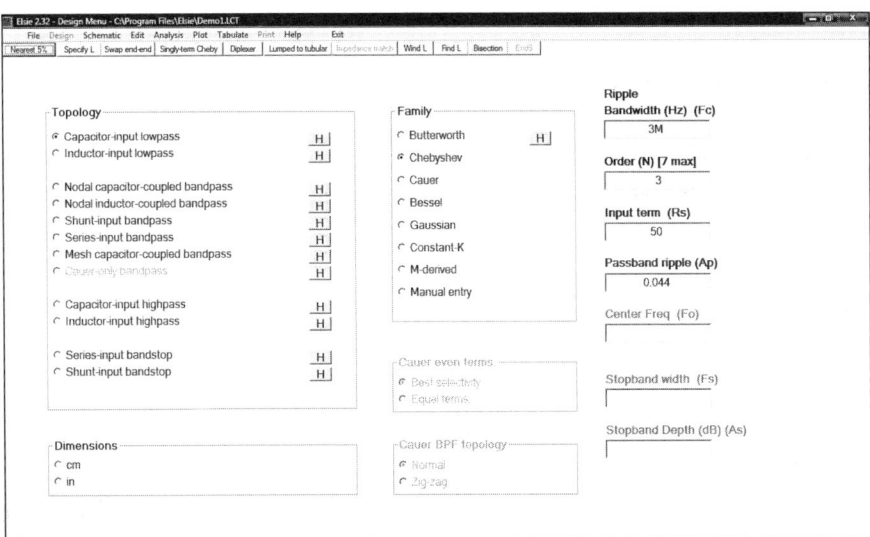

Figure 2 — The DESIGN window is used to select a filter's design parameters, such as the filter topology, the family of design equations that will be used and several performance requirements.

---

[1]*The ARRL Handbook for Radio Communications,* 2010 Edition. Available from your ARRL dealer or the ARRL Bookstore, ARRL order no. 1448 (Hardcover 1462). Telephone 860-594-0355, or toll-free in the US 888-277-5289; www.arrl.org/shop; pubsales@arrl.org.

Conventions and Controls

ELSIE uses the same number entry conventions as *LTspice*. For very large and very small numbers, use scientific notation, such as 1E6 for 1,000,000 or 1E–11 for ten pico somethings. Upper or lower case k represents kilo. Capital M represents mega and lower case m represents milli — don't get them mixed up!

Frequency values are always entered in hertz — use scientific notation or prefixes such as k, M or G. Capacitance values are always in farads, so use scientific notation or prefixes such as u for micro (1u is 1 µF), n for nano and p for pico. Similarly, inductor values are always in henrys; m for millihenrys, u for µH and n for nH. Resistance values are always in ohms and the same prefix conventions are used. Ripple and attenuation values are in dB.

ELSIE has seven different windows that display information or collect design information to operate the program. The different window choices are listed along the top of the screen: Design, Schematic, Edit, Analysis, Plot, Tabulate and Help. File is a menu for saving or loading a filter design or for exporting a *netlist* or schematic of the filter in a format that can be read by SPICE simulators (including *LTspice*). A netlist is a list of connections between the filter components and can be read by many simulation and circuit-board layout programs.

In addition, *ELSIE* includes an extensive "walk-through" under the HELP menu. Click on HELP, make sure the CONTENTS tab is selected, then click WALKTHROUGH. Once you're beginning to use *ELSIE*, the information in each section of Help will explain the design process and how to interpret the program's results.

Driving Miss *ELSIE*

Click the OPEN OLD button and from the list of designs (the software includes dozens of sample designs) select *Demo1.LCT*. The software will load the filter design named Demo1 and you'll see the screen shown in Figure 1. This is the program's *schematic view*. Filter input is on the left and the output is on the right. The section of the schematic in green is the part that *ELSIE* designs.

At the bottom left of the schematic are three lines of text — these describe the filter's design parameters. The bandwidth of the filter is 3 MHz, the filter's design is from the Chebyshev family and the maximum ripple in the filter's passband is 0.044 dB.

Click on the DESIGN menu label (next to FILE at the top of the screen) and the screen in Figure 2 will appear. The buttons labeled H indicate that "instant help" is available to explain that function or selection. Click on them and see! This screen is used to define the requirements for a filter and here you can see more information about filter Demo1. In addition to the information on the schematic view the *topology* is "Capacitor-Input lowpass." A filter's topology describes how the components are arranged. Since the filter is a low-pass type from the Chebyshev family, the Ripple Bandwidth (Fc) means that the filter's transmission will be no lower than twice the maximum ripple (–0.088 dB) at 3 MHz. The order of the filter is 3, meaning that it consists of three individual energy-storing reactances. The filter's input termination impedance has been set to 50 Ω. Select the TABULATE window and browse through the list of values to determine the frequency at which the response is closest to –0.088 dB (between 2.9935 and 3.0025 MHz).

Now open the PLOT window using the list of windows across the top of the screen. This window can display many different types of useful graphs. The default is the TRANSMISSION PLOT showing filter response versus frequency. (Click the top plot selection button at the upper left of the screen if some other graph is displayed.) Response in dB is shown on the Y axis with larger negative values meaning more attenuation of the input signal and frequency is on the X axis.

Switch between the different types of plots by clicking on the different plot selection buttons. Holding the cursor over the button without clicking (called "mousing over" a control) will display a label describing what data the plot contains. For example, the second button (TRANSMISSION AND RETURN) adds return loss to the graph. The third button (TRANSMISSION AND ITS ANGLE) shows the familiar Bode style plot of both amplitude and phase response as seen in Figure 3. The graph scales are determined automatically by the program, although some graph parameters can be adjusted by selecting the Analysis window.

Editing and Adjusting

You can adjust the component values, too. You can do this manually or use some of the program's design options. For example, return to the SCHEMATIC window — the component values of 907.131 pF and 2.92962 µH aren't exactly common in vendor catalogs! *ELSIE* allows you to change the values to the nearest standard value in the 5% series for capacitors and inductors.

Select the DESIGN window and click the 5% VALUES button underneath the FILE menu label at the upper left-hand part of the screen. The NEAREST VALUES ROUTINE window will open showing the closest standard values; in this case 910 pF and 3 µH. Select the TRANSFER ALL NEAREST-VALUE PARTS... option then click END NEAREST-VALUE ROUTINE. We could just as easily elect to only modify the capacitors or inductors. If only one type of component is changed to a standard value, the program can also recalculate ("retune") the remaining components so that the filter meets the original design requirements.

Return to the SCHEMATIC window and confirm that the parts values are now standardized. What effect has this had on filter performance? Checking the PLOT window doesn't show a lot of change, but if you look for Fc in the TABULATE window, you'll see that Fc is now between 2.9226 and 2.9314 MHz — a shift of about 70 kHz or 2.3%. Feel free to experiment with the design, changing the component values and getting comfortable with some of the controls.

Further Reading

Filter design involves knowing some terminology. You can catch up on most of *ELSIE*'s terms by doing a little reading. Hands-On Radio experiments #50 and #51 on Filter Design define a lot of the terms and processes used to design filters.[2]

Next Month

Let's build a filter or two next month, shall we? The goal will be to design something useful for your shack!

[2]All previous Hands-On Radio experiments are available to ARRL members as downloadable PDF files at **www.arrl.org/hands-on-radio-experiments-resources-and-faq**.

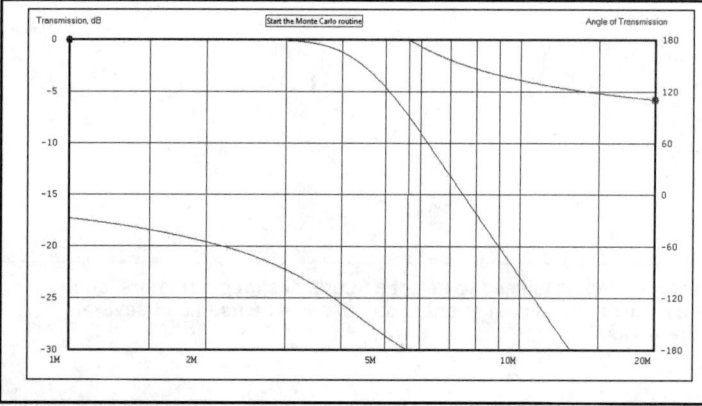

Figure 3 — The filter's response can be displayed in a number of different ways, from the familiar transmission and phase response plot shown here to a Smith chart view of return loss.

Experiment #88 — ELSIE Filter Design — Part 2

HF operators who live near an AM broadcast transmitting facility often find the strong signals overload the input amplifiers and filters of a receiver, particularly on the lower HF amateur bands. Let's use *ELSIE* to design a *broadcast reject filter* to reduce the amplitude of those strong signals.

Specifying Filters

In order to use a filter design tool such as ELSIE, however, we need to speak the language of the filter designer. Figure 1 shows some of the terms used to describe a high-pass filter response. The graph is of *transmission* (the ratio of output to input signal strength in dB) versus frequency. For a passive filter, the maximum value of transmission is 0 dB. As less of the input signal is transmitted to the output, the value of transmission, in dB, becomes increasingly negative. Negative transmission is the same as attenuation.

On the frequency axis, F_C is the cutoff frequency at which the output signal power is ½ that of the input signal. F_S defines the *stopband width*. The passband for a high-pass filter consists of all frequencies above F_C. The stopband for a high-pass filter consists of all frequencies below FS. The *transition region* is the range of frequencies between the stopband and passband. (For a low-pass filter, FS and FC are reversed, along with the gray boxes they bound.)

The colored boxes leave a space between them through which the filter's amplitude response curve must pass. The borders of the colored boxes establish the required performance for the filter. The colored box defining the stopband is bounded on the bottom by the stopband depth. The filter's transmission in the stopband must be equal to or below the stopband depth. The colored box defining the passband is bounded on the top by the *passband ripple*. The filter's transmission in the passband must be equal to or greater than the passband ripple. (Why it's called ripple and not passband depth will be explained shortly.) Between FS and FC is the transition region in which the response curve passes between the stopband and passband. Any response curve that passes through the space between the colored boxes meets the performance requirements for the filter. (There are many other filter parameters beside amplitude response, but that's all we're going to consider now.)

What's passband ripple? Why is it called ripple? The filter response curve in Figure 1 is nice and smooth, typical of the common Butterworth family of filters often referred to as *maximally flat passband*. In the passband of other filter responses, such as the Chebyshev and Cauer, the response curve varies up and down before finally entering the transition region below F_C. This variation is referred to as ripple and it occurs in both the stopband and the passband. Allowing ripple results in a steeper response in the transition region. Speaking graphically, if we move the bottom of the stopband gray box up (reduce stopband depth or increase stopband ripple) and the top of the passband gray box down (increase passband ripple), the gray boxes can be moved closer together for steeper rolloff.

Broadcast Reject Filter

We need to attenuate all signals below the lowest amateur band, 160 meters with a low end of 1800 kHz, which will be our cut off frequency, F_C. The AM broadcast band covers from 550 kHz to 1700 kHz, so our stopband width, F_S, is 1700 kHz.

How much attenuation do we need? Generally speaking, the higher the change in response with frequency, the higher the required order of the filter and the more inductors and capacitors that are needed. In order to keep the filter simple, we should specify the smallest amount of stopband depth that allows our receiver to act properly. Unless you live very close to an AM antenna, a stopband depth of 20 dB should be sufficient.

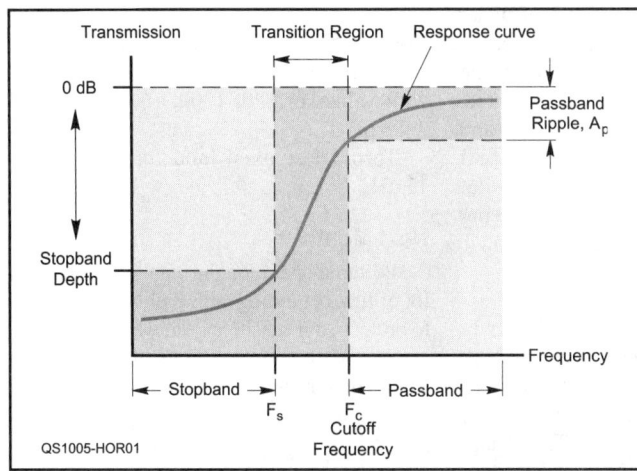

Figure 1 — The key specifications for a filter's amplitude response can be sketched on a transmission vs frequency graph. As long as the filter response curve stays between the boxes, it meets the design specifications.

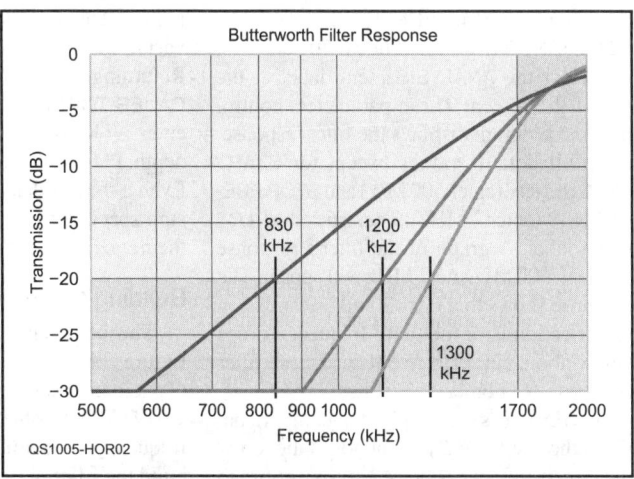

Figure 2 — Three Butterworth filter responses for third order (red curve), fifth order (blue curve) and seventh order (green curve). None provide the required 20 dB of attenuation across the AM broadcast band.

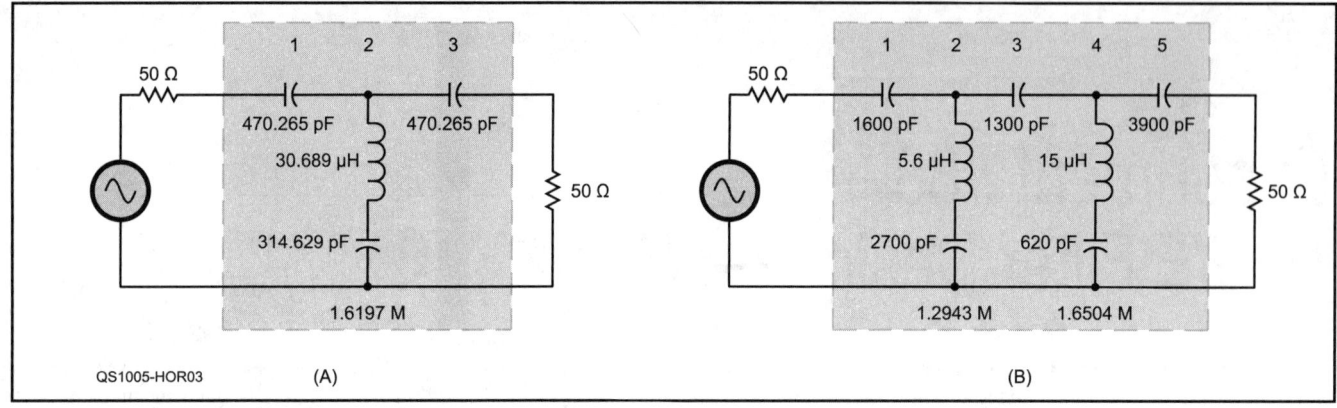

Figure 3 — At (A), the third order Cauer filter meets the specifications but not if the values are replaced with the nearest 5% series standard value. At (B), the fifth order filter using standard 5% series component values meets the performance requirements.

Run *ELSIE*, clicking NEW DESIGN on the opening screen. You're now in the DESIGN screen. Start on the left with TOPOLOGY, which refers to the arrangement of components. There are two choices for highpass filters, CAPACITOR INPUT and INDUCTOR INPUT. Select CAPACITOR INPUT, then click the H button to the right. A screen will pop up showing the filter's component arrangement or topology. Click RETURN, then select INDUCTOR INPUT and click H. Compare the two topologies. You'll note the difference in the blue response curves and also that the capacitor input topology uses fewer inductors. Since inductors are generally lossier and more expensive than capacitors, we'll use the capacitor input topology, so leave this selection checked.

Comparing Families

Next, select the BUTTERWORTH family and click H to compare the responses. You can see the ripple in the Chebyshev and Cauer families. Click RETURN TO DESIGN and enter values for FC (1800 kHz), ORDER (let's begin with 3), and INPUT TERMINATION RESISTANCE (R_S should be 50 Ω).

Select the ANALYSIS menu label at the top of the screen. These parameters control how the program displays the filter response. Keep all default values except for START FREQUENCY (enter 500 kHz) and STOP FREQUENCY (enter 2 MHz). Then click the PLOT menu label. A graph of the filter's response between 500 kHz and 2 MHz will appear. The response shows that our –20 dB transmission specification isn't met until frequency drops below about 830 kHz. We'll need more filter sections — a higher order filter.

ELSIE can save graphs for comparison. Click the SAVE OVERLAY button at the top of the screen and then SAVE AS 1. Now return to the DESIGN screen and increase the order to five. Click PLOT again — the filter response is still not good enough, but the –20 dB frequency has increased to about 1200 kHz. Save the graph as OVERLAY 2, then change the ORDER to 7 and plot again. Click GET OVERLAY and then GET 1 and GET 2 so that you see a graph similar to Figure 2. While all three filters have less than 3 dB of attenuation at 1.8 MHz, even though we've increased the order to seven (click SCHEMATIC to see the filter components) we're still not close to attenuating all of the AM broadcast band signals by 20 dB.

Return to the DESIGN screen and change the filter family to CHEBYSHEV. The passband ripple (AP) window will change to black as this is a design input for this family. With an order of three and a value of 1 dB for AP, plot the filter response and find FS for this filter (close to 900 kHz). It's an improvement, but we're not there yet! Increase the order to five and then to seven, saving the overlays for comparison as before. The seventh order Chebyshev comes pretty close, with an FS between 1.6 and 1.7 MHz, but is there a better filter?

The Cauer filter design increases filter complexity by substituting tuned circuits for some of the components. The tuned circuit creates a very steep response in the transition region. This comes at the expense of wider variations in attenuation in the stopband. Returning to the DESIGN screen, select the CAUER TOPOLOGY with an order of three, enter 1.7 MHz for FS and 20 dB for stopband depth. Plot the response — what a difference! Even a third order Cauer filter meets our design specification. (Verify this by reviewing the numeric data in the TABULATE window.)

Building the Filter

The schematic for the filter shown in Figure 3A has reasonable values for both capacitance and inductance. (The number 1.6197 M below the schematic is the resonant frequency of the tuned circuit.) To actually build the filter, we should try to use standard values.[1] Returning to the DESIGN screen, click NEAREST 5% and then TRANSFER ALL NEAREST-VALUE PARTS TO PARTS LIST. After clicking END NEAREST-VALUE ROUTINE, plot the response again. You'll see that the filter no longer meets the stopband depth requirement by a dB or so. Changing the filter component values deoptimized the design. Increase order to five in the DESIGN screen and repeat the process of changing components to the nearest 5% value as shown in Figure 3B. Plot the response and you'll see that it easily meets the filter specifications.

You can build this filter by tack soldering the components onto a scrap of printed circuit board as shown in Figure 2 of Experiment #44.[2] The 5.6 and 15 μH inductors can be wound as toroids (see Experiment #47) or purchased as axial inductors. Use plastic film or silvered mica capacitors. Connections into and out of the filter can be made with short pieces of RG-58 or RG-174 and a metal candy box will serve nicely as an enclosure. Test the performance of your filter by tuning in AM stations on your receiver and recording the signal level and frequency. Then insert the filter in the antenna feed line (don't transmit through it!) and remeasure the signal levels. The S-meter reading should be at least 3 S units lower with the filter in-line.

Parts List

Plastic film or silvered mica capacitors with values of 620, 1300, 1600, 2700 and 3900 pF.

Toroidal or axial inductors of 5.6 and 15 μH.

Next Month

Portable operation during the summer is a lot of fun, beginning with Field Day. In order to protect your sensitive gear, we'll take a look at a pair of over-voltage protection circuits for ac and dc power.

---

[1] A good discussion of component standard values is online at **www.ebyte.it/library/educards/ee/EE_StandardValues.html**.

[2] All previous Hands-On Radio experiments are available to ARRL members as downloadable PDF files at **www.arrl.org/hands-on-radio**.

Experiment 107
PCB Layout — *Part 1*

We've done quite a few simple tasks with the *LTspice* circuit simulation package and I hope many of my readers have added it to their ham radio toolset.[1] This month we are going to embark on a multicolumn adventure of learning to use the next set of computer based electronic tools, the schematic editor and PC board layout program. The result will be a practical gadget for the mobile or portable operator!

Ready, Fire, Aim!

Not so fast — don't fire up the web browser just yet! If you'd like an overview of the schematic-entry and PCB CAD (printed circuit board/computer aided design) process, check out the Computer-Aided Circuit Design chapter of *The ARRL Handbook* in which Dale Grover, KD8KYZ, lays out a comprehensive overview of the use of PCB CAD tools.[2] This particular set of columns will teach with a specific example. We'll take a circuit from concept, through simulation, all the way to the schematic editor and PCB layout software. Ordering circuit boards and then populating them with components will result in a working model of our "product." Ready? Okay, let's begin.

The Product Definition

The product will be a means of detecting when battery voltage drops below some predetermined level, disconnecting the load and preventing running down the battery any further. Those form our three *functional requirements* for the product. Let's go further and come up with electrical *performance specifications* for those requirements.

Most automotive batteries are considered discharged when terminal voltage falls to 10.5 V at some level of current. We'd like our device to disconnect the load before we reach that level since it would be nice to be able to start the car. Since not all of us have the same comfort level about discharging our battery nor do we want to use expensive precision components, we'll make the *setpoint* for the circuit adjustable. Let's design for a range of 10.5 to 11.5 V.

Another characteristic of batteries is that their terminal voltage changes with load current due to internal resistance: the higher the load current, the lower the terminal voltage. If we simply sense when battery voltage falls below a certain threshold, that might happen on a single transmitted code element or spoken syllable. The battery would still have plenty of charge, even though its voltage might drop for short periods. To prevent prematurely disconnecting the load, the circuit should wait until battery voltage falls below the selected threshold for some minimum amount of time. Because of the wide variety of batteries and loads, let's take a relaxed approach and allow 30 seconds before disconnecting the load.

If the circuit continues to draw current, it might discharge the battery on its own! We want the circuit to operate only when activated by the operator, and once it disconnects the load we want the circuit to de-activate and stay that way until the operator turns it back on. This is called a *latching* function — it turns off and stays off!

Voltage Sensing

Sensing voltage can be performed by a *comparator* circuit as described in Hands-On Radio Experiment 11. A simple voltage-sensing comparator circuit is shown in Figure 1 after you enter it into *LTspice*.

We'll use the LT1841 dual, general-purpose comparator for this circuit. With LTspice running, you can find that part by clicking on the COMPONENT symbol in the tool bar. Scroll to the left in the parts list and you'll see [COMPARATORS]. Double click this label to access the list of comparators provided with *LTspice* and select the LT1841.

With the schematic created, be sure to add labels to the various connections or "nets" as shown by right clicking the wire line, select LABEL NET, enter the label in the provided window, then press ENTER. The text label box will appear and can be assigned to a wire or net by positioning the connection dot of the box on the wire and clicking once.

Battery voltage is represented by the voltage source, V1, which is set to 12 V. A voltage reference for the comparator, V_{REF}, is established by a 6.2 V Zener diode, D1. Select ZENER from *LTspice*'s list of components, place it on the schematic, then right click and select PICK NEW DIODE to find the exact part in the figure. (We're going to change this diode to a different part number when we design the final circuit, but this part was in the *LTspice* component library.) R3 limits the current through D1 to (12 – 6.2 V) / 10 kΩ = 0.58 mA. C2 acts as a noise and RFI filter.

R1 and R2 divide the battery voltage with C1 acting as a noise and RFI filter. (Note

[1]*LTspice* is discussed in Hands-On Radio experiments 83-86 and 105-106. All previous experiments are available to ARRL members at **www.arrl.org/hands-on-radio**.
[2]*The ARRL Handbook for Radio Communications,* 2012 Edition. Available from your ARRL dealer or the ARRL Bookstore, ARRL order no. 6672 (Hardcover 6634). Telephone 860-594-0355, or toll-free in the US 888-277-5289; **www.arrl.org/shop**; pubsales@arrl.org.

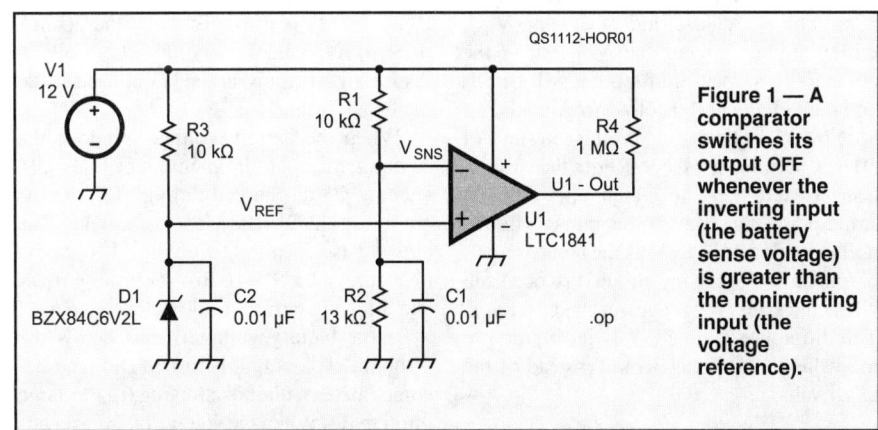

Figure 1 — A comparator switches its output OFF whenever the inverting input (the battery sense voltage) is greater than the noninverting input (the voltage reference).

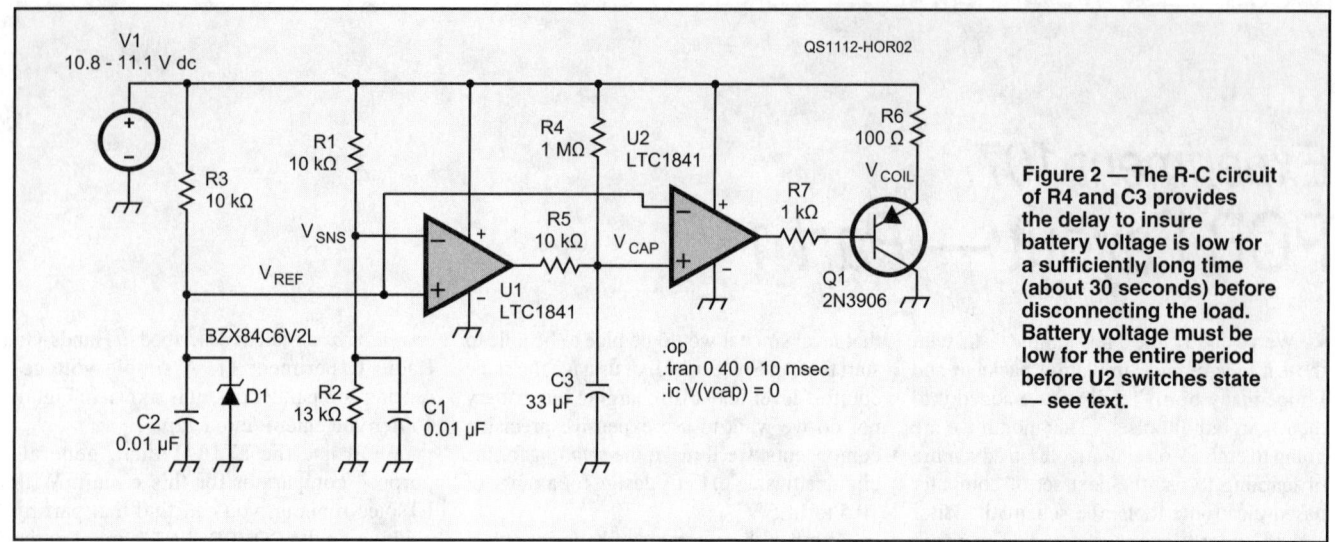

Figure 2 — The R-C circuit of R4 and C3 provides the delay to insure battery voltage is low for a sufficiently long time (about 30 seconds) before disconnecting the load. Battery voltage must be low for the entire period before U2 switches state — see text.

that adding these small capacitors costs little and can save a lot of grief if using sensitive circuits in environments where there is likely to be strong RF.) I arbitrarily set R1 to 10 kΩ and solved for R2 so that the battery sense voltage, V_{SNS}, will equal V_{REF} when the battery voltage is 11 V, exactly in the middle of our sensing range of 10.5 to 11.5 V.

R2 = 6.2 V × 10 kΩ / (11 − 6.2)
= 62 / 4.8 = 12.92 kΩ

A standard value of 13 kΩ will be fine.

Because V_{REF} is connected to the comparator's noninverting (+) input, whenever V_{SNS} is greater than V_{REF}, U1's open-drain output circuit will turn ON and pull the voltage at R4 low, labeled U1-OUT.

Simulate this circuit by selecting EDIT SIMULATION CMD from the SIMULATE menu. Select the DC OP PNT tab and click OK. Place the simulation symbol anywhere on the schematic and op. will appear when you click. Select RUN from the SIMULATE menu and the OPERATING POINT LIST will appear in a small window. In the list of node voltages, you'll see V(V_{REF}): 6.15858 VOLTAGE. This is the voltage reference value, V_{REF}. Also look for V($_{VSNS}$) and V(u1−out). V_{SNS} will be slightly greater than V_{REF} and so the comparator's open-drain output will be ON and the voltage will be close to ground.

Now lower the battery voltage in steps of 50 mV, starting at 11.1 V. Rerun the simulation at each step and watch V(u1−out). At some point, the comparator output voltage will jump to nearly 11 V as the comparator's output turns OFF. In my circuit, that occurred when the battery voltage reached 10.85 V. The difference from 11 V is mostly due to the value of R2 being 13 kΩ instead of the exact value required.

Delays, Delays

Now we need a circuit that will detect when U1's output is OFF for the prescribed

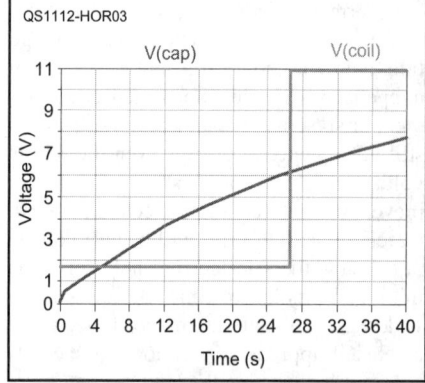

Figure 3 — As the R-C circuit charges, the voltage on C3 eventually exceeds the voltage reference, V_{REF}, at 6.2 V. At that point, U2 switches its output to an open circuit, turning off Q1 and the current through R6, which simulates a relay coil.

period of time. A comparator will do this job, too, with U2 and the basic R-C circuit of Figure 2. The simplest method of switching large currents is an electromechanical relay, but since *LTspice* doesn't include a model for a relay, I've simulated the relay's coil with R6, a 100 Ω resistor. To carry the 50 mA or more typical of 12 V dc relay coils, a 2N3906 PNP transistor (Q1) is used to do the heavy lifting.

When the battery voltage is above the setpoint, the open-drain output of U1 is ON, keeping C3 discharged through R5 and the output of U2 ON and close to ground. This allows base current to flow in Q1 — about (11 − 0.7) / 1 kΩ ≈ 10 mA — turning it ON and allowing current to flow through R6.

If the battery voltage drops below the setpoint, U1's output is turned OFF and becomes an open circuit, allowing C3 to charge through R4. When the voltage on C3 exceeds V_{REF}, U2 switches its output to an open circuit. This also turns OFF Q1 as shown by Figure 3, in which you can see C3 charging

and U2's output turning OFF.

It's All About Timing

How large must C3 be to charge from 0 V to V_{REF} in 30 seconds? To find out, we'll need to solve the equation for capacitor voltage in our R-C circuit:

$$V_{REF} = V_{BATT}\left[1 - e^{\frac{-t}{RC}}\right] \quad [Eq\ 1]$$

where V_{BATT} is the battery voltage. Having selected R = 1 MΩ, the solution is:

$$C3 = \frac{-t}{R4 \times \ln\left(\frac{V_{BATT} - V_{REF}}{V_{BATT}}\right)}$$

$$= \frac{-30s}{1M\Omega \times \ln\left(\frac{4.8}{11}\right)} = 36.2\mu F \quad [Eq\ 2]$$

where ln is the natural logarithm, $\log_e$. The closest standard value for C3 is 33 μF and that will result in a charging time of about 10% less than 30 seconds, which is acceptable.

If battery voltage climbs back above the setpoint while C3 is charging, U1 will turn ON again, discharging C3 through R5. R5 is required to limit the current through the open-drain output to no more than V_{BATT} / 10 kΩ = 12 mA which is a safe value for IC outputs.

To simulate the entire circuit, you'll need to configure the TRANSIENT SIMULATION command. Set the STOP TIME to 40 seconds and the MAXIMUM TIMESTEP to 10 ms. The timing capacitor should also be discharged at the start. This can be done with a *Spice* directive to SET INITIAL CONDITIONS for the voltage across C3. Click the SPICE DIRECTIVE symbol on the toolbar (op.) or select SPICE DIRECTIVE from the EDIT menu and enter IC. V(VCAP)=0, then click OK and place the text on the schematic. Now, when you run the transient simulation, it will begin with C3 discharged.

Experiment 108
PCB Layout — *Part 2*

Last month, we designed a circuit that would disconnect a load when battery voltage dropped below a certain level for a specific length of time.[1] There are still a couple of features to add — adjustable disconnect voltage, a means of switching load current and the circuit must latch off in a power down state when the load is disconnected. In addition, I'm going to introduce a new schematic editor as we start down the road toward producing a circuit board. Let's get going by transferring our circuit to the new CAD package.

Fabrication Services

Over the past few years, numerous online PCB *fabricators* have sprung up. One of the best known is ExpressPCB (**expresspcb.com**), which also provides free schematic editor and PCB layout software. You use the ExpressPCB tools to design your circuit and lay out the PCB then submit the resulting files to ExpressPCB for fabrication. You get the boards back by mail in a few days!

ExpressPCB is unique in that the format of the files it uses is proprietary — if you use their software, you'll have to use their fabrication service. Nevertheless, Express-PCB is an easy way to get used to schematic entry and PCB layout. Once you learn how, you'll be more informed about the basic steps and will be in a better position to use a more advanced package such as *Eagle* (**www.cadsoftusa.com**) or *Kicad* (**www.lis.inpg.fr/realise_au_lis/kicad**), both of which have free versions. They both can generate standard format output files, but are more complex to learn and use. We'll start with the simple and inexpensive *MiniBoard* service from ExpressPCB and you can take it from there!

ExpressPCB Software

Download and install the ExpressPCB schematic editor and layout program as one file (**ExpressPCBSetup.exe**) after clicking DOWNLOAD EXPRESSPCB from the ExpressPCB home page. There is one version for *Windows XP*, *Windows 2000* and *Windows NT* and another for *Windows Vista* and *Windows 7*. Run the file and two programs will be installed: *ExpressSCH* (the schematic editor) and *ExpressPCB* (the PCB layout program).

The first time you run either package, you'll be prompted to read a QUICK START GUIDE. Keep the guide open while you work with the programs or print it out on paper. You'll be able to use it as a handy reminder of how to perform basic functions until you've gotten used to the program.

Entering the Schematic

Run *ExpressSCH*, creating a blank schematic. Maximize the window to full screen and begin by editing the title block at the lower right. Double-click on SCHEMATIC NAME and enter "Low Battery Detector" in the text window that opens, followed by ENTER to place the text on the schematic. You can click and drag to move the text around. See Figure 1 for ideas of what to enter into the other fields of the title block. Now use the FILE menu to save the file with a filename of "Low Battery Detector" or whatever you prefer.

Next, recreate the *LTspice* schematic from last time, leaving out the voltage source, V1, and R6, the resistor simulating the relay coil. Start with the resistors. Click the PLACE A COMPONENT tool button at the left hand side of the screen and click the left-most stick figure button in the toolbar. From the pull down menu at the right side of the upper toolbar, select PASSIVE — RESISTOR. The cursor will change to a double-lined cross. Move it into the schematic area. Each time you click, a resistor symbol will be placed on the schematic. Place five resistors on the schematic then right click anywhere. The cursor will change back to an arrowhead. Select the component tool again and click the second stick figure toolbar button from the left. Place two more resistors on the schematic, noting how they are oriented. These will be the resistors in series with the comparator outputs.

Repeat this general process to create two non-polarized capacitors, one polarized capacitor, a Zener diode and a PNP transistor. (The diode and transistor can be found in the SEMICONDUCTOR group of components.) Select the SYMBOL or LABEL tool and place seven power ground symbols on the schematic. Arrange the components and symbols approximately as in Figure 1.

We're going to change the dual comparator to a more common type, the LM393. The LTC1841 was convenient for simulating in *LTspice* but is not rated for automotive voltages. The LM393 can operate at voltages as high as 36 V and is available as a through hole DIP package. Download the LM393 data sheet for future reference at **www.national.com/ds/LM/LM393.pdf**. Select the symbol IC — NATIONAL — LM393 — COMPARATOR — DIP-8 and place it on the schematic. Both sections of the dual comparator will be placed on the schematic as a pair. Click on one of the sections and it can be moved independently. The Zener diode we'll use here is the more common 1N5234B.

Use the WIRE tool to connect the compo-

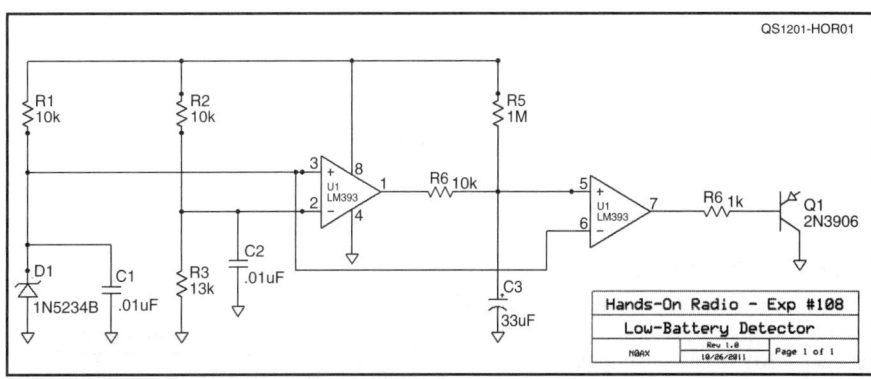

Figure 1 — A partial schematic that results from transferring the *LTspice* design into *ExpressSCH*. [The schematics have been rendered to match the software. — *Ed.*]

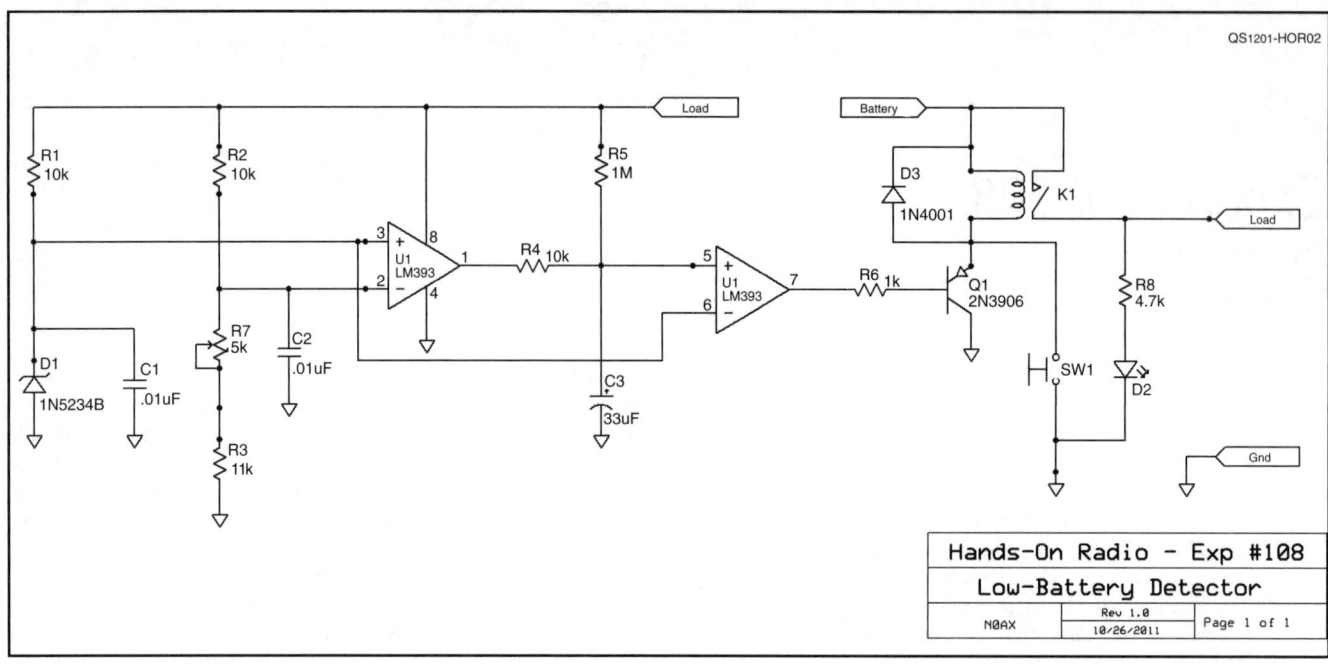

Figure 2 — The final schematic after adding the relay and latching circuits. Ports labeled Battery, Load and Gnd are used to indicate external power connections.

nents together without worrying about their placement on the schematic. The wires will *rubber band* to maintain the connection as the components move. It is important that you make connections between pins of components and not to the lines between components. If you're not sure whether you've made a connection to a pin, select and move the component. If the connection was made properly, the connection will stay connected to the component. Be careful to connect the comparator inputs correctly as the *LTspice* and *ExpressSCH* conventions for orienting the inverting (–) and non-inverting (+) inputs are different. Use the INSERT A CORNER IN A WIRE tool to reroute the wires in the conventional right angle fashion.

To assign component values, right click on the component then select SET COMPONENT PROPERTIES. In the window that appears, enter the component designator (such as R1, C1, D1, U1 and so forth) and the value or part type (1N5234B, 0.01 µF and so on), then click OK. Move the components around so that the labels are in the clear. The result should look similar to Figure 1. Both comparator sections should be designated U1 to avoid problems when linking the schematic and layout.

Adding the Features

The threshold voltage should be made adjustable by replacing R3 with a series combination of a 10 kΩ resistor and a potentiometer. Using a spreadsheet, the switching threshold, $V_{THR} = 6.2 (1 + R2/R3)$ was evaluated for several combinations of standard fixed and potentiometer values. I selected an 11 kΩ fixed resistor and a 5 kΩ potentiometer, giving a range of 10.1 to 11.8 V centered close to 11 V. Add the potentiometer to the schematic and change the value of R3. C2 should remain connected to the inverting input of U1A.[1]

An SPST relay will switch the load current. The stock library of schematic symbols that comes with *ExpressPCB* does not include relays so you'll have to add the symbol "SPST — Relay.s" provided on the Hands-On Radio website, obtain one off the Internet (many designers have published collections of symbols for *ExpressPCB* online), or create your own. Assuming you use the Hands-On symbol Relay-SPST.s, download it to the SCHCOMPONENTS_LIBRARY folder where *ExpressPCB* was installed — usually C:*ProgramFiles\ExpressPCB*. You can now select the component and place it on the schematic. Connect one side of the coil to the collector of Q1. Add a 1N4001 kickback diode, D3, across the coil as shown in Figure 2 to protect Q1 when it suddenly interrupts the coil current.

To represent the battery and load connections, create left and right pointing port symbols. From the COMPONENTS menu, select

[1]All previous Hands-On Radio experiments are available to ARRL members at **www.arrl.org/ hands-on-radio**. The design spreadsheet is available in the section for this experiment.

COMPONENT & SYMBOL MANAGER. Click the LIBRARY SYMBOLS check box and select PORT — RIGHT POINTING — 8 LETTERS WIDE, then click INSERT INTO SCHEMATIC. Move the port symbol to a clear spot, then right click it and edit the component properties so that the label is Battery. Connect the symbol to both the coil and the NO (normally open) contact as shown in the figure. Create two PORT — LEFT POINTING — 8 LETTERS WIDE symbols with a label of Load. Connect one port to the relay's remaining contact and the other to the power connection of the comparator electronics. Create one more left pointing port with a label of Gnd and connect it to a ground symbol.

To create the latching function, we've just taken the first step — supplying the sensing circuit from the load voltage means that if the load is disconnected, the sensing circuit is as well, turning off Q1 and keeping the relay contacts open. How does the circuit turn on in the first place? That's where SW1 comes in.

To turn the circuit ON and close the relay contacts, close SW1. This bypasses Q1 and closes the relay, applying power to the sensing circuit. If battery voltage is sufficiently high, Q1 will turn ON and keep the relay closed after SW1 is released. Once battery voltage drops below the threshold, the relay will turn OFF and the circuit will have to be manually activated again. R8 and D2 provide a visual indication of whether the load is powered or not. We are now ready to proceed with circuit board layout!

Experiment 109
PCB Layout — *Part 3*

After last month you have a complete schematic that is ready for layout. Now we'll switch over to the PCB layout software and turn the circuit design into a circuit board that can be ordered from ExpressPCB. I assume that the reader is somewhat familiar with circuit board structure. If not, additional background is available in the CAD chapter of the 2011 or 2012 edition of *The ARRL Handbook*.[1]

Checking and Fixing Errors

First, the schematic should be checked to see if there are any connection or configuration errors. From the FILE menu select CHECK SCHEMATIC FOR NETLIST ERRORS. My first pass was not happy with Q1 — there were no pin numbers assigned on the component because I was using a general-purpose transistor symbol. I followed the instructions given by the error checker and assigned the pins as 1 — Emitter, 2 — Base and 3 — Collector, the standard order for a 2N3906 in a TO-92 package. Rerunning the error check turned up one not quite connected wire that I would have never spotted by eye. I fixed it and was ready for layout.

ExpressPCB Layout

Run the *ExpressPCB* software and read the *Quick Start Guide* as before. Because the layout software will be unfamiliar, open the HELP file, too. The default size of the circuit board is the same as for the Miniboard service we'll be using, but check anyway by moving the cursor between opposite corners and checking the X-Y coordinates in the lower left. It should be 3.8 inches wide and 2.5 inches high. Assign a name to the file and save it using the SAVE AS function.

Instead of component symbols, the layout software has component outlines with a pad for each pin of the component. Operation is similar to the schematic editor — click the PLACE A COMPONENT tool (look for the IC symbol), select a component from the list, and click to place it on the layout.

To obtain a list of the components you'll need, go back to the schematic and from the EDIT menu, select the COPY BILL OF MATERIALS TO CLIPBOARD function. Open a text editor, paste the list into a blank document and print it. Place the required quantity of each component onto the layout:

- (C1, 2) Cap — Lead spacing 0.1 inch (2.5 mm)
- (C3) Cap — Radial electrolytic — Lead spacing 0.1 inch (2.5 mm)
- (D1) Diode — Zener 0.5 W (lead spacing 0.3 inch, hole 0.029 inch)
- (D2) LED — T1
- (D3) Diode — DO-41 (lead spacing 0.4 inch, hole 0.040 inch)
- (Q1) Semiconductor — TO-92
- (U1) DIP — 8 pin
- (R1-7) Resistor — 0.25 W (lead spacing 0.4 inch)
- (R8) Potentiometer — Bourns series 3386F
- (SW1) Switch — 6 mm push button

Place the components more or less as the schematic has them arranged as in Figure 1 which shows the *silk screen* layer (yellow on the display). The silk screen shows the shape of each component and the placement of all pads along with the designator of the component.

Right-click on each component and edit the component properties to assign designators (R1, C1, D1, etc) to each component corresponding to the designators on the schematic. This tells *ExpressPCB* which layout pins and pads should be connected together. If you were using an *auto-router* program, the software could then create the traces just by looking at the *net list* from the schematic and matching up designators and pin numbers (*net* is short for network). *ExpressPCB* has a manual routing tool so you'll be doing that chore yourself.

Connecting Components

Laying out the traces is a lot of fun and can be an interesting puzzle, as you will see. Print out a copy of the schematic and get a highlighter pen or colored pencil. As you make connections, highlight or trace them on the paper schematic to show the connection has been made. That way, it will be easier to

---

[1]*The ARRL Handbook for Radio Communications*, 2012 Edition. Available from your ARRL dealer or the ARRL Bookstore, ARRL order no. 6672 (Hardcover 6634). Telephone 860-594-0355, or toll-free in the US 888-277-5289; **www.arrl.org/shop; pubsales@arrl.org**.

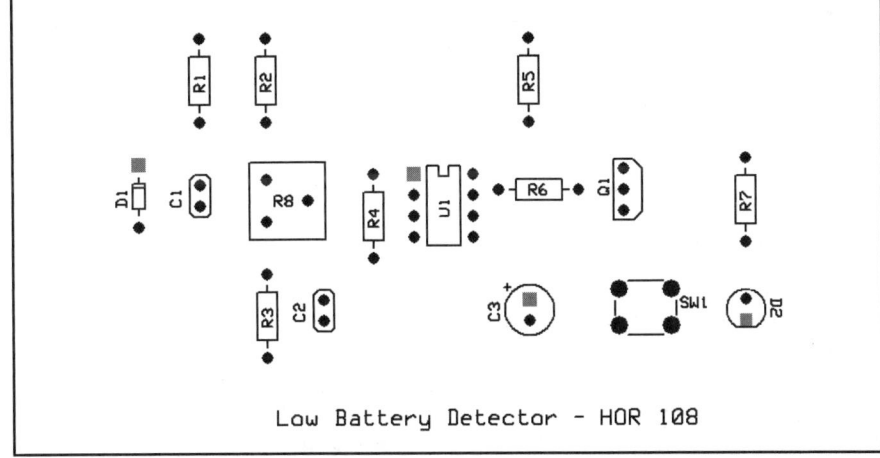

Figure 1 — PCB layout begins by selecting a component outline for each required component on the layout. At this stage exact placement is not important.

be sure you've made all necessary connections.

The next step is to link the schematic and layout so the software can be sure you are making the necessary connections. From the FILE menu, use the LINK TO SCHEMATIC function and select the schematic file.

Click the HIGHLIGHT NET CONNECTIONS tool button (at the bottom of the column of buttons) then click on a component pin. All of the other pins that should be connected to that pin will be shown in blue. Select the TRACE tool or type T and connect all of the highlighted pins using traces on the copper (bottom) side of the board shown in green. Don't worry about making corners or choosing board layers — just make straight traces for now. (It's no surprise that this stage of the layout is called a *rat's nest*.) Keep double checking against the schematic. Complete a section of the circuit then move the parts around or rotate them if you want to make the traces less snarled.

Now use the CORNER tool to bend the traces around and between pins. You can change the trace from the component side of the board (red) to the copper side of the board (green) by selecting it and clicking on the toolbar MOVE TO LAYER CONTROLS. When a segment of a trace is moved from one layer to another, a new pad at one or both ends of the segment is created. This is called a *via* and consists of a small hole and pad on each layer. I enjoy the puzzle of "solving the board" with the fewest vias — or even none, on a simple board like this one.

Creating the Relay

The relay we're going to use is a T90N1D12-12 SPST (Form 1A) with contacts rated for 30 A. It is available from Digi-Key (**www.digikey.com**), part number PB104-ND. Download the data sheet for reference. Page 3 of the data sheet is where you'll find mechanical drawings showing the pin connections. This is a bottom view of the pins that will have to be reversed since we view the board from the component side. Also note that the data sheet's top pin is not present on the SPST relay and the metal frame of the relay is "live" — connected to the moving pole of the relay — with its own pin.

Because the software library doesn't include this relay, we'll have to create the component. In the HELP file, read the instructions for creating custom components. The procedure involves using the PLACE A PAD tool to create and arrange pads for each of the relay pins in the pattern specified on the data sheet (remembering that the data sheet pattern is looking from the bottom, not the top).

Once you have the pads placed, a component outline is then drawn on the yellow silkscreen layer. The outline and pads are grouped into the single component in Figure 2 and saved in the custom component folder. I've done all of that work for you and placed the component file Relay - T90N1.p on the Hands-On Radio website.[2] Download that component and save it in the **C:\Program Files\ExpressPCB\PCBComponents_Custom** folder. (If you want to make the component yourself, save the layout and work using a new blank layout.)

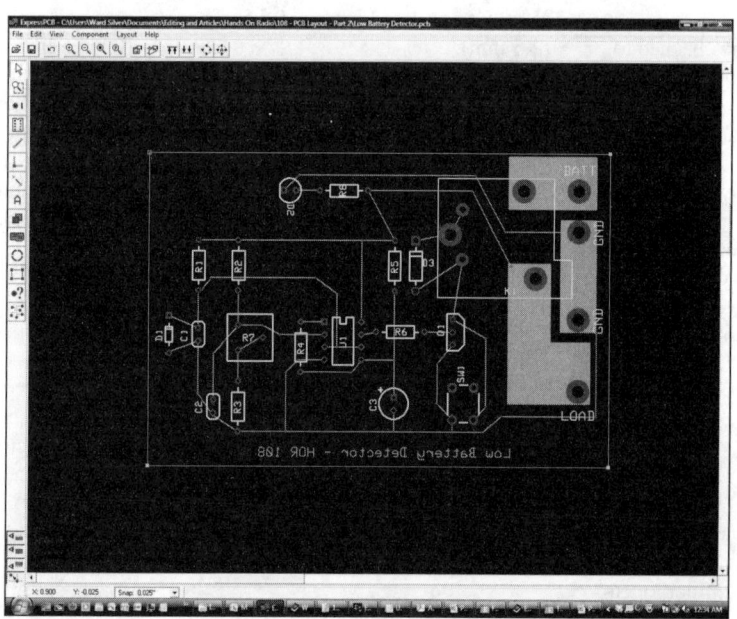

Figure 2 — A custom component is created for the SPST relay used for K1. The component is created by selecting and arranging pads, then drawing a component outline on the silkscreen layer.

Figure 3 — The final PCB layout with all traces on the board's copper side shown in green. Large rectangular areas of copper are used to carry the high load current.

Using the COMPONENT menu's COMPONENT MANAGER function, you'll find the relay in the CUSTOM COMPONENTS group. Click INSERT INTO SCHEMATIC then move and rotate the component, place it in the desired location, moving other components if necessary. Complete the coil connections, leaving the high-current connections for last.

The BATTERY, LOAD and GROUND connections can carry up to 25 A, so wide copper traces are required. (See the *Circuit Board Design Tips* document in the HELP menu for information on trace width and current capacity.) I began by placing 0.2 inch pads with 0.1 inch holes for #12 AWG wire at the edge of the board. I then used the PLACE A RECTANGLE tool to create large blocks of copper connecting the wire and relay pads. Power and ground connections to the rest of the circuit can then be made to the LOAD and GROUND pads.

Figure 3 shows the layout I came up with. The components are placed roughly as the schematic shows them. I managed to route all of the connecting traces on the bottom layer but that's just a matter of personal taste. Clear areas are left in three corners of the board for mounting holes to be drilled if desired. The next step is to order boards and get ready to build.

In "PCB Layout — *Part 4*" we will build our board.

---

[2]All previous Hands-On Radio experiments are available to ARRL members at **www.arrl.org/hands-on-radio**. The design spreadsheet is available in the section for this experiment.

Experiment 110
PCB Layout — *Part 4*

It's time for the rubber to meet the road — the boards designed previously have been delivered and so have all the components.[1] Let's build the circuit and compare it to our requirements and expectations. It's time for *Incoming Quality Inspection*.

Collect the parts and boards as shown in Figure 1. The PC boards are delivered individually shrink-wrapped so cut one out of its package and verify that it's the right size and that the traces and silk screen are oriented correctly. Go through your parts and confirm that all of them are the correct value or part number by placing them on the schematic as in Figure 2.

Check the fit of the relay into the holes on the board. This is where we find out if our new component was created properly. Next, check the pushbutton switch and make sure the contact closures connect to the expected pairs of pads. Because the exact part number was listed in the *ExpressPCB* components I was confident of proper fit but generic components may vary.

Now perform a visual and electronic check against the printed schematic — called *buzzing out* the board. This simple board can be checked manually by using your voltmeter's continuity test function. If everything is ready, you're ready to start building.

Testing the Board

Even for simple circuits, it's important to perform the initial prototype testing in a step-by-step fashion, building each part of the circuit independently. In this way, you can find and correct problems while they are isolated. If everything is assembled before testing it can be much harder to find problems because they often affect the entire circuit.

Before you start working with the board, if you live in a dry location consider taking steps to minimize electrostatic discharge (ESD) by working on a grounded metal surface or using a static-discharge mat or cuff. (I use a cuff connected to the workbench safety ground.) None of the parts in this circuit are particularly sensitive but a finger-to-IC spark can damage the LM393.

Until the relay is installed, use the LOAD and GND terminals for power connection. The power supply should be adjustable. Remove power from the board between each step.

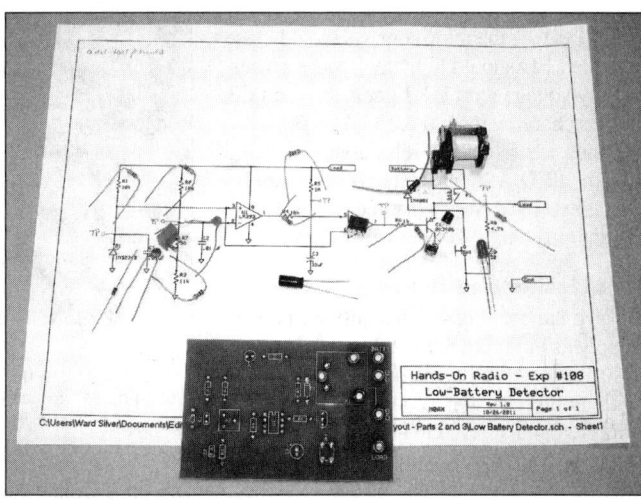

Figure 2 — The easiest way to confirm that you have all of the parts and that they have the right value is to place them on the schematic.

Step 1, Voltage Reference — Place and solder R1, D1 and C1. Apply 12 V and use a voltmeter to verify that the voltage at U1-3 (pin 3 of U1) is 6.2 V ±5%.

Step 2, Battery Voltage Sensing — Place and solder R2, R3, R7 and C2. Apply 12 V and adjust R7 through its range to verify that voltage at U1-2 varies from approximately 10.1 to 11.8 V. Now set the power supply voltage to the threshold voltage at which you want to disconnect the battery. Verify that you can set the voltage at U1-2 to 6.2 V by adjusting R7. Leave R7 set at this position.

Step 3, Comparator Switching and Timing — Place and solder U1 (be sure to get pin 1 oriented correctly), R4, R5, R6 and C3 (pay attention to the marking of the capacitor's negative lead). Apply 12 V and verify that the output of both comparators (U1-1 and U1-7) is LOW (near 0 V). This indicates that the power supply voltage is above the minimum so that the relay would be supplying power to the load. Now reduce the power supply voltage until the minimum threshold is reached to verify that C3 begins charging. The voltage at U1-5 will begin to increase slowly. Verify that it takes approximately 30 seconds for the voltage at U1-5 to reach

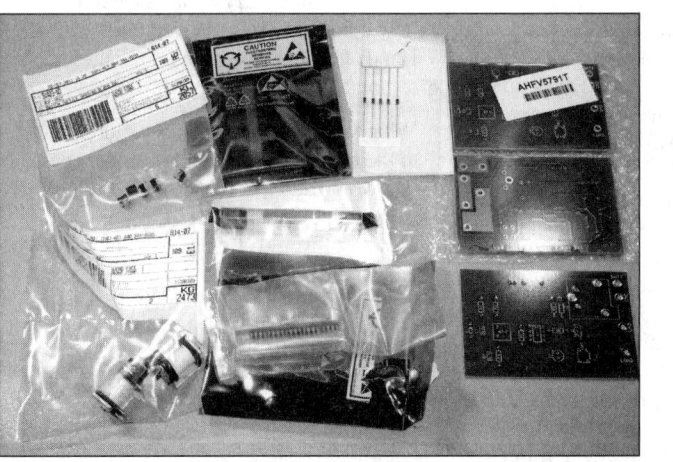

Figure 1 — Before beginning assembly make a thorough inspection of all parts and the finished PC boards. Be sure custom components fit their hole pattern properly.

[1]All previous Hands-On Radio experiments are available to ARRL members at www.arrl.org/hands-on-radio.

6.2 V and that U1-7 goes HIGH when that voltage is reached.

Step 4, Relay Drive — Place and solder K1, Q1, D2, D3, R8 and SW1. Use a rubber band or two to hold K1 securely against the board while it's being soldered. Play close attention to the cathode markings or package shapes of Q1, D2 and D3. Apply 12 V to the board — now to the BATT terminal. The relay should remain de-energized and the LED should stay OFF.

Press SW1 — the relay contacts should close and the LED should light up. Reduce power supply voltage until U1-1 goes HIGH and C3 begins charging toward 6.2 V. After 30 seconds, verify that the relay contacts open and the LED goes out. Return power supply voltage to 12 V. Verify that the relay stays de-energized until you press SW1 again.

Troubleshooting

I tested the board using an adjustable power supply as in Figure 3. First, I tested with a resistive load (12 Ω, 1 A draw) and then with mobile rigs. On a couple of the rigs, I was greatly dismayed to hear the unmistakable buzz of rapid relay cycling when low voltage was detected! The voltage reference and voltage at U1-5 both looked stable and clean — on a voltmeter. Connecting an oscilloscope to U1-5 and U1-7 quickly showed the problem. As power was removed from the LOAD circuit, the comparator output turned back ON just long enough to close the relay for a few milliseconds before turning OFF again and creating an oscillating cycle.

To make a long story short, this problem turned out to be caused by filter capacitors inside the radio. If sufficiently large capacitors are present on the LOAD circuit, enough voltage remains to keep the comparator circuit alive for several milliseconds after opening the relay contacts instead of voltage being removed abruptly. As V_{CC} dropped, the comparator output switched back ON very briefly — long enough to energize the relay and cause the cycling.

After several attempts at modifying the comparator circuit, it became clear that the easiest solution was to operate the comparator circuit from the BATT circuit instead of the LOAD circuit. This kept the comparator in control at all times and eliminated the cycling. To make this modification on the PC board, use a razor blade or hobby knife to cut the trace connecting U1-8 to the LOAD circuit. Use a short piece of insulated wire such as wire-wrap or light hookup wire to connect U1-8 to the BATT circuit connection on the relay.

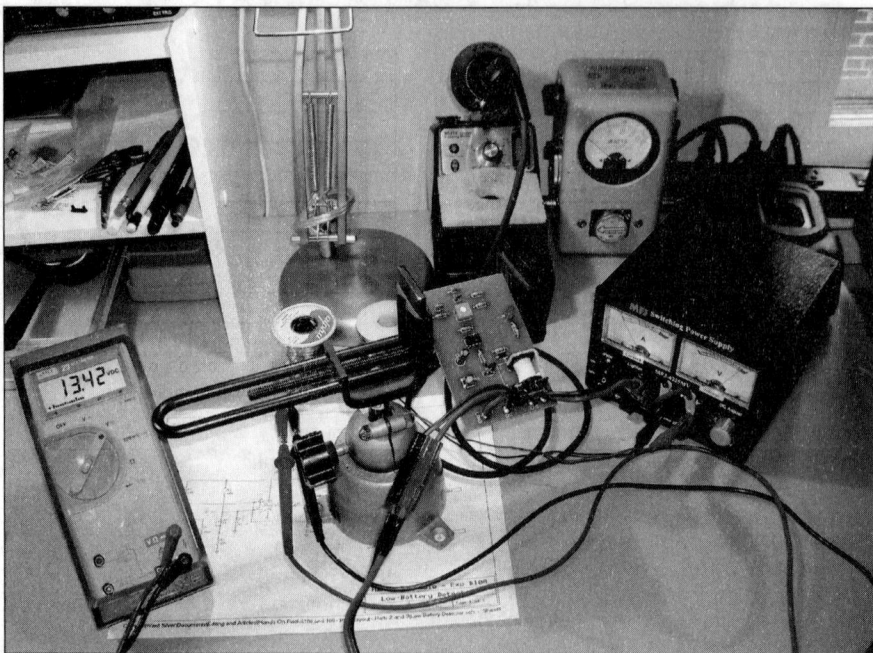

Figure 3 — The completed board being tested on the workbench. Anderson Powerpole connections were used to connect to the adjustable power supply and the ICOM IC-207H mobile rig serving as a test load.

Does this violate the "no OFF power" requirement by allowing battery current to flow after the load is disconnected? To a degree, yes, although the LM393 does not draw more than 0.4 mA of supply current according to the datasheet. This is less current drain than most batteries experience from self-discharge and other vehicle loads so I consider it an acceptable trade-off. I'll be experimenting with using another transistor to switch power to the comparator. Updates to the circuit will be posted on the Hands-On Radio website.

This problem illustrates an important lesson about simulation — it rarely includes all of the system in which the circuit must operate. In this case, a slower than expected removal of power from the LM393 circuit resulted in unexpected behavior. Perhaps a more complete simulation would have picked this up but my experiences in product development tell me that there is almost always at least one SDT (Some Darned Thing) that the computer or designer doesn't know about. Plan for your initial testing to turn up behavior that will require some adjustment to the design. This is why the fourth part of the design cycle in Experiment #83 must always be "compare": Design, Simulate, Build, Compare.

Final Testing and Improvements

The final test was to connect the board to a portable VHF/UHF radio powered by a lawn tractor lead-acid battery. I inserted the low battery detector in the radio's power supply line and spent the day listening to and talking on the repeater while keeping an eye on the voltmeter. Late in the afternoon during a contact…click! The voltage had dropped low enough for long enough to trigger the disconnect. Even though the battery voltage recovered as predicted, the relay contacts stayed open until I pressed the RESET switch.

Obviously, it will take more testing before the detector can be considered "finished" but it is now working as expected. I'll mount it in an enclosure and take it on the road to see how it performs in a real vehicle. Watch the Hands-On Radio website for updates about testing and circuit changes.

What kind of improvements would I make to the board layout? There is lots of space, so I would add a grid of isolated pads to support additional prototyping and circuit changes. I would also add test points (isolated pads) for connecting voltmeters and 'scope probes, including ground points. Once I've used the circuit for a while, I may find that I need transient suppressors or additional filtering capacitors. An updated schematic and layout are available on the Hands-On Radio website.

I hope you've enjoyed this foray into laying out your own circuit boards. With the inexpensive software tools and the ability to submit designs directly to fabricators, it is easier than ever to build professional, reliable projects for your shack.

Antennas and Transmission Lines

Experiment #82 — Antenna Height

Ask an experienced ham about antenna height and the answer will often begin, "It depends…" On what? Doesn't an antenna work better higher above ground? Well, it depends! There is such a thing as "too high," believe it or not, and we'll do a quick modeling experiment this month to illustrate why.

Antenna Modeling

In the "good old days" antenna design involved a lot of "cut and try" consisting of cutting the wire or tubing, assembling the antenna and hoisting it into position. Those with access to a test range could actually measure the antenna's pattern. But most amateurs could only evaluate the antenna by patiently conducting comparisons and trying to draw conclusions despite the changing conditions. Today, the personal computer has given amateurs access to sophisticated modeling programs that take a lot of the cutting and trying out of the process — if you were careful in applying the modeling tool.

We're going to make use of one of the most popular modeling programs, *EZNEC* by W7EL, to observe the effects of antenna height above ground.[1] If you already own *EZNEC*, skip ahead to the next section. If not, download the free demo copy of *EZNEC 5.0* from the program's Web site. It's a full featured version of the program, including the help information, but limited in the precision with which it can analyze an antenna. The demo version will suffice for this experiment. If you develop a taste for modeling larger or more complex antennas, you can buy a package with additional capabilities from the same Web site. The ARRL also offers an online antenna modeling course.[2]

The Backyard Dipole

EZNEC comes with a number of prepared antenna models. If you run the program and click the OPEN button (or select FILE, then OPEN), you'll see a list of available model files. The one we're going to work with is

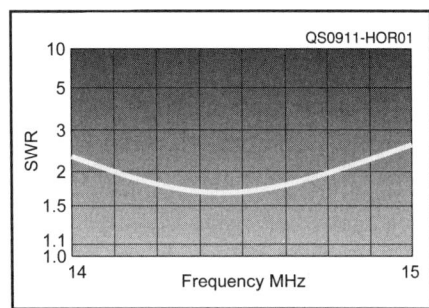

Figure 1 — The SWR curve for the back yard dipole at a height of 30 feet over *EZNEC's* Real Ground. The resonant frequency of 14.46 MHz indicates the dipole is too short.

called BYDIPOLE (*backyard dipole*). It's an electrical half wave (λ/2) long, center fed dipole, cut for the 20 meter band. Open this file, review the antenna structure and the program's configuration. By clicking the right-arrow (>) next to WIRES, you can see that the dipole is made from 33.43 feet of #12 AWG bare wire, fed in the middle and suspended at a height of 30 feet. Click VIEW ANT to see the antenna and keep this window open as you make changes as a check.

Click the SWR button and plot SWR from 14 to 15 MHz in steps of 0.02 MHz. By moving the cursor along the curve, you can find the frequency of resonance at which feed point reactance is zero — 14.46 MHz — out of the band! According to the well-known formula for dipole length, l (in feet) = 468/f (in MHz), the length checks out as 33.43 feet. What happened? The constant 468 in the length formula is an approximation based on a dipole that is higher above ground in terms of its wavelength. Multiplying 33.43 × 14.46 shows that 484 would be more appropriate. Recalculating the wire length as 484 / 14 = 34.57 feet results in resonance at 14 MHz. Find the length for which resonance occurs at 14.175 MHz, the center of the band. (34.14 feet) What is the feed point impedance at resonance? (87 Ω)

There's Always an Angle

Go back to the main *EZNEC* screen, click > next to GROUND TYPE, select FREE SPACE and run the SWR curve again. You'll have to extend the SWR curve below 14 MHz because the resonant frequency drops to 13.8 MHz at which the feed point impedance is 72 Ω. The absence of ground reflection causes the resonant frequency to be lower. The reflection creates an *image* antenna as far below the surface as the real antenna is above ground. This is the same model for the vertical ground-plane antenna, except that the dipole is horizontal and above ground.

The image antenna has two primary effects. The first, as you have observed, is to change the resonant frequency and feed point impedance from the free space values. The second is to create a two element phased array with one element being the real antenna and the image forming the second element. The combination of the fields from both result in the radiation pattern of the antenna.

Figures 2A through 2G show the elevation radiation pattern of the dipole at six electrical heights; from ⅛ through 2 λ. Table 1 shows several measurements for each height. First, the dipole length is shown for resonance at 14.175 MHz along with the l × f constant. The next column shows feed point impedance and SWR at resonance. Finally, maximum gain with respect to an isotropic radiator in free space (dBi) is shown along with the vertical angle at which it occurs. Our simple backyard

Table 1
Variation in Dipole Performance with Height

Height in Wavelengths at 14.175 MHz (feet)	Resonant Length in Feet (l × f)	Feed-point Impedance in Ω (SWR)	Max Gain (dBi) at Angle (Degrees)
⅛ (8.8)	33.0 (467.8)	31.5 (1.59)	8.3 @ 90
¼ (17.4)	32.9 (466.4)	81.7 (1.63)	6.5 @ 62
½ (34.7)	34.1 (483.4)	69.6 (1.39)	7.9 @ 28
¾ (52.0)	33.4 (473.4)	73.4 (1.47)	7.3 @ 18
1 (69.4)	33.9 (480.5)	71.9 (1.44)	7.7 @ 14
1½ (104.1)	33.8 (479.1)	72.0 (1.44)	7.8 @ 9
2 (138.8)	33.8 (479.1)	72.3 (1.45)	7.9 @ 7

---

[1]Several versions of *EZNEC* antenna modeling software are available from developer Roy Lewallen, W7EL, at **www.eznec.com**.

[2]ARRL online course EC-004 is no longer available.

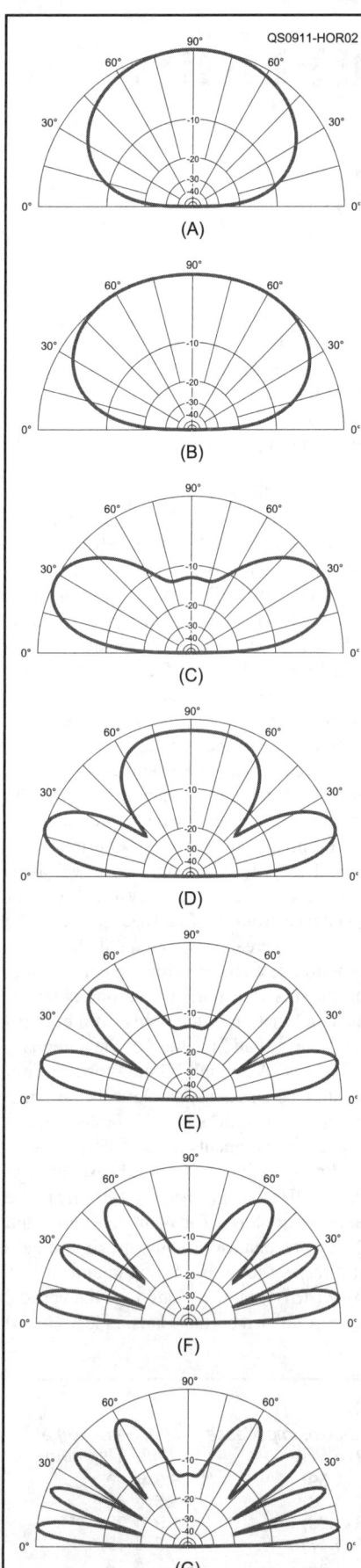

Figure 2 — Six radiation patterns for the dipole at different heights: (A) ⅛ λ, (B) ¼ λ, (C) ½ λ, (D) ¾ λ, (E) 1 λ, (F) 1½ λ, (G) 2 λ.

Antenna Modeling: Boon or Bane?

As anyone who has ever used a computer simulation program knows, pitfalls abound. In fact, while creating Figure 2, the author misentered some parameters, leading to some serious head-scratching about the dipole's surprising behavior! Nevertheless, no one seems to be clamoring for the days before modeling software.

It's important to pay attention to the limitations of any program, usually discussed in the users manuals. Results too good to be true, very different than expected, or that change dramatically with small changes in the design often indicate a problem. Because of modeling programs, antenna design has blossomed in Amateur Radio, one of the most active areas of experimentation and discovery. Skeptics need not fear that the computer will completely displace construction and on the air tests. As W7IUV once said, "I don't think anyone has made an *EZNEC* to *EZNEC* contact yet!"

dipole is a fairly complex antenna!

You can see why using the formula 468/f can lead to a lot of extra antenna trimming — the variation from 466.4 at ¼ λ high to 483.4 at ½ λ high results in a difference of 1.2 feet! The most rapid change takes place between heights of ¼ and ½ λ — about where most hams install a 20 meter dipole. Impedance also varies from 31.5 to 81.7 Ω. As the antenna is raised beyond 1 λ, resonant length and feed point impedance stabilize near their free-space values.

The radiation pattern, however, does not stabilize — quite the opposite! Starting from a low height and radiating mostly straight up, the pattern flattens out until at ½ λ it is radiating mostly at low vertical angles because the signal reflected from the ground cancels the direct radiation in the vertical direction. This would be a good DX antenna, but poor for short skip and regional coverage.

As the antenna is raised to ¾ λ, the reflected energy no longer cancels, creating a large vertical lobe. Even though the main lobe has essentially the same gain and is at a lower angle (18°) than at ½ λ, short-skip high-angle signals would be stronger, creating interference to low angle DX signals. The pattern continues to break up into more and more lobes as the antenna is raised. Even though the maximum gain and vertical angle are very similar above ½ λ, the on air performance of these antennas, especially on a crowded band, would be very different!

Gaining Gain

If you are struggling to hear and be heard through the pileups on 20 meters and are won-

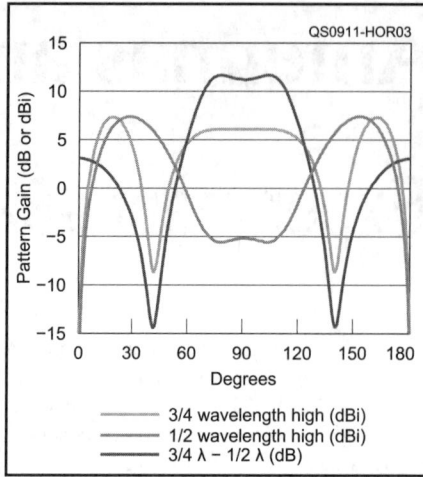

Figure 3 — The difference in gain (red) between the dipole at ¾ λ high (green) and ½ λ (red). The higher dipole has more gain at low angles, but also at high angles.

dering if your dipole would do any better if raised from its present 35 feet to about 50 feet, take a look at Figure 3. The red line shows the difference in gain between the dipole at ¾ λ (50 ft) and ½ λ (35 ft). If the red line is above 0, the higher dipole has more gain.

The higher antenna does have a bit more gain below 22° — a maximum of 3 dB near the horizon. However, in the large range of vertical angles between 22 and 56°, the lower dipole is stronger. And what of the angles above 56°? If you are having trouble with interference from nearby and regional stations, the higher dipole would be a poor choice because it will hear those stations 10 to 12 dB better than the low antenna. Higher is not always better — you would have to raise the dipole to 1 λ or higher before it begins rejecting high angle signals.

Recommended Reading

The best antenna reference for hams is *The ARRL Antenna Book*.[3] For those who want an easier to digest introduction to the subject, the ARRL also offers *Basic Antennas*.[4] There are a number of other excellent books that focus on specific types of antennas or offer collections of designs. Hams love reading about antennas — almost as much as using them!

---

[3]R. D. Straw, Editor, *The ARRL Antenna Book*, 21st Edition. Available from your ARRL dealer or the ARRL Bookstore, ARRL order no. 9876. Telephone 860-594-0355, or toll-free in the US 888-277-5289; **www.arrl.org/shop; pubsales@arrl.org**.

[4]J. Hallas, W1ZR, *Basic Antennas — Understanding Practical Antennas and Designs*, available from your ARRL dealer or the ARRL Bookstore, ARRL order no. 9994. Telephone 860-594-0355, or toll-free in the US 888-277-5289; **www.arrl.org/shop/; pubsales@arrl.org**.

Experiment 113
Radiation Patterns

While most readers are quite familiar with antenna radiation patterns, there are several less familiar features and alternate styles. This month's article shows a few examples generated by the *EZNEC* free demo package (**www.eznec.com**).

Angle Refresher

If you plot two dimensional radiation patterns you are really plotting the data from a "slice" through the complete three dimensional (3D) pattern. For example, the 3D pattern for a wire dipole in free space looks a lot like a bagel with the axis of the antenna running through the small hole at the center. What about at other angles or over ground? It's important to understand just where the slice is taking place.

Figure 1 shows the 3D elevation pattern for a 20 meter dipole, ½ wavelength above ground. The axis of the dipole is aligned as shown with its center at the origin under the Z axis label. Ground reflections have "squashed the bagel" into a beanbag shaped pattern with radiation in all horizontal directions and very little straight up.

The elevation pattern is shown in red on the 3D pattern. It shows the surface of the 3D pattern cut by the plane shown as the tilted rectangle. The rectangle is perpendicular to the ground and rotated around the center of the dipole. The azimuth angle specifies the orientation of the slicing plane with respect to the X axis — 20° in this case. (0° is usually, but not always, aligned with the antenna's direction of maximum gain.) You might imagine the slicing plane spinning around and generating a series of elevation patterns that taken together form the 3D pattern surface.

Azimuth patterns are not quite so simply envisioned, and Figure 2 shows why. In the case of an azimuth pattern, the slice is no longer a plane — it's a cone! This is the very same 3D pattern as in Figure 1 with the dipole oriented in the same way. Staying with the bagel metaphor, now imagine sticking a very sharp knife into the bagel's center at the specified elevation angle above the cutting board, er… horizontal plane. Now twirl the knife around its point, keeping the angle constant so that you cut out a cone of bagel. The resulting edge on the surface of the bagel is the azimuth pattern shown by the red line on the surface of the 3D pattern in the figure.

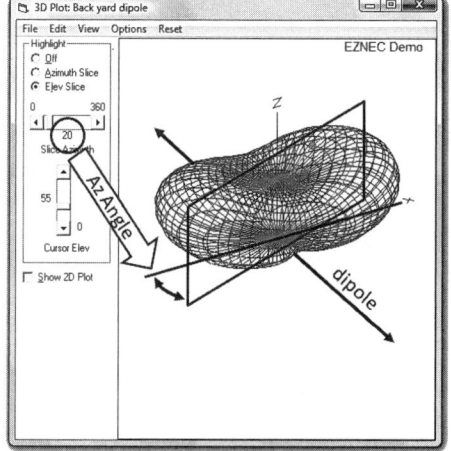

Figure 1 — An elevation pattern is taken in a plane containing the center of the antenna and at a specified azimuth angle from the X axis.

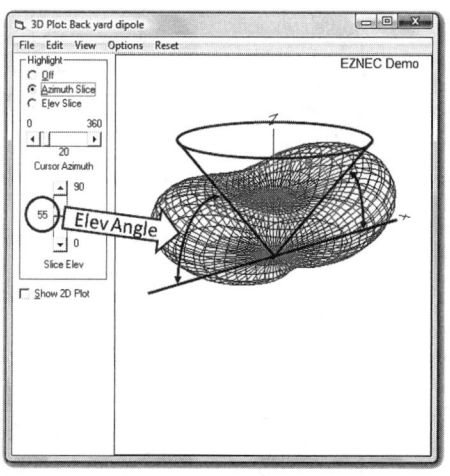

Figure 2 — An azimuth pattern is taken at a constant elevation angle around an antenna. It can be thought of as the line where the cone and 3D pattern intersect.

As the elevation angle is lowered, the cone gets wider and wider. As the elevation angle is increased, the cone gets narrower and narrower. It is customary to show an antenna's azimuth pattern for the elevation angle at which gain is maximum. From Figure 1 that would be an elevation angle of about 30°. Similarly, an elevation pattern is shown at the azimuth angle for which gain is maximum — 0° for the dipole.

You can see that for even this simple antenna a single pair of azimuth and elevation patterns doesn't begin to tell the whole story. If a 3D pattern is available, it gives you clues as to the antenna's general behavior. Then you can use the specific azimuth and elevation patterns to zero in on performance at specific bearings and elevations.

Some antenna modeling programs such as *4nec2* can generate color coded 3D patterns (see Figure 3) that are a bit easier to read than mesh surfaces.[1] It's pretty easy to overdo it with color but being able to visualize the 3D pattern clearly is important.

Map or Math?

If you looked at the usual azimuth pattern of Figure 4 you might assume that the antenna was pointing straight east and that rotating it

---

[1]*4nec2* is published by Arie Voors (**home.ict.nl/~arivoors**).

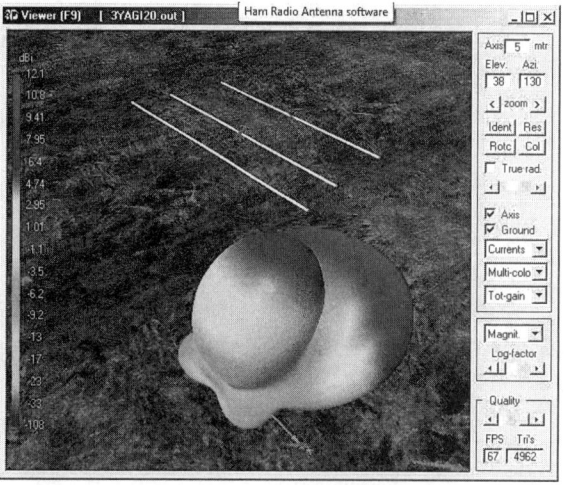

Figure 3 — Colorful 3D patterns, such as this one generated by the free modeling program *4nec2*, help the designer visualize antenna behavior.

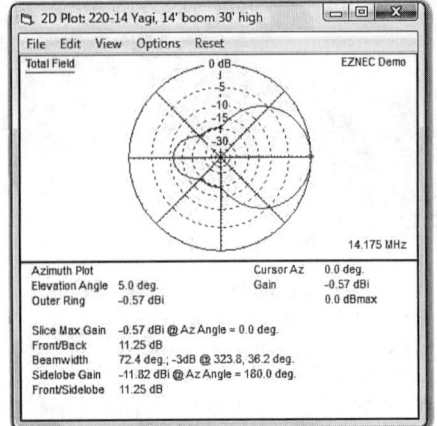

Figure 4 — Azimuth patterns can be drawn with 0° at the right and angle increasing counterclockwise (as shown here) or according to compass bearings with north at the top and angle increasing clockwise. In the second case, the main lobe of the antenna would be pointing east.

counterclockwise (CCW) to point north would also rotate the pattern CCW. You would be wrong. The default for nearly all azimuth patterns is the mathematical standard for drawing polar plots in which zero degrees (0°) is at the right and angle increases in the CCW direction — the opposite of a map. For compass bearing scales, north is at the top and angle increases in the clockwise (CW) direction.

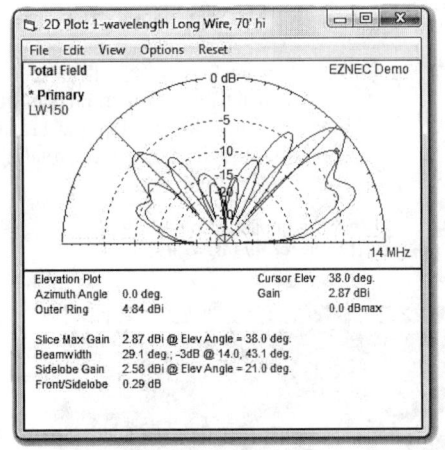

(A)

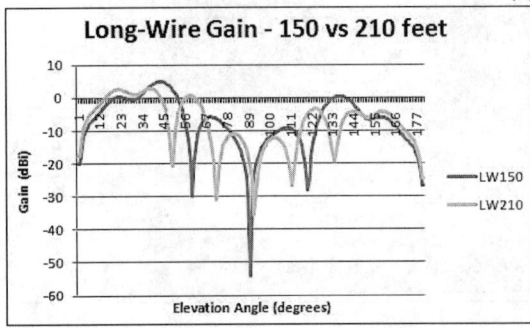

(B)

Figure 5 —The standard polar plot format at A compares the elevation patterns of two long wire antennas. The same information is presented in Linear dB format at B.

direction so that east (90°) is at the right and so forth.

This criss-cross of conventions is not an issue for most amateur antennas because they are symmetrical along the line of maximum forward gain. Theoretically, a symmetrical antenna such as a Yagi has a pattern that is the same to either side of the forward direction. More complex antennas, such as electrically steered or asymmetric arrays, have asymmetric patterns. For these antennas, it very much does matter which angle convention is used.

The way to change the angle convention in *EZNEC* is from the main window's OPTIONS | ANGLE CONVENTION menu selection. Pick either COMPASS BEARING or CCW FROM X AXIS as desired.

Elevation patterns can take the unusual angular assignment of 0° to both the left and right hand horizontal axes with straight up being 90°. *EZNEC* uses a convention of 0° being to the right and 180° to the left regardless of the selected azimuth convention.

E and H Plane Patterns

Another frequent area of confusion is in the relationship of E and H plane patterns to azimuth and elevation patterns and to vertical and horizontal directions. Let's start by defining what is meant by the terms E and H plane. The electromagnetic field radiated by an antenna is a combination of an E field (or electric field) and an H field (or magnetic field). In the far field of the antenna, the E and H fields are perpendicular. So far, so good, right?

The orientation of the E and H fields is determined by the orientation of the currents in the antenna. An ac current flowing in a straight line generates both an E field parallel to the line and an H field that curls around the line according to the right-hand rule assuming we are talking about conventional current that flows from positive to negative.[2] Further assuming we are talking about antennas constructed of thin wire or tubing elements, the orientation of the elements also determines the orientation of the E and H fields because the current direction and element direction are the same. Again, so far, so good.

How are a dipole's E and H fields oriented? Since the current in the dipole flows in the same direction as the axis of the dipole, the E field is oriented parallel to the dipole and the H field forms

[2]**en.wikipedia.org/wiki/Right-hand_rule**

circles around the dipole at right angles to the E field. If a radiation pattern is generated in a plane that contains the dipole, it will show the strength of the E field around the antenna. Similarly, a pattern that cuts through the dipole at right angles shows the H field. An E field (or H field) radiation pattern not only shows the strength of the E field (or H field) but is *also* oriented in the direction of maximum gain.

The orientation of E and H fields and patterns depends solely on the orientation of the antenna currents, which is almost always the same as for the elements of amateur antennas. Here's an example: If a Yagi has horizontal elements, the azimuth pattern with an elevation angle containing the peak of the main lobe is an E field pattern. An elevation pattern at right angles to the Yagi elements through the main lobe is an H field pattern. Whether the E or H field patterns are azimuth, elevation, horizontal or vertical patterns depends on the orientation of the antenna with respect to the surface of the Earth.

Linear dB Pattern Plots

The "linear dB" method of graphing an antenna's radiation pattern is not common in amateur use (yet) but is very common in professional and academic publications. In this type of graph, the vertical axis is gain in dB and the azimuth or elevation angle is the horizontal axis.

Figure 5 compares a regular polar radiation pattern to a linear dB plot to illustrate why the linear graph is particularly useful when comparing two antenna designs. Figure 5A shows the usual polar elevation pattern for a terminated long wire antenna on 20 meters. A pattern is shown for a 210 foot wire (black) and a 150 foot wire (blue). Figure 5B shows the same information plotted in linear dB format.

Which type of plot makes it easier to compare the performance of the two antennas? Around the main lobe, the polar plot does well. Away from the main lobe and particularly at low gains and nulls, the linear graph is much easier to read both in angle and amplitude. In addition, it's far easier to see the difference in gain on the linear graph at any angle. Linear dB patterns are used to compare antennas and to present a clearer picture of the antenna pattern away from the main lobe. In *EZNEC*, the Linear dB scale is available by using the OPTIONS | 2D, 3D PLOT SCALE menu selection.

A Pattern of Options

We have just scratched the surface of all the many pattern options available to antenna designers. Hopefully, though, this column has made you more aware of the importance of the more common options and parameters for the patterns you use to select and buy antennas.

Experiment 92 — The 468 Factor

This month, I'm going to revisit a topic originally touched upon in Experiment #82 — Antenna Height.[1] We've all seen this number over and over again — the "magic number" that gives us the length of a half wavelength dipole in feet from the dipole's resonant frequency: L = 468/f. In free space the wavelength in feet is 492/f, but a practical half wavelength antenna is shorter so the constant is smaller. After publishing the column showing what happens as an antenna gets raised and lowered, my interest was piqued. [This column is based on **eham.net** article #23802, "Where Does 468 Come From?"]

The number 468 is on the license exams and in the literature. It's been there ever since I started reading about ham radio in the mid 1960s. It's a pillar of amateur antenna theory. Every ham is expected to memorize it. And it's wrong.

It would be more accurate to say that it's rarely correct. There are certain instances where it's close, but using it often leads to wasted wire. The usual instructions to a new ham are, "Calculate how much wire you need using 468/f and then add a couple of feet." What that really means is the value 468 is too small and we compensate for the error by adding a couple of feet. If 468 isn't right, why do we use it? Answering that question required me to take a meandering trip along the paths of history.

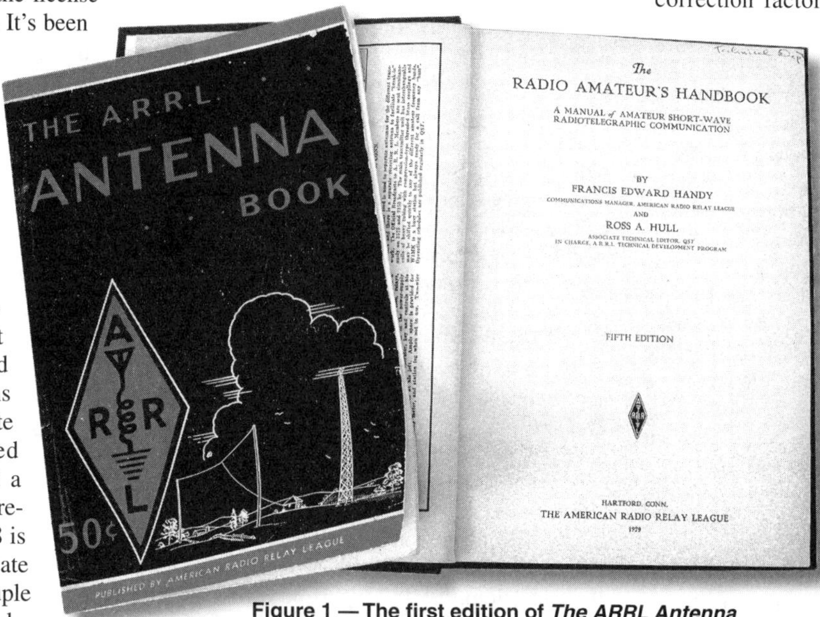

The Search for 468

Recently, I had the opportunity to spend a few days at ARRL Headquarters to plan upcoming writing and editing projects. The ARRL has a great technical library with every edition of ARRL publications and technical publications going back decades. (If you ever get close to Connecticut, it's well worth dropping in on the ARRL for a tour.) I had some time one afternoon and decided to find out when and how the number 468 first appeared in the literature of Amateur Radio.

My first stop was *The ARRL Antenna Book*'s initial edition in 1939. Sure enough, on page 13 in the chapter on "Antenna Properties," the familiar formula 468/f appears. The *Antenna Book* states that the *end effect* caused by the attachment of insulators at the ends of the antenna results in the approximately 5% reduction in length from the free space 492/f to 468/f. The text goes on to state that the percentage "varies slightly with different

Figure 1 — The first edition of *The ARRL Antenna Book* from 1939 and the 1929 edition of *The ARRL Handbook*, in which the source of 468 was found.

installations," but doesn't say how, nor is a citation provided to identify how the value of 468 was obtained.

Since it is unlikely that the value of 468 appeared in the *Antenna Book* without any "prior art," I next turned to *The ARRL Handbook*'s first edition in 1926. That turned out to be a dry hole — no formula for antenna length and nothing in 1927 or 1928 either. Then, in the 1929 edition's "Antennas" chapter on page 128, I hit pay dirt. The text defines natural wavelength as the highest wavelength (the lowest frequency) at which the Hertz antenna (a half wavelength dipole) will resonate. It is stated that "The natural wavelength of the wire…will be its length in meters multiplied by 2.1." Hmmm…2.1 is 5% longer than would be the free space value of 2. (Remember, the text is discussing wavelength, not frequency.) Farther down the page I saw, "Speaking in terms of feet, the natural wavelength of the antenna will be its length in feet divided by 1.56." That equation translates to L = (300 × 1.56)/f and 300 × 1.56 is 468. Here were the headwaters of the mighty River 468!

Still, no background for the correction was given. Where does the use of a correction factor originate? Back to the stacks. Did I really want to go through all of the *QST* magazines until I found my answer? Well, not really, but inspiration struck in the form of the online *QST* archives.[2] I logged into the ARRL Web site, brought up the *QST* archive search page and…hit another roadblock. I couldn't very well search for "468" because it was unlikely to be a keyword. "Dipole" would return hundreds of hits. Then I realized that in the early days, a half wavelength dipole would have been referred to as a "Hertz" antenna or "Hertzian" antenna. I entered the former and scrolled down to the very earliest entries.

The oldest article on Hertz antennas was in the July 1925 issue by 9BXQ and titled "The Hertz Antenna at 20 and 40 Meters" but it didn't discuss a formula for length. The next oldest article, October 1926's "The Length of the Hertz Antenna" by G. William Lang, turned out to be what I was looking for.

In the article, Lang (who was apparently

---

[1]All previous Hands-On Radio experiments are available to ARRL members as downloadable PDF files at **www.arrl.org/hands-on-radio**.

[2]Archives of ARRL publications are available online to ARRL members at **www.arrl.org/arrl-periodicals-archive-search**.

Antennas and Transmission Lines 57

ANT NO	WIRE LENGTH TIP TO TIP		FUNDAMENTAL WAVELENGTH	K BY WHICH λ IS MULTIPLIED TO GET WIRE LENGTH		SHAPE
	Feet	Meters	Meters	Feet	Meters	
1	30	9.1	21.3	1.409	.4275	/
2	31	9.4	22.0	1.410	.4275	/
3	44	12.5	28.0	1.57	.4460	‾\
4	56	17.1	39.5	1.418	.4300	‾\
5	61	18.6	43.0	1.415	.4320	L
6	76	23.1	51.5	1.47	.4480	—
7	57.5	17.5 plus	40.0	1.435	.4380	‾\
8	110.	33.5	78.0	1.410	.4300	L
9	55	16.7	37.0	1.465	.4450	—

TABLE SHOWING MEASUREMENTS MADE AND CONSTANTS FOR EACH ANTENNA

Figure 2 — Table from page 16, Oct 1926 *QST*, reproduced as a figure to show the old style lettering. Do you see the error in the wire length for antenna No. 3? It should be 13.4 rather than 12.5 meters.

Table 1
Variation in Length × Feet with Height of a 20 Meter Dipole

Height in Wavelengths at 14.175 MHz (feet)	Length × Feet
⅛ (8.8)	467.8
¼ (17.4)	466.4
½ (34.7)	483.4
¾ (52.0)	473.4
1 (69.4)	480.5
1½ (104.1)	479.1
2 (138.8)	479.1

not a ham, but worked in the Department of Radio Operations for Radio Station WBZ in Boston) set up some Hertz antennas at amateur station 1KA and also measured antennas at station 1CK and 1KF. He used an oscillator and a wavemeter to determine the frequency at which the antenna resonated then measured the entire antenna — tip-to-tip, including the counterpoise.

A table of correction values was derived, with the free space wavelength in meters multiplied by an average value of 1.46 to get the antenna's resonant wavelength in feet. This corresponds to an equation of L = 438/f. This is the first suggestion that the actual resonant length of a practical amateur antenna can be predicted by applying a correction factor to a free space wavelength.

The table is reproduced in Figure 2, neatly lettered in the style of the day. Take a look at the right hand column, though. It shows the *shape* of the antenna. Antenna enthusiasts will recognize an early inverted L along with a vertical and a "bent vertical." In those days, it was not at all clear (at least to amateurs) what effect antenna shape had on resonant wavelength or on radiating capability. Thus, it was important to note everything when making measurements. These antennas also included a *counterpoise* — another wire laid out on or near the ground.

The experiments of 1925 and 1926 took place on or near 40 meters. In those days, CW operation on what we now call the "low bands" of 80 and 40 meters was the norm. At these wavelengths, a half wavelength dipole was of a reasonable length. It could be made of ordinary copper wire, probably #8 to #14 AWG, and installed in the back yard at heights of 20 to 40 feet. For these antennas, ⅛ to ¼ wavelength above ground, a value of 468 is about right, resulting in the equation printed in the *ARRL Handbook* in 1929.

Discovering this article was a little spooky, since one of my first "serious" ham radio publications was a report on some antenna testing I performed with Steve Morris, K7LXC.[3] Here I was, discovering that my answer was to be found in the report of another set of antenna tests nearly 85 years before!

Why Doesn't 468 Mean 468?

In truth, many variables affect the resonant frequency of a half wavelength dipole. The two primary factors are the length to diameter ratio of the antenna conductor and, most strongly, the antenna's height above ground. These can combine to change the actual correction factor quite a bit! (Insulation can also affect an antenna's electrical length.) In Experiment #82, I modeled a typical 20 meter dipole made of #12 AWG uninsulated wire at heights from ⅛ to 2 wavelengths over realistic ground and calculated the correction factor when its length was adjusted to resonance at each height. (The resonant frequency is the frequency at which the feed-point impedance is purely resistive, i.e. X=0) It varied from 466 to 481 over that range as you can see in Table 1.[4] Clearly, using 468/f would lead to an antenna being too short most of the time.

If 468 is too small and rarely correct, what should you do? Realistically, you should expect to trim your dipole to get the resonant frequency you want. Instead of being frustrated that the calculations aren't exact, learn to adjust the antenna's length efficiently by using an instrument such as an antenna analyzer. Start with an estimated value based on a more realistic formula such as 490/f that results in a small amount of extra wire for attaching insulators. During tuning, twist the wire connections together or use clamps, then raise the antenna into position and measure. When it's right, only then solder and weatherproof the connections. Recognize that every antenna's circumstances are slightly different — height, ground conductivity, thickness of wire, nearby conductors and so forth.

Magic Numbers

Another lesson to learn from this exploration is to realize that "magic numbers" in formulas have often been determined through experimentation under specific circumstances. As such, they likely depend on a variety of factors that you may not be able to replicate. They will only approximate what you actually encounter. If the assumptions behind the value are given — you can use that information by comparing it to your situation. If the assumptions are not known — you should allow for variations or try to find a more accurate model representative of your own circumstances.

I hope you've enjoyed reading about this journey as much as I enjoyed taking it, opening the covers of books nearly 80 years old and mapping the stream of knowledge back to its sources — finding there the footprints of wireless pioneers that set ham radio on the course we travel today.

For Further Reading

This would be a great opportunity for you to use the *QST* archives and find that October 1926 article. When you're through, enjoy a trip through ham history in the pages of that old *QST*. The only thing you'll be missing online will be the wonderful smell of an ancient magazine!

---

[3]W. Silver, NØAX, and S. Morris, K7LXC, "HF Tribander Performance — Test Methods and Results, Second Edition," Champion Radio Products, www.championradio.com, 2000.

[4]See also J. Hallas, W1ZR, "The Real World Meets Your Real Antenna," *QST*, Apr 2010, p 47.

Experiment 94 — SWR and Transmission Line Loss

SWR (standing wave ratio) is something everybody measures — it's one of the most widely used numbers in ham radio. But not everybody understands what SWR is and what affects it. In this month's experiment, we'll have a quick refresher on SWR basics and then examine the effect of losses in the transmission line on the SWR you measure back at the shack.

SWR Basics

If power traveling through the transmission line encounters a load impedance, Z_L, different from the transmission line's characteristic impedance, Z_0, some of the power is reflected back along the transmission line. This creates a stationary pattern of voltage (and current) along the line. The stationary pattern is called a *standing wave* and the ratio of the peak to minimum voltage or current is called the *standing wave ratio* or *SWR*. (A complete treatment of SWR is available in *The ARRL Antenna Book*.[1])

SWR is a numeric way of describing the relationship between the transmission line's Z_0 to the load impedance, Z_L, such as the impedance of an antenna. The simplest SWR calculation is SWR = Z_L/Z_0 or Z_0/Z_L, whichever is greater than 1. (SWR is never less than 1:1, pronounced "one to one.") If a 50 Ω transmission line is connected to a 75 Ω load, the SWR = 75/50 = 1.5:1. If the load is 25 Ω, the SWR = 50/25 = 2.0:1.

In order to work with the incident or forward power, P_f, traveling toward the load and reflected power, P_r, traveling away from the load, it is useful to define a *reflection coefficient*, denoted by the symbol ρ (rho). Sometimes the symbol Γ (gamma) is used.

$$|\rho| = \frac{SWR-1}{SWR+1} \quad \text{and} \quad SWR = \frac{1+\rho}{1-\rho}$$

This equation only calculates the magnitude of the reflection coefficient, which is a complex number when either the line or load impedance contains reactance.

The amount of forward and reflected power can be used to calculate the SWR (and vice versa):

$$\rho = \sqrt{\frac{P_r}{P_f}} \quad \text{and} \quad SWR = \frac{1+\sqrt{P_r/P_f}}{1-\sqrt{P_r/P_f}}$$

Thus, forward and reflected power, SWR and the reflection coefficient are all conveniently related.

Line Loss

No transmission line is perfect — some of the power input to the line will be dissipated as heat. The heat is the result of either resistive loss (proportional to the square of the current) in the conductors or from losses in the dielectric (proportional to the square of the voltage). This is called *line loss*. The amount of loss in the transmission line as power flows through it to a load with $Z_L = Z_0$ is called the *matched-line loss*, abbreviated ML and measured in dB per unit of length, such as dB/ft.[2]

ML increases with frequency and is usually specified in dB/100 feet at several frequencies, often 1-10-100-1000 MHz. For example, the table of transmission line characteristics on page 24-18 of *The ARRL Antenna Book* shows that RG-58C/U (Belden part number 8262) has an ML of 0.4 dB/100 feet at 1 MHz, 1.4 dB/100 feet at 10 MHz, 4.9 dB/100 feet at 100 MHz, and 21.5 dB/100 feet at 1 GHz. (This would not be a good transmission line for use on the 23 cm band!)

Loss Due to SWR

If the load impedance is not equal to that of the transmission line, causing some of the power to be reflected from the load, not all of it is absorbed by the load. If the load is an antenna, that means not all of the power is radiated. The reflected power travels back to the other end of the line where all or part of it may be reflected again for another trip to the load. Eventually, all power input to a transmission line is either dissipated as heat in the line or absorbed by the load or generating source impedances.

Is this extra loss a problem to worry about? On HF, for SWR of less than 2:1, the additional loss over and above ML due to part of the power being reflected is usually less than 0.5 dB. For lines of reasonable length, this is an insignificant amount in nearly all circumstances. Once the SWR reaches 3:1 or higher, however, the additional loss can be noticeable, or worse. Figure 1 is a graph of additional loss due to SWR for varying values of ML.

To use Figure 1, first determine ML for the length of transmission line being used. (Table 1 lists data for a few common cables and Figure 23 on page 24-20 of *The ARRL Antenna Book* shows ML for 100 feet of many types of common transmission lines over the frequency range from 1 to 1000 MHz.) Then use Figure 1 to determine the additional loss and add it to the ML to obtain the total loss.

For example, what is the total loss of a 200 foot piece of RG-58C/U at 100 MHz if the SWR is 4:1? The manufacturer specifies ML for RG-58C/U as 4.9 dB/100 ft at 100 MHz, so ML for a 200 foot length is

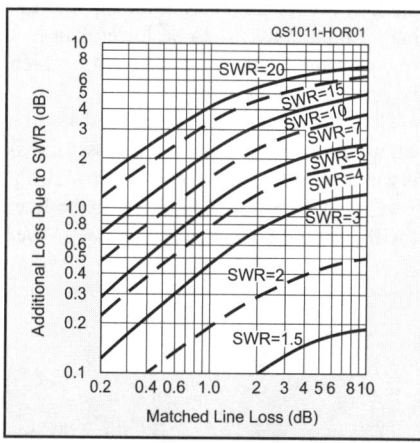

Figure 1 — Additional line loss due to SWR as measured at the load. Total loss is equal to ML plus the additional loss.

---

[1]R. D. Straw, Editor, *The ARRL Antenna Book*, 21st Edition. Available from your ARRL dealer or the ARRL Bookstore, ARRL order no. 9876. Telephone 860-594-0355, or toll-free in the US 888-277-5289; www.arrl.org/shop; pubsales@arrl.org.

[2]S. Ophanidis, *Electromagnetic Waves and Antennas*, ECE Department, Rutgers University, www.ece.rutgers.edu/~orfanidi/ewa/, Chapter 10.

Table 1
Matched Loss (ML) of Common Transmission Lines as a Function of Frequency

Line Type	ML for 100 feet at Frequency (MHz)			
	1	10	100	1000
RG-213	0.2	0.6	1.9	8.0
RG-8X	0.2	0.7	2.3	7.4
RG-58C/U	0.4	1.4	4.9	21.5
RG-59	0.3	0.9	2.6	8.5
RG-174	1.9	3.3	8.4	34.0

Cables with the same RG designation from different manufacturers may differ in characteristics. Consult manufacturers' data sheets.

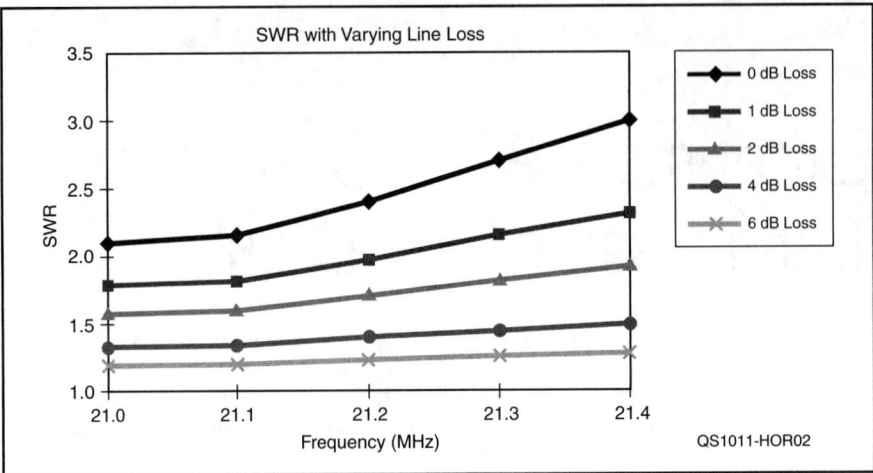

Figure 2 — Increasing line loss (ML) reduces SWR of a 15 meter quad measured at the input to the transmission line. The top curve (0 dB line loss) shows SWR at the antenna that is the same as input SWR for a lossless line.

$2 \times 4.9 = 9.8$ dB. From Figure 1, start at ML ≈ 10 dB on the horizontal axis and follow a vertical line up to where it intersects the SWR = 4 curve. On the graph's left hand vertical axis, that intersection is just less than 2 dB, so additional loss is about 1.9 dB. Total loss is then $9.8 + 1.9 = 11.7$ dB. Again, this would not be a very good transmission line choice, since only about $\frac{1}{15}$ of the input power would make it into whatever load is attached.

Input SWR

Note that the caption for Figure 1 refers to *SWR measured at the load*. When first studying SWR, say for a license exam, the study material often states that "SWR does not change along a transmission line." For this statement to be true, the transmission line must be lossless. Why? Because some of the reflected power is lost due to ML on its return trip from the load back to the source. As we get closer to the source, less of the original forward power has been lost, so its value is increasing! From the second equation, you can see that if P_r decreases and P_f increases, then SWR gets closer to 1:1. In other words, line loss causes the SWR at the input to the line — *input SWR* — to be lower than if it were measured at the load, the point at which the ratio of P_r to P_f is lowest.

Let's take this to an extreme and assume that we have a very, very long and lossy transmission line and apply some power to it, P_f. If we measure SWR at the input to the line, it will always be 1:1 because no power ever returns from the load and $P_r = 0$. It doesn't matter what the load impedance is at all! This is an important lesson: line loss reduces input SWR.

Most transmission lines are not terribly lossy in the frequency ranges they're designed for, but how much do they really affect the typical SWR curve? Figure 2 shows the effects of loss on the input SWR curve for a 15 meter quad antenna for ML values of 0, 1, 2, 4 and 6 dB. As you can see, the lossier the line, the better the antenna looks — in terms of SWR, anyway. With 6 dB of line loss, you might wonder why your received signal is weak, even though SWR never exceeds 1.3:1. (The formula for input SWR was obtained from Reference 2 and an *Excel* spreadsheet is provided on the Hands-On Radio Web site for making these calculations and graphs.[3])

Observing the Effects of Line Loss

You can observe this effect yourself with the help of an antenna analyzer that can operate at and above 100 MHz, such as an MFJ-259 or similar instrument. Acquire a long piece of coaxial cable and short the far end of the cable with a short piece of wire,

[3]All previous Hands-On Radio experiments are available to ARRL members as downloadable PDF files at **www.arrl.org/hands-on-radio**.

creating a load SWR equal to infinity. (A short circuit is easier to create than an open circuit due to stray capacitance.) Connect the antenna analyzer to the other end, set to a frequency of a few MHz. The SWR indication should be very high or infinite. Now increase the frequency while watching SWR on the analyzer. As frequency increases, you will see SWR begin to decrease. For most cable types, above 100 MHz SWR will decrease rapidly. Compare different types of cable to see the effects of line loss. The lower the input SWR reading (at the analyzer), the lossier the line at that frequency because load SWR is infinite!

For Further Reading

Read the Transmission Lines chapter of either the *ARRL Handbook* or the *ARRL Antenna Book* for more information about these amazing cables that we take for granted.[4] The mathematically inclined reader may enjoy browsing through Reference 2 or any engineering textbook that covers transmission lines.

[4]*The ARRL Handbook for Radio Communications*, 2011 Edition. Available from your ARRL dealer or the ARRL Bookstore, ARRL order no. 0953 (Hardcover 0960). Telephone 860-594-0355, or toll-free in the US 888-277-5289; **www.arrl.org/shop; pubsales@arrl.org**.

Experiment #81 — Synchronous Transformers

Why are these transmission line sections *synchronous*? Synchronous has multiple definitions, but the one that applies here is: *going on at the same rate and exactly together; recurring together* — from *syn* (together with) + *chron* (time) + *ous* (possessing or having the quality of). In this case, time refers to phase. Specifically, the phase of waves in the transmission line used to make up this month's project!

Quarter Wave Transformer (Q section)

The best-known example of a synchronous transformer in Amateur Radio is the *quarter wave transformer*, also known as a *Q section*, shown in Figure 1. The Q section creates an impedance match between two impedances, Z_1 and Z_2, by inserting between them a quarter-wavelength of transmission line with a characteristic impedance that is the geometric mean of the impedances to be matched, $Z_Q = \sqrt{(Z_1 \times Z_2)}$.

That deceptively simple equation represents what happens as the result of an infinite series of reflections occurring at the junctions of the three sections of transmission line. Figure 2 illustrates the first few steps. Let's follow along. Beginning in Step 1, as the electromagnetic wave in Line 1 encounters transmission lines with impedance different than that in which it's traveling.

From transmission line theory, we know that some of the incident wave's energy will be reflected at the impedance discontinuity (Step 2), generating a reflected wave. Viewing the waves in terms of voltage, the incident wave is E_{I1} and the reflected wave is E_{R1}. The ratio between the incident and reflected voltages is the *reflection coefficient*, $\rho_1 = E_{R1} / E_{I1}$. That means $E_{R1} = \rho_1 \times E_{I1}$.

If the wave encounters an infinite or zero impedance (an open or short circuit, respectively), $\rho = 1$. If the impedance the wave encounters is the same as the impedance of the line it's traveling through, such as a matched load, there is no reflection and $\rho = 0$. For any other value of impedance encountered, ρ is between 0 and 1. (The full representation of ρ includes phase, meaning that ρ is really a complex number of the form $|\rho| \angle \theta$, but for the purposes of this discussion, we will consider the magnitude and phase of ρ separately.)

The reflected wave, E_{R1}, travels back along Line 1 in the opposite direction from the incident wave. The remaining energy that wasn't reflected continues on in Line 2 with a new voltage, E_{I2}. Whatever generated the original wave, E_{I1}, will eventually see the reflected wave, E_{R1}, return.

That might be the end of the story, but there is another change in impedance a little further along where Line 2 meets Line 3, and another set of reflections is generated (Step 3), with E_{I3} continuing on in Line 3. If the reflection coefficient at the second discontinuity is ρ_2, then another reflected wave, E_{R2}, is generated with a voltage of $\rho_2 E_{I2}$, traveling in the same direction as E_{R1}. It encounters the initial discontinuity

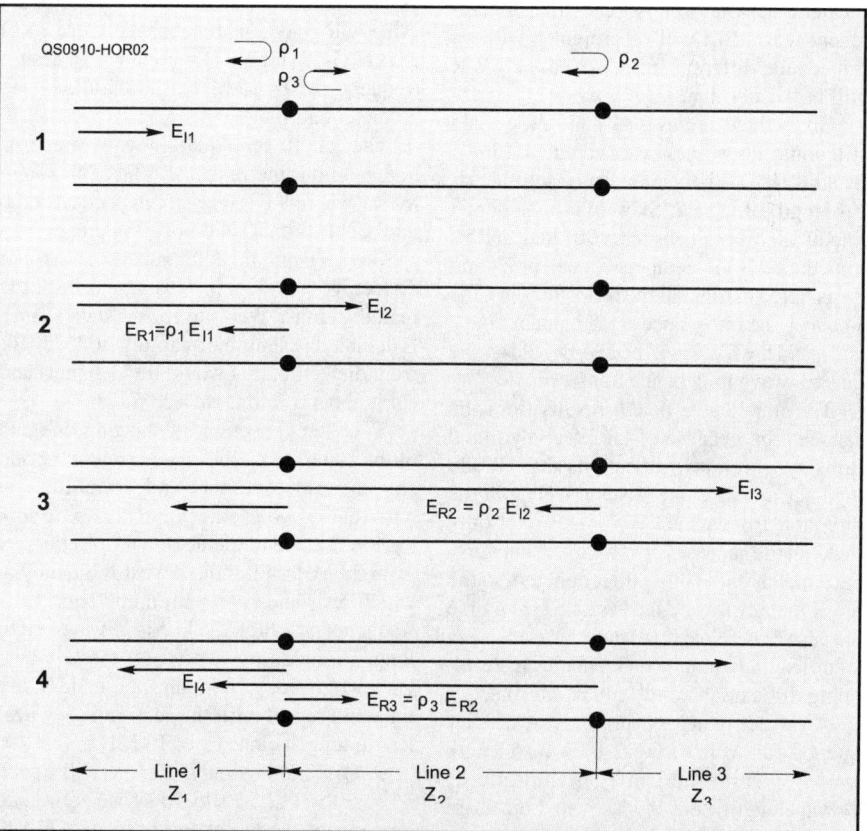

Figure 1 — The ¼ wave and ¹⁄₁₂ wave synchronous transformers. A series of carefully phased reflections create an impedance match.

Figure 2 — The initial sequence of reflections generated by a ¼ wavelength section of transmission line inserted between lines of different characteristic impedances.

(Step 4) while traveling in the opposite direction as the original incident wave, E_I, and generates another pair of waves, E_{I4} and E_{R3}, according to reflection coefficient ρ_3.

We now have three waves adding together in Line 1 — E_{I1}, E_{R1} and E_{I4}. E_{R1} is smaller than E_{I1} and E_{I4} is smaller than E_{R1}. The resulting voltage sum, however, depends on the relative phase of all three waves. The phase difference between E_{I1} and E_{R1} depends on the relationship between Z_1 and Z_2: if $Z_2 > Z_1$, then E_{R1} is in phase with E_{I1}, and out of phase if $Z_2 < Z_1$. (We are assuming all three impedances are purely resistive, so no additional phase shift due to reactance occurs.) E_{I4} differs in phase from E_{R1} by twice the electrical length of Line 2 because the wave has to travel through Line 2 once in each direction before returning to the junction of Lines 1 and 2.

You can see that more reflections will occur as E_{R3} travels back to the junction of Line 2 and Line 3 and generates another pair of waves. This happens forever, until the incident wave E_{I1} ceases. (We are also ignoring any reflections created at the unseen ends of Line 1 and Line 3.) As each set of increasingly smaller reflected and incident waves are combined, the result eventually converges on a steady-state value for the voltages of the waves traveling in each direction in all three segments of transmission line.

In Line 1, this combination of voltages means that whatever is generating the incident wave, E_{I1}, will be presented with an impedance different from Z_1 and the SWR will be greater than 1:1. Or will it?

Given fixed values for Z_1 and Z_3, we can still control how the waves add up in Line 1 by adjusting both the impedance and electrical length of Line 2. Skipping to the punch line, if Line 2 happens to be 90° long and its impedance is the geometric mean of Z_1 and Z_2 as noted earlier, all the reflections in Line 1 cancel, the impedance at the input to Line 1 is Z_1, and the SWR will be 1:1. Behold — the quarter-wave impedance transformer!

The impedance match results from the cancellation because of the precisely timed (thus synchronous) reflections that all add together to cancel all waves traveling in the direction from which E_{I1} comes. Not only that, but the same set of wave mechanics create a match in the other direction, too, so the SWR looking toward the Q section from Line 3 is also 1:1! (Note that I have glossed over a significant amount of mathematics in developing this explanation.[1] You're welcome.)

An exact match occurs only at the frequency for which Line 2 is ¼ wavelength long (or some odd integer multiple of ¼ wavelength) and if losses in Line 2 are

[1]R. Lay, W9DMK, "A Transient Analysis of an Impedance Transforming Device (The Quarter-Wave Transformer)," **www.qsl.net/w9dmk/qrtrwav4.pdf**.

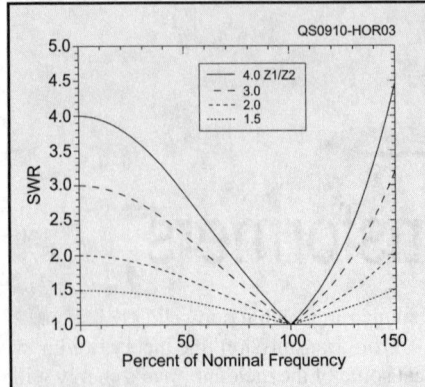

Figure 3 — The bandwidth of the ¹⁄₁₂ wavelength transformer. The resulting SWR for frequencies from dc to 1.5 times the design frequency and resistive impedance mismatches up to 4:1 (from Note 3).

low enough that they have an insignificant effect. Nevertheless, the Q section provides an excellent match over several percent of bandwidth — good enough to use for a whole band at and above 7 MHz. Let's make one!

Making a Q Section

The hardest part about making a Q section is often just coming up with the odd impedance transmission line! However, there's no need to obsess over getting an exact match — any impedance within 10% of the exact value will give good results. We'll use 75 Ω coax (RG-59) for our Q section with a design frequency of 14.15 MHz, the middle of the 20 meter band.

The length of a quarter-wave piece of transmission line in feet is VF × 246 / f (in MHz), where VF is the *velocity factor* for the line. Cable with a solid polyethylene dielectric has a VF of 0.66. If your cable has a foam dielectric, the VF is 0.78 to 0.83 (check the manufacturer's Web site for the data sheet). Calculate the length of cable you need (for solid dielectric, that would be 11.5 feet) and cut it about 5% long, or 12.05 feet.

Tune the Q section by shorting one end of the cable (twist the center conductor and shield braid together) and attaching the other end to your antenna analyzer. Ignore the SWR value and tune down in frequency from 28 MHz. Find the lowest frequency at which resistance is a minimum. This is the frequency at which the line is ½ wavelength long, so divide by two to get the ¼ wavelength frequency. Trim an inch or less at a time and repeat until the section is ¼ wavelength long at close to 14.15 MHz.

A Q section using 75 Ω line will match $75^2 / 50 = 112.5$ Ω to 50 Ω, so terminate one end of the Q section with two 220 Ω resistors in parallel using reasonably short leads. Verify that the SWR at the other end of the Q section is close to 1:1. Vary the frequency and find the Q section's *SWR bandwidth* — the range of frequencies over which SWR is 2:1 or less. Substitute different values of terminating resistance to find out how much the termination can vary in either direction before SWR at the other end exceeds 2:1. Try the Q section at a frequency where it is ¾ wavelength long and again at ⁵⁄₄ wavelength.

Twelfth Wave Transformers

The Q section is really a special case of series section matching.[2] There's no restriction (other than complexity) that there be just one matching section. In fact, the two section variation shown in Figure 1B is quite handy for matching two different impedances of transmission line, such as 50 Ω coax and 75 Ω hardline.[3] Best of all, it doesn't require any special transmission line impedances, only sections of line with the same impedances that are to be matched!

This configuration is referred to as a *twelfth wave transformer* because when the ratio of the impedances to be matched is 1.5:1, as is the case with 50 and 75 Ω cables, the electrical length of the two matching sections between the lines to be matched is 0.0815 λ, quite close to λ/12 (0.0833 λ). As the lowest line in Figure 3 shows, the SWR bandwidth of the transformer is quite broad, but decreases as the ratio of impedances to be matched increases. You can use this trick to make good use of surplus low loss 75 Ω CATV hardline between 50 Ω antennas and radios!

Parts List

- Coaxial cable, 20 feet of RG-59.
- Resistors, 2 each, 220 Ω, ¼ W carbon composition (non-inductive).

Recommended Reading

Quarter-wave transformers are the basis for non-reflective optical coatings. As you read through **en.wikipedia.org/wiki/Optical_coating** you may recognize several similarities between optics and transmission lines. ARRL members should also download the referenced *QST* articles from the *QST* Archive.

Next Month

How high should it go? Isn't higher better? Find out next month as we do a virtual experiment using antenna modeling software to observe the effects of height above ground on antennas.[4]

[2]F. Regier, OD5CG, "Series-Section Transmission-Line Impedance Matching," *QST*, Jul 1978, pp 14-16.
[3]D. Emerson, AA7FV, "Try a Twelfth-Wave Transformer," *QST*, Jun 1997, pp 43-44.
[4]Previous Hands-On Radio columns and a complete parts list for all experiments are available to ARRL members at **www.arrl.org/tis/info/HTML/Hands-OnRadio**.

Experiment 111
Coiled-Coax Chokes

After several months laboring on circuit boards it's time to get back to something a little more "radio-y," don't you think? Back in Experiment #91 the topic was an overview of *common-mode chokes* that are used primarily to fight RFI, isolate antennas from feed lines and reduce RF current on the outside surface of coax shields.[1] This month, we'll take a closer look at one of the most popular types of common-mode choke, the *coiled-coax choke*. By building one in three different ways you'll get a better idea of how they work.

Functions of a Balun

The coiled-coax choke is often referred to as a *choke balun*. You probably already know that *balun* is an abbreviation for **bal**anced to **un**balanced. What may not be clear is that balun is a general term referring to *any* device that transfers differential-mode signals between balanced and unbalanced systems while maintaining symmetrical signals in the balanced system. That's a mouthful!

You'll encounter both *voltage* and *current* baluns. Voltage baluns develop equal voltages at the balanced terminals and current baluns develop equal currents. If the balanced terminals both have equal impedances, such as a properly terminated parallel conductor feed line, equal voltage and equal currents will result with either type. If the balanced system is an antenna system, however, the impedances of the two terminals are often quite different. That means equal currents won't flow into the antenna terminals and since current is what results in radiating signals, the antenna won't be radiating symmetrically. Thus, a current balun is preferred for use with an antenna to force the currents to be equal.

Why wouldn't an antenna have roughly equal impedances at its terminals? You might have noticed the word *system* in the previous paragraph. That's a clue that there might be more than just the antenna itself getting involved. Sure enough, there is —

[1] All previous "Hands-On Radio" experiments are available to ARRL members at **www.arrl.org/hands-on-radio**.

the outside of the coax shield that is also connected to one side of the antenna either directly or by coupling to the antenna through proximity. The function of the current balun is to disconnect that "third wire" while forcing equal currents into the actual antenna terminals. (This is described in detail by Roy Lewellan, W7EL, in his article "Baluns: What They Do and How They Do It."[2])

There are three common methods of performing the current balun function. The first is to turn the outside surface of the coax shield into a quarter-wavelength transmission line that has high impedance at the point where the balanced and unbalanced systems meet. This is a *sleeve balun*, often used at VHF and UHF where wavelengths are short. The second method creates high impedances with ferrite beads and the third method by coiling the feed line to form an inductor. (All three techniques are discussed in *The ARRL Handbook* and *The ARRL Antenna Book*.[3,4])

The inductive reactance of the coiled feed line acts as an *RF choke*, reducing current flow and forcing the equal currents inside the feed line to divide equally between the antenna terminals. Similarly, received signal currents on the antenna have no other path and so combine in equal amounts inside the feed line.

Making a Coiled-Coax Choke Balun

While coiling the cable does create an inductor, you must also consider the effects of inter-winding capacitance that creates a

[2] R. Lewallen, W7EL, "Baluns: What They Do and How They Do It," **www.eznec.com/Amateur/Articles/Baluns.pdf**.
[3] *The ARRL Handbook for Radio Communications,* 2012 Edition. Available from your ARRL dealer or the ARRL Bookstore, ARRL order no. 6672 (Hardcover 6634). Telephone 860-594-0355, or toll-free in the US 888-277-5289; **www.arrl.org/shop**; **pubsales@arrl.org**.
[4] *The ARRL Antenna Book,* 22nd Edition. Available from your ARRL dealer or the ARRL Bookstore, ARRL order no. 6948. Telephone 860-594-0355, or toll-free in the US 888-277-5289; **www.arrl.org/shop**; **pubsales@arrl.org**.

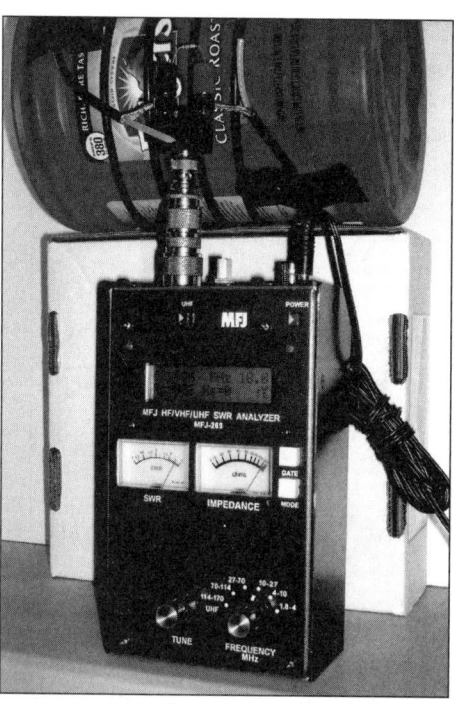

Figure 1 — The coiled-coax choke balun is wound on a plastic form and attached to the analyzer using a binding-post adaptor. The analyzer then reads R_S and X_S. A 1 kΩ swamping resistor across the binding posts limits the variations in R_S and X_S to displayable values.

parallel-resonant circuit with the coil so that it has a *self-resonant frequency*, f_0: Below f_0, the balun acts like an inductor and above f_0 it acts like a capacitor.

This is fairly easy to demonstrate by using an antenna analyzer that can display impedance (or a combination of resistance and reactance) such as the MFJ-269 shown in Figure 1. All you'll need is a few feet of spare coax and a 1 kΩ resistor. For the baluns tested in this experiment, I used 9 feet of RG-58 coax. Some plastic forms are also required — I used a 6 inch diameter empty coffee container and a 3 inch diameter almost-empty peanut butter jar. Any plastic or glass form will do — just be sure not to use metal.

Strip about 1½ inches of jacket off each end of the coax and create pigtails by twisting the braid strands together. Wind the coax onto the larger form with about 1 inch spacing between the turns (also known as *pitch*).

Antennas and Transmission Lines 63

Arrange the coax so that the pigtails are close enough to attach to the analyzer and secure the coax to the form with electrical tape.

If you have a binding-post adapter as shown in Figure 1, connect one pigtail to each post. If you don't have an adapter, solder the pigtails to pieces of wire and clamp them to the analyzer's RF connector or stick them in the center socket. Support the coil with something non-conducting — I used a small cardboard box. The entire setup should be assembled on a wooden or plastic table away from large metal surfaces.

Turn on the analyzer and set it to display series resistance (R_S) and reactance (X_S) — this is the default on most analyzers. Sweep the analyzer across the upper HF bands (the MFJ-269 10-27 MHz range is about right) and look for a range of frequencies where R_S peaks and X_S becomes a very small value. The middle frequency in this range is f_0. For my larger coil, f_0 was 17.96 MHz.

If the values exceed the analyzer range (this is likely), add a 1 kΩ resistor in parallel with the coil (not in series). This *swamping resistor* limits the variation of resistance and reactance to displayable values without affecting f_0.

Measuring the Balun

For reliable impedance measurement there are many things wrong with the setup in Figure 1. Measuring absolute impedance at RF is very tricky and susceptible to many sources of error. We are only going to measure f_0 and observe how impedance changes with coil shape — that's a lot easier.

Tune the analyzer to resonance and watch the displayed values for R_S and X_S as you touch the jacket of the coil, the case of the analyzer, and the power cord for the analyzer. Even bringing your hands *close* to the coil or analyzer should make the values change a little. With the sensitivity of the test in mind, as you take data try not to disturb the setup and don't touch the analyzer case while adjusting the controls or reading the display.

Record R_S and X_S at 1 MHz intervals throughout your chosen range — from well below f_0 to well above f_0. Record the actual value of f_0. Use a spreadsheet to capture all the values and make a graph as shown in Figure 2A through 2C. (An *Excel* spreadsheet for plotting all graphs is available on the Hands-On Radio website for this experiment.) It should look something like the graphs of Figure 2.

The spreadsheet also calculates the approximate inductance of the coil and plots its reactance ($X_L = 2\pi fL$) as the straight dashed line on the graph. (The spreadsheet also calculates and graphs the magnitude of the polar form of the impedance, $|Z|\angle\theta$.) Obviously, our real-world coil behaves a lot differently than an ideal inductor. Remember that by adding the swamping resistor, the analyzer is reading the parallel combination of it and the coil so the values will probably not be close to the ideal inductor's reactance. We are only interested in the shape of the curve, though.

Assuming you see a graph like that in Figure 2A, self-resonance is clearly shown. Impedance of the coil will increase up to f_0 and then drop off. The balun is most effective at reducing shield current from about ½ to ⅔ f_0 to a few MHz above f_0. You'll also see X_S dive toward zero and then rise again. Most analyzers don't display the sign of reactance, but by understanding resonant circuits you'll realize that the sign reverses at f_0, changing from inductive to capacitance reactance.

Now loosen the coil on the form and move the turns together so the jackets of adjacent turns are touching. Secure the coil to the form and repeat the measurements. You'll see f_0 shift downward quite a bit in Figure 2B as the inductance of the coil and the inter-turn capacitance increase.

Transfer the coil to the smaller form and repeat the measurements. For the peanut butter jar coil shown in Figure 3, f_0 increased by about 1 MHz and the response curve broadened as shown in Figure 2C, reflecting the lower Q of long, narrow coils. You can continue to alter the shape of the coil to observe how it affects f_0.

You now know how to make a coiled-coax choke balun and can even tune it for the highest impedance right where you need it. Don't be too precise as the sensitivity test showed you that lots of things can affect the coil, including attaching it to an antenna. There are several pretested designs for coiled-coax choke baluns in *The ARRL Handbook* and *The ARRL Antenna Book*, as well.

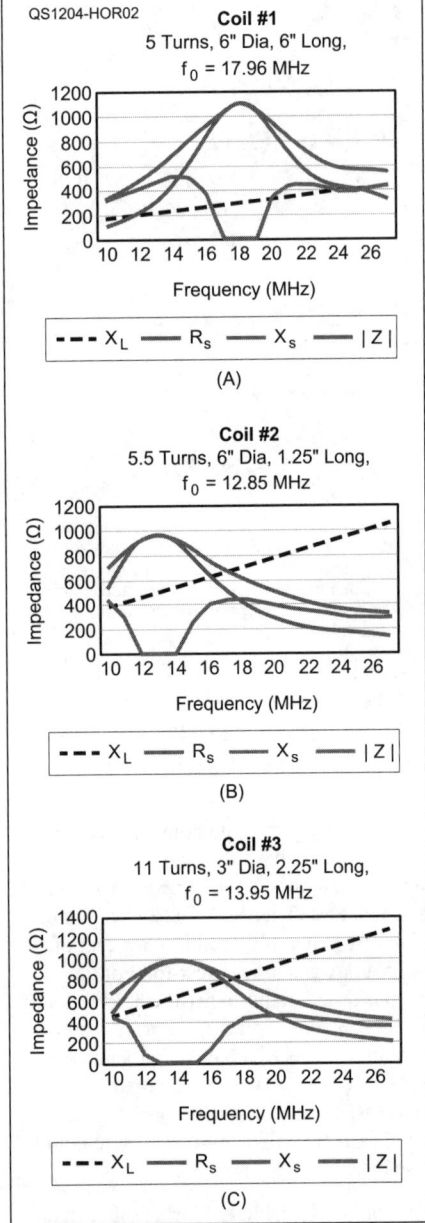

Figure 2 — Three different baluns, all using the same length of coax, are wound on different diameter forms and with different turn spacing (pitch). The resulting shift in self-resonant frequency, f_0, is dramatic.

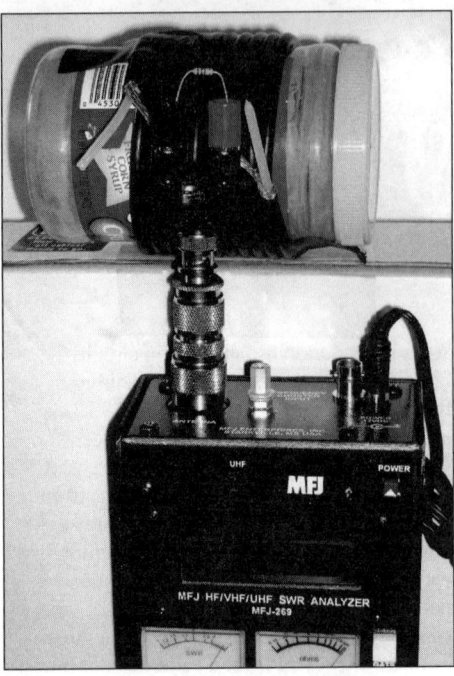

Figure 3 — A smaller diameter coil may have similar inductance to a larger coil but its lower Q causes its impedance curve to be broader. All coils in this article are made from the same length of coax feed line.

Experiment 96
Open Wire Transmission Lines

Over the past few years, the open wire transmission line has enjoyed something of a rebirth in ham radio. Before coaxial feed lines were still impractical for the ordinary ham, open wire line was often homebrewed. Insulators, ceramic or waterproofed wood, separated solid copper wire by up to 6 inches. Link or transformer coupling was used to connect the transmitter output circuits to the balanced feed line. The science of open wire lines was a hot topic as in an excellent January 1934 *QST* article.[1] The back of *QST* contained numerous ads for insulators and feed-through insulators and other necessary parts for "rolling your own."

Figure 1 shows a familiar transmission line configurations (two wire) and three not so familiar configurations (single wire, four wire and five wire). The balanced four wire and unbalanced five wire lines were widely used for high power HF applications long after World War II and are still found at some shortwave broadcast stations. The five wire configuration is a sort of "skeleton coax."

While coaxial cable is far more convenient to use, there are applications in which it is not a good choice for an antenna system. The lower loss of open wire line compared to coaxial cable makes it an effective choice with high SWR, such as from non-resonant antennas or those with a high feed point impedance. For very long feed lines, such as to a distant tower or to antennas in the upper HF/lower VHF spectrum, open wire line may be a very cost effective solution compared to hardline. Open wire line can also be used to make two wire switched direction Beverage antennas.[2]

Why Don't Open Wire Lines Radiate?

With properly terminated coaxial cable, the field of a high frequency signal is completely contained between the inner surface of the outer shield and the outer surface of the center conductor. If there are no breaks in the shield, the field cannot escape. Neither can a high frequency signal from outside the cable cause current to flow inside the cable. This is fairly intuitive. Open wire lines, however, have both (or all) conductors exposed. Why don't they radiate?

Well, actually, they *do* radiate a little bit as given in the following formula for two wire lines with a wire to wire spacing S of less than $1/10$ of a wavelength:

Radiated power = $160 \times I^2 \times (\pi \times S/\lambda)^2$

where I is the line current and S/λ is the *electrical line spacing*.[3] (This equation also requires the line to be non-resonant, meaning not an integral number of ¼ wavelengths long.) For example, if a 600 Ω line with wire to wire separation of 2 inches is carrying 1 kW at a frequency of 14 MHz, the line

[3] F. Terman, *Radio Engineer's Handbook*, First Edition, 1943, McGraw-Hill.

current, I = $\sqrt{(1000/600)}$ = 1.29 A and the line spacing is 0.0024 λ. The radiated power is then $160 (1.29)^2 (3.14 \times 0.0024)^2$ = 0.015 W, which is 48 dB below the power being carried in the line.

The real question is why don't they radiate *more*? Each individual conductor in the line does radiate an electromagnetic field as would a single, isolated wire. In a symmetric, balanced line, such as our two wire example, however, the currents in the two wires have opposite polarities and radiate fields that cancel almost completely at distances more than a few line spacings away. As the electrical line spacing, S/λ, increases you can see from the formula that radiated power also increases, either because physical distance between wires increases or as the frequency of the signal in the line increases — or both! A good rule of thumb is not to use open wire lines with line spacings greater than $1/10$ of a

[1] R. Glover, "A Practical Transmission-Line System for the Doublet Antenna," *QST*, Jan 1934, pp 17-22. *QST* articles more than four years old can be viewed by ARRL members by using the ARRL Periodicals Archive and Search, www.arrl.org/arrl-periodicals-archive-search.

[2] W. Silver, NØAX, "A Cool Beverage Four-Pack," *QST*, Apr 2006, pp 33-36.

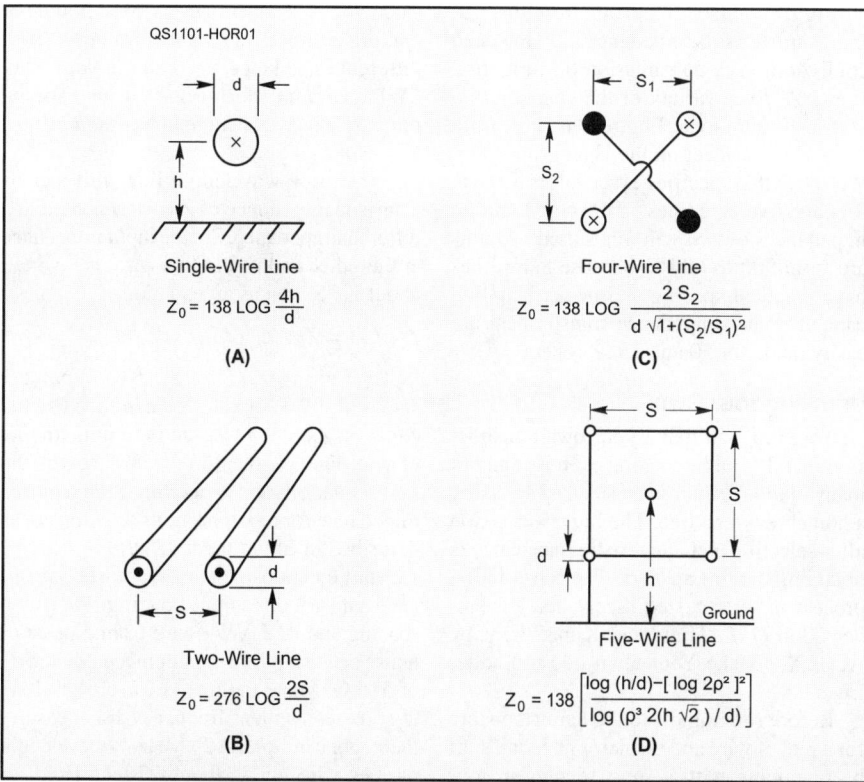

Figure 1 — Several types of transmission lines and the formulas that determine their characteristic impedance. Note in (D) ρ = zh/s.

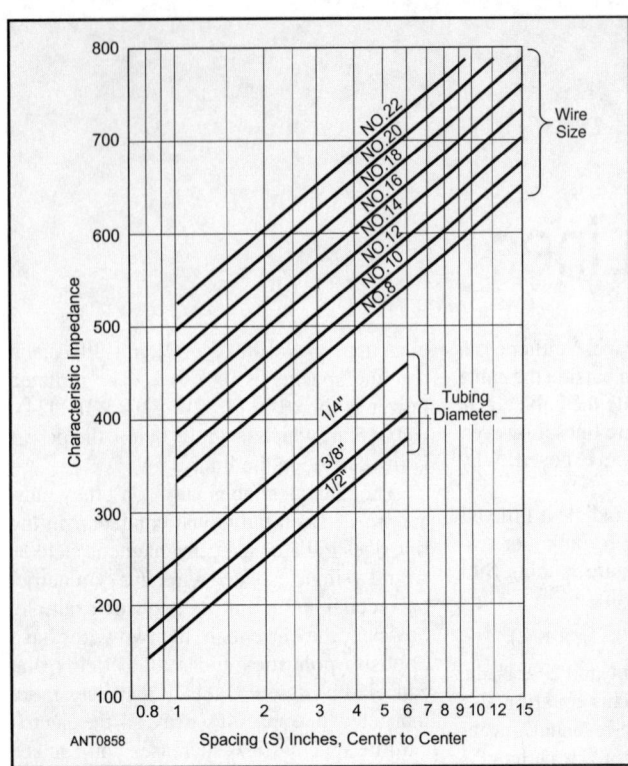

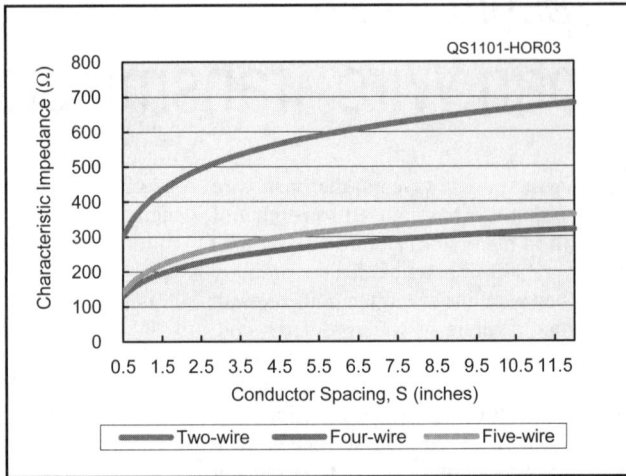

Figure 2 — Characteristic impedance as a function of conductor spacing and size for two wire lines.

Figure 3 — Characteristic impedance for two, four and five wire lines made from #12 AWG wire 60 inches above ground. Height above ground affects the unbalanced five wire configuration. The spreadsheet is available for download from the Hands-On Radio Web page.

wavelength (0.1 λ), at which point radiation from the line approaches 20 dB below the carried signal.

Characteristic Impedances

Figure 1 also gives the formulas for the characteristic impedance, Z_0, for each configuration of transmission line. Figure 2 shows Z_0 for a variety of the common two wire configurations. Figure 3 is the output from a spreadsheet on the Hands-On Radio Web page that specifies exact values for two, four and five wire lines.[4] The most common impedances of two wire lines used by hams are from 300 to 600 Ω because those lines have practical spacing and power handling capability and impedance transformers are easily made for 50 and 75 Ω systems.

Unrolling Your Own

You, too, can make your own transmission line. It's an interesting exercise and not many of today's hams can say they are using a homebrew feed line. The biggest decision after selecting a characteristic impedance is obtaining the line spacers. The construction process is summarized nicely in a December 2006 *QST* "Hints and Kinks" item by AC0AX, "Make Your Own 600 Ω Ladder Line."

Before running off to the hardware store for a reel of wire and insulator material, start by practicing with a short section of your two wire line. Constructing the line on the workbench is not particularly difficult — strip some wire and appropriate some paper or plastic drinking straws from the kitchen for insulators. Make two cuts halfway through the straw at the desired spacing, stretch the wire out straight, and press it into the cuts. Voila — transmission line!

How can you measure Z_0 to confirm your calculations? Rather than go to the trouble of making an impedance transformer for each different impedance, you can use your 50 Ω SWR analyzer to do the job by using a special property of ¼ wavelength lines — *impedance inversion*.

When a ¼ wavelength line with a characteristic impedance of Z_0 is terminated in a different impedance, Z_L, the input impedance at the other end of the line, Z_{IN}, is inverted about Z_0. Stated as an equation:

$$Z_{IN} / Z_0 = Z_0 / Z_L \text{ or } Z_{IN} = Z_0^2 / Z_L$$

By varying Z_L until the SWR analyzer shows an SWR of 1:1, you can then use the equation $Z_0 = \sqrt{(50 Z_L)}$ to determine the Z_0 of your line. (If this looks familiar, it is the same equation used to determine the required impedance for a synchronous transformer as described in Experiment #81.)

Start by choosing a characteristic impedance and a convenient spacing. A 3 inch spacing and #12 AWG wire from a piece of house ac wiring cable result in a predicted Z_0 of 516 Ω. Make the line ¼ wavelength long at some convenient frequency for measurement. For example, at 29 MHz, ¼ wavelength of open wire line is about 7.7 feet. Hold the line above a nonmetallic work surface on nonconductive supports so that it is as straight as possible and with no kinks or abrupt bends. Use a binding post adapter such as a Pomona 1699 (UHF) to connect the one end of the line to the SWR analyzer or homebrew an adapter, keeping all connections short and direct. This is the input end. Leave the other end open with nothing contacting the unconnected wires.

Determine the frequency at which the line is exactly ¼ wavelength long by adjusting the analyzer frequency and watching the resistance value. SWR will remain infinite, but when R reaches a minimum, the line is ¼ wavelength long.

Without changing frequency, tack solder a noninductive resistor (carbon composition or carbon film will do) with a value approximately $Z_L = Z_0^2 / Z_{IN}$ across the open end of the line, In our example with a 516 Ω characteristic impedance, Z_L should be 5325 Ω. Using the closest standard value of 5.1 kΩ, the input impedance should be approximately Z_0^2 / Z_L = 52 Ω. If the analyzer reads a higher value, Z_0 is lower than 516 Ω and vice versa.

You can also use this technique to determine Z_0 of an unknown piece of open wire line. Start by shorting the load end of the line and finding the *lowest* frequency at which the value of R reaches a minimum value. The line is ½ wavelength long at this frequency, so it will be ¼ wavelength long at ½ that frequency. Now terminate the line in varying values of Z_L until you get an input impedance close to 50 Ω and the characteristic impedance of the line is $Z_0 = \sqrt{(50 Z_L)}$.

[4]All previous Hands-On Radio experiments are available to ARRL members at **www.arrl.org/Hands-On-Radio**.

Experiment 116
The Quarter-Three-Quarter Wave Balun

As the 22nd edition of *The ARRL Antenna Book* was being prepared, Frank Donovan, W3LPL, suggested a balun design for the book.[1] "It's one of my favorites," he said enthusiastically, "and nobody seems to know about it. It's easy to construct and works great!"

The design doesn't have a catchy name, so perhaps that's why it's relatively unknown. I'll give it one — the *quarter-three-quarter wave (Q3Q) balun* — and maybe it will waltz into your antenna system. Shown in Figure 1, the professional literature shows the Q3Q balun to be a variation of the *hybrid ring* balun in Figure 2.[2] In both, $\lambda/4$-long transmission lines sections perform both impedance matching and insure equal and opposite currents in the load. But first — a word about baluns.

What is a Balun?
The word *balun* (pronounced "BAL-un") is a contraction of <u>bal</u>anced to <u>un</u>balanced and refers to any device that transfers differential-mode signals between a balanced and unbalanced system while maintaining symmetrical currents or voltages in the balanced system. The term applies only to that *function* — it doesn't matter whether the balun is made of transmission lines, flux-coupled transformers, or some other structure that simply chokes off unbalanced current.

An impedance transformer may or may not perform the balun function. There are balanced to balanced and unbalanced to unbalanced impedance transformers. Multiple devices are often combined and called a "balun." For example, a "4:1 balun" can be a 4:1 impedance transformer in series with ferrite beads forming a choke balun.

Other names for baluns are common, such as "line isolator" for a choke balun or they are named for their construction such as bead

---

[1] *The ARRL Antenna Book,* 22nd Edition. Available from your ARRL dealer or the ARRL Bookstore, ARRL order no. 6948. Telephone 860-594-0355, or toll-free in the US 888-277-5289; **www.arrl.org/shop**; **pubsales@arrl.org**.
[2] Johnson & Jasik, *Antenna Engineering Handbook,* 2nd edition, section 43-6.

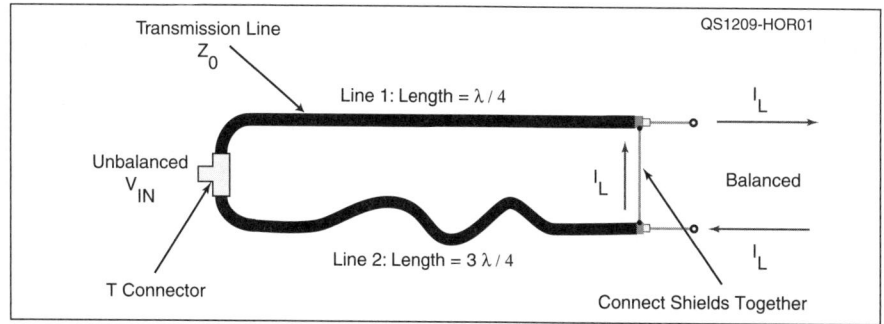

Figure 1 — The quarter-three-quarter wave balun uses the current-forcing function of odd $\lambda/4$ feed lines and the $\lambda/2$ delay of the longer line to cause equal and opposite currents to flow in the antenna terminals.

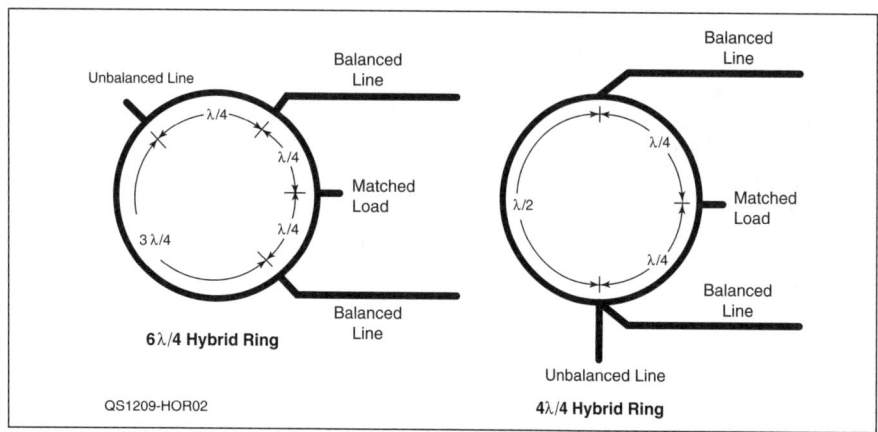

Figure 2 — Hybrid ring baluns use different combinations of $\lambda/4$ lines and delay sections to match impedances and balance load and line currents.

balun, coiled-coax balun and sleeve balun. What is important is to separate the function from the construction. Got it? Here we go!

Transmission Line Baluns
In the Q3Q balun, the $\lambda/4$ current-forcing property of transmission lines is used to perform the balun function. (For more uses of $\lambda/4$ transmission lines, see Hands-On Radio Experiment #81.[3]) If a transmission line is an odd number of quarter wavelengths long ($\lambda/4$, $3\lambda/4$, $5\lambda/4$ or more) the output current, I_L in Figure 1, is independent of the load impedance and equal to the line input voltage, V_{IN},

[3] All previous Hands-On Radio experiments are available to ARRL members at **www.arrl.org/hands-on-radio**.

divided by the line's characteristic impedance, Z_0.

$$I_L = V_{IN} / Z_0$$

In effect, a voltage source at the $\lambda/4$ line's input is turned into a current source at the output. The same amount of power is transferred from the input to output — no power is created — less the line's loss at that frequency. Since line 1 is $\lambda/4$ long and line 2 is $3\lambda/4$ long and they have the same input voltage at the T connector, their output currents will also be the same.

Since line 2 is $\lambda/2$ longer than line 1, its current will have a 180° phase difference from the current at the output of line 1. This results in equal-and-opposite currents flowing in each terminal of the balanced connection. Looking closely at Figure 1, you can see the same current flowing right from the center conductor in line 1 as flows left to the center conductor of line 2. Since the currents on the inside of the shield are equal and opposite to the center conductor current, the shield currents balance, too. Thus, the balanced system sees equal and opposite currents flowing in each terminal without regard to their impedance (within reason).

Disconnecting the Shield

The Q3Q balun also solves another important problem — common-mode current on the outside of the transmission line. Connecting a coaxial feed line directly to an antenna connects the shield to one terminal of the antenna. This isn't a problem for current inside the transmission line — it flows into the antenna terminal just as it's supposed to. The problem comes from the *outside* surface of the transmission line.

Due to the skin effect, the outer surface of a coaxial shield is completely independent of the inner surface at RF.[4] The effect is to attach a "third wire" formed by the outer surface of the shield to the antenna. This creates a path for common-mode currents that can have a big effect on antenna system performance.[5]

A close look at Figure 1 shows the Q3Q balun does not have a "third wire" path created by the outside of the shield. Since the shield currents balance, the shields can be connected directly together and the current flowing out of the shield of line 2 flows into the shield of line 1. No connection of the shields to the balanced system is required — the unbalancing third-wire path is not present and no current from the antenna connection flows on the outside of either line 1 or line 2.

The feed line and balun may still have current induced on the outer shield surface from the antenna's radiated signal. Additional chokes may be needed along the line to minimize common-mode current. Because the shield connection is independent of the balanced connections, it may be connected (but doesn't have to be) to some other ground or common reference without upsetting the balanced system. For example, it can be connected to the boom of a Yagi that is insulated from a balanced driven element.

The Q3Q is a *monoband* design — that is, it only acts as a balun at the design frequency for which the transmission lines are the required number of λ/4 long. (The Q3Q design can be used over a bandwidth of about 10% centered on the design frequency.) That

[4]en.wikipedia.org/wiki/Skin_effect
[5]www.eznec.com/amateur/articles/baluns.pdf

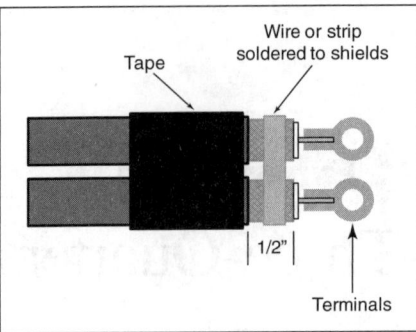

Figure 3 — The shield connection between lines 1 and 2 should be kept short for good high frequency performance. Precise adjustment of final line length for use at VHF+ should be performed with the terminals attached. At HF the connection length is not critical and the center conductors can be extended beyond the shields for easier feed point connection..

means this balun should not be used on triband or multi-band antennas. Nevertheless, many useful single-band applications exist around even a modest station — verticals, dipoles, J-poles and monoband Yagis including VHF+ antennas. The cost is modest: one T connector, two cable plugs and an extra 3λ/4 of coax. I think you should build one.

Building a Q3Q balun

Building a *synchronous balun* — meaning that its function is performed by precise timing of waves in the transmission line — requires that the line sections be precisely the right length at the design frequency. This can be done using an antenna or SWR analyzer. You can either measure the line's *velocity factor (VF)* and calculate how long the sections should be or you can use the analyzer to trim a section of line to the right length by estimating and trimming.

Whichever method you choose, do not assume that the VF for your coax is exactly what is published in tables for that type of coax. Variations of a few percent are common, particularly if the part number is not a current MIL SPEC designation, such as a manufacturer's proprietary cable or an obsolete designator such as RG-8.

The manual supplied with your analyzer may have instructions for determining the VF of transmission line or cutting lines to specific electrical lengths. If so, follow those instructions. In any case begin by installing the required cable plug on one end of the cable and attaching it to the T connector. Make all measurements through the T connector so that its length is included in both line 1 and line 2. If you do not use the T connector when trimming the lines to length, the assembled length will be too long for the desired frequency. The frequency at which the balun works best will end up too low. At HF and the lower VHF ranges, this will result in a small shift in frequency but at higher frequencies the error can be significant.

If you do not have the instructions for creating specific electrical lengths of cable, here is the basic procedure for making a Q3Q balun with a design frequency of f_0:

- Find the nominal VF for your cable from a table or handbook.

- Beginning with the λ/4 line (line 1 in Figure 1), install a cable plug on one end of the line and attach the T connector.

- Estimate the λ/4 length of cable as ¼ × VF × 983.6 / f_0 (MHz) and cut the cable a few inches long. Make a clean, square cut with no loose strands on the end.

- Using the antenna analyzer, find the *lowest* frequency at which the reactance (X) value reaches a minimum value. Use the digital display value and not the analog meter indication for the most accurate reading. This is the frequency at which the line is λ/4 long: f_{TEST}.

- Calculate the correct length as follows: correct length = current length × f_{TEST} / f_0.

- Cut the line to the proper length in a couple of steps, making measurements at each length. Attach terminals, if needed and make a final length trim if necessary.

- Repeat the trimming steps for a 3λ/4 line that is three times as long. The frequency f_{TEST} will be ⅓ the λ/4 frequency for the λ/4 line.

- Remove ½ inch from the jackets of the lines. Wrap a wire or a narrow metal strip around the exposed shields and solder it to both shields as in Figure 3 so that the connection between the cables is as short as possible.

- Waterproof the exposed ends and tape the line sections together to prevent the connection from being pulled apart. The balun lines can be coiled together as one package and you have made a Q3Q balun.

And Now for Something You'll Really Like

The noise reported as Jovian in Experiment #114 turns out to have considerably more mundane terrestrial origins. The noise from Jupiter looks rather different as it happens — tune to the Hands-On Radio website (**www.arrl.org/hands-on-radio**) for more information and links to sound files of the *real thing*. Thanks to Whitman Reeve and Dave Typinski, AJ4CO, for the correction.

Experiment #119
The Q3Q Balun Redux

Experiment #116, which presented the Quarter-Three-Quarter Wave Balun really captured the interest of readers, generating quite a bit of e-mail! When I receive correspondence that I think might be of interest to readers, I usually post it on the Hands-On Radio web page.[1] This time, the comments needed their own column.

As with so many other techniques and designs, the Q3Q balun was invented long ago and has been applied in many different applications. It first appeared in the amateur literature in a May 1977 *Ham Radio* article, "A New Coaxial Balun" by Jim Dietrich, WAØRDX.[2] As noted in Experiment #116, it is a variation of the hybrid-ring or "rat race" mixers more commonly found in microwave applications.

Going Inside the Q3Q

As discussed in Experiment #81 which is about synchronous transformers, the Q3Q's ¼ wavelength (and ¾ wavelength) sections also perform impedance transformation. Each Q3Q section is terminated in half of the load impedance, $Z_L/2 = Z_0/2$ so the SWR inside each section is 2:1. (For simplicity, we are assuming the load impedance is purely resistive.) At the other end of the ¼ and ¾ wavelength Q3Q sections, because of impedance inverting, the input impedance is $2Z_L = 2Z_0$. Impedance at the input to the section containing the extra ½ wavelength is still $2Z_0$ because impedance repeats every ½ wavelength inside a transmission line. At the T connector, the two input impedances are connected in parallel: $2Z_0$ in parallel with $2Z_0$ is equal to Z_0. This results in a 1:1 SWR in the main feed line.

So, when everything — the main feed line and the two balun sections — is made out of cable with the same Z_0 and the load impedance is the same as Z_0, the Q3Q just acts as a 1:1 balun to make sure each side of the load (usually an antenna's driven element) has equal current.

But wait — how can a pair of feed lines with an SWR of 2:1 create an SWR of 1:1? Remember that SWR is created by reflections and that reflections have both an amplitude and phase. The reflected wave in the longer of the Q3Q's sections has to travel an extra ½ wavelength or 180° before it reaches the T connector. That means the reflections in the two sections are out of phase where they connect together and cancel — even though they have the same amplitude. As far as the main feed line is concerned, there are no reflected waves from the Q3Q and the SWR is 1:1.

Subtracting One Q

What happens if the ¼ wavelength sections are removed from each feed line of the Q3Q? This leaves the main feed line connected directly to one side of the load and the other side connected through the remaining ½ wavelength section. This is the classic ½ wavelength 4:1 voltage balun shown in Figure 1.

If the load impedance is $4Z_0$, each section of the balun is connected to ½ of the load impedance or $2Z_0$. This impedance also appears at the input to the ½ wavelength section so at the junction of the two feed lines, the impedances, each $2Z_0$, are connected in parallel. The result is the main feed line "sees" an impedance of Z_0 for an SWR of 1:1. Without the parallel connection, the SWR inside either line would be 2:1 but again, the reflected waves combine to cancel in the main feed line.

This description assumes that the load impedance is perfectly balanced so that the equal and out of phase voltages create equal and out of phase currents. If the load is not balanced, unequal currents will flow in each half of the load. Since current — not voltage — is what causes antennas to radiate, the antenna's radiation pattern will also be unbalanced. Voltage baluns do not ensure equal currents in the load terminals.

Using the Q3Q for Impedance Matching

Back to the Q3Q design — what happens when Z_L is *not* equal to Z_0? In that case, the impedance ½ wavelength away from the load in each section is no longer $2Z_0$ and the paralleled impedances at the T connector no longer combine to give Z_0 and an SWR of 1:1. I hate when that happens.

Never fear, there are more tricks up the sleeve of this balun as WAØRDX's original article demonstrates. The load impedance can still be matched by using transmission

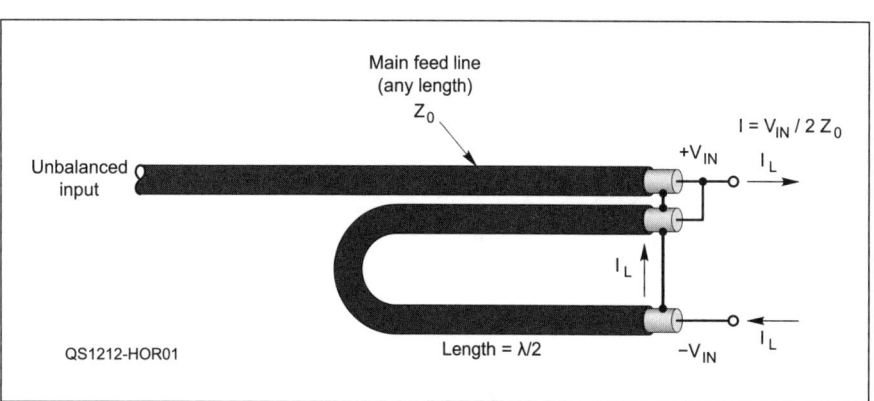

Figure 1 — The 4:1 voltage balun, also known as the ½ wave loop balun, presents out of phase voltages to each half of the balanced load. If the load is imbalanced, current in the load will also be imbalanced.

---

[1] All previous Hands-On Radio experiments and supplemental information are available to ARRL members at **www.arrl.org/hands-on-radio**.
[2] J. Dietrich, WAØRDX, "A New Coaxial Balun," *Ham Radio*, May 1977, pp 26-28.

line sections with different characteristic impedances, just as if a single ¼ wavelength synchronous transformer were being used.

For example, the WAØRDX article shows that the low driven element impedance of a Yagi — perhaps 10 to 15 Ω — can be matched by constructing the feed line sections from coaxial cable with a characteristic impedance of $\sqrt{(Z_L \times Z_0)} \approx 25$ Ω. Paralleling two 50 Ω feed lines creates the required 25 Ω feed line. Similarly, matching a loop with a feed point impedance of 100 Ω requires Q3Q sections with a characteristic impedance of $\sqrt{(Z_L \times Z_0)} \approx 70$ Ω. In this case, RG-59 or RG-11 with a $Z_0 = 75$ Ω would do the job.

Quarter Wave Voltage to Current Transformation

Reviewing just a bit, the current that flows in the Q3Q load terminals, I_L, is equal to the voltage at the T connector, V_{IN}, divided by the transmission line's characteristic impedance, Z_0, so that $I_L = V_{IN} / Z_0$. In this somewhat simplified view, the actual impedance, Z_L, of the load between the two terminals does not affect I_L. The current-forcing function of the quarter-wave line depends only on the input voltage and the line's characteristic impedance.

But just how does the line "force" any value of current? There is no source of power at the load or in the line — what if the load impedance is very high? Wouldn't voltage also skyrocket? In a perfect world, yes, but in our real world, no. I consulted Joe Reisert, W1JR, on the subject and he suggested that the limiting factor would be loss in the feed line.

Ah, ha! This is touched on in the classic *Reflections* by Walt Maxwell, W2DU (SK), in which he discusses reflected power and standing waves.[3] When power is reflected from a termination, the reflected waves experience loss just as waves in the forward direction do. With each trip back and forth in the feed line, more energy is dissipated as

[3]W. Maxwell, W2DU, *Reflections*, 2nd edition, WorldRadio Books, 2001, Chapters 8 and 9.

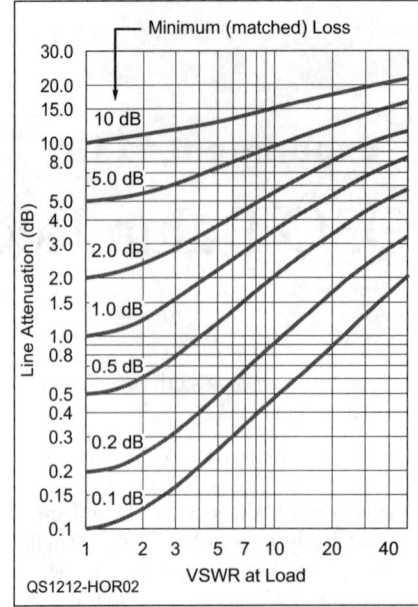

Figure 2 — Total insertion loss in a transmission line terminated in a mismatch. [Courtesy of Joe Reisert, W1JR]

heat. This puts a limit on how high a voltage can be built up by the resulting standing waves. If the maximum voltage is limited, so too is the maximum current that can be developed at the load.

To find out just how much loss is involved, we need to calculate the *total loss* of the line with a given termination. There are charts that show the additional loss which must be added when the SWR is greater than 1:1 but Joe created a single graph that shows total loss for a given SWR and matched feed line loss as shown in Figure 2.[4] To find the total loss, first determine the SWR at the load. From the SWR value on the horizontal axis, proceed vertically to the curve representing the feed line's matched loss (loss with SWR of 1:1). At the intersection, the total loss can be read on the vertical axis. This is less error prone than using two loss

[4]J. Reisert, W1JR, "VHF/UHF World," *Ham Radio*, Oct 1987, pp. 27-38.

charts and adding values together.

The ARRL's *Transmission Lines for Windows* (*TLW*) software by Dean Straw, N6BV (included with *The ARRL Antenna Book*), can also be used to calculate the loss and allows the user to select the type of cable.[5] Length of the feed line can be specified in either physical or electrical length. Let's take a look at typical loss values.

For a 100 foot length of RG-213 terminated in a 50 Ω load, at 14.0 MHz, *TLW* calculates the matched loss to be 0.780 dB. If the load is increased to 100 Ω (SWR = 2:1), loss increases to 0.924 dB. For loads of 1000 Ω (SWR = 20:1) and 5000 Ω (100:1), loss increases to 4.417 dB and 9.884 dB, respectively. The loss from high SWR leaves less and less power to develop high voltages anywhere along the line.

Our ½ wavelength line adds a small additional loss: assuming a 100:1 SWR for an open-circuit termination, the ½ wavelength section of RG-213 has a total loss of 3.088 dB according to *TLW*. Because the SWR at the input to the ½ wavelength section will also be very high, there will be high losses in the main feed line as well — about 9.8 dB for a 100 foot length of RG-213 at 14.0 MHz. The available output power in this case (assuming the transmitter is still pumping 100 W into the line) with 9.8 + 3.088 = 12.9 dB of loss is only 5 W, reducing the available current in the 5000 Ω load to $\sqrt{(5/5000)} = 32$ mA. So there are strict limits on how much current can be "forced" into loads that create high SWR.

Summary

Combining a ½ wavelength voltage balun with a pair of ¼ wavelength matching sections was one of today's best kept ham radio secrets. No more! Make sure you add this handy feed line gadget to your toolbox of radio know-how.

[5]*The ARRL Antenna Book*, 22nd Edition. Available from your ARRL dealer or the ARRL Bookstore, ARRL order no. 6948. Telephone 860-594-0355, or toll-free in the US 888-277-5289; **www.arrl.org/shop**; **pubsales@arrl.org**.

Electronic Circuits and Components

Experiment #62 — About Resistors

When is an ohm not an ohm? It depends on the home of the ohm! There are many different types of resistors and all of them have different characteristics that make them suitable or unsuitable for certain types of uses around radio circuits. Let's have a look and even have a demonstration.

Terms to Learn

Parasitic — a secondary characteristic of a component that depends on its physical construction.

Ideal — a component without parasitic characteristics.

Resistor Fundamentals

Every conductor resists the flow of current (except for superconductors). Georg Ohm defined the exact relationship between voltage (E), current (I) and resistance (R), formulating his famous law learned by every Technician class licensee:

$E = I \times R, I = E / R$ or $R = E / I$

Electrons collide with the atoms that make up the material as they flow through it. The collision transfers some of the electron's energy to the atoms, which vibrate in response. These vibrations increase the temperature of the material and result in power dissipation calculated as:

Power (P) = $I^2 \times R$ or $P = E^2 / R$

An *ideal* resistor does not care whether the current flowing through it is ac or dc. The electrons collide with atoms going in either direction. Nevertheless, the shape and construction of the material affects the electrical characteristics of the resistor. Practical resistors that can be used in electronics require a pair of electrodes or leads. The resistive material must be coated with an insulating material to protect it from the surrounding environment and vice versa. Construction details of the real resistors, shown in Figure 1, result in the electrical model for the resistor shown in Figure 2. The model includes *parasitic* characteristics created as a consequence of how the resistor is constructed. Look for the ideal resistor, R, buried deep inside the model! R_S is the equivalent resistance of the real resistor.

The series inductance, L_S, is created both by the leads attached to the resistor and the structure of the resistive material. Wire wound and spiral cut film resistors have the highest parasitic inductances, carbon composition resistors the least. Notice that for a surface-mount resistor, there are no leads. This greatly reduces L_S.

A resistor's electrodes also form a very small capacitor, C_P, modifying the resistor's behavior at very high frequencies. Although the resistor's coating is a very good insulator, current can still flow in very small amounts across its surface as a leakage current, represented by R_P. This becomes important when the resistor has a very high value or is used in a high voltage circuit.

Resistor Types

There are several different resistor construction methods and body styles or packages that are designed for a certain range of applied voltage, power dissipation or other considerations.

Carbon Composition Resistors

Composition means that the resistive material is a mix of carbon and stabilizing compounds. The amount of carbon in the mix determines the resistance of the material. A small cylinder, much like a pencil lead, is held between the two electrodes and coated with resin or phenolic, making a resistor with low L_S that is often used in RF circuits.

Carbon composition resistors are typically available with power ratings of ¼ to 2 W. They can also handle temporary overloads much better than film resistors because the heat is distributed evenly throughout the cylinder of resistive material. That makes them a good choice for circuits that protect against and absorb pulses and transients. Unfortunately, these resistors are also strongly influenced by temperature and humidity and so are not good for circuits that depend on precise, stable resistance values.

Cermet or Metal Oxide Resistors

Cermet (ceramic-metal mix) or metal oxide resistors are replacing carbon composition resistors in many radio applications

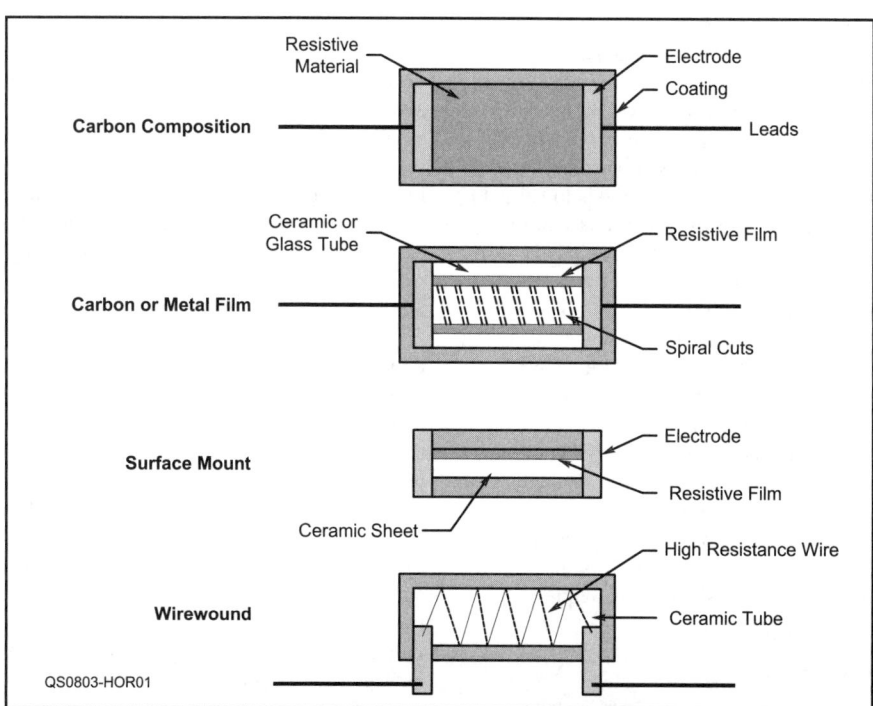

Figure 1 — The electrical characteristics of different resistor types are strongly affected by their construction. Reactance from parasitic inductance and capacitance strongly impacts the resistor's behavior at RF.

because their value is more stable than that of carbon composition resistors.[1] These are constructed much like carbon composition resistors, substituting the cermet or metal oxide for the carbon composition material.

Film Resistors

A film resistor uses a very thin coating of carbon or metal on an insulating substrate, such as ceramic or glass as the resistive material. The value of the resistance is determined by the thickness of the film and the amount of carbon or metal it contains. These resistors are available with very accurate and stable values, but they are unable to handle large amounts of power because the film is so thin. Overloads can also damage the film by creating *hot spots* inside the resistor, changing its value permanently. The value of film resistors is sometimes adjusted before sealing by cutting away some of the film with a laser in a process called *trimming*. If the film is deposited on the inside of a tube, the trimming process creates a coil-like current path that raises the L_S of the resistor. If your circuit operates at high frequencies, be sure the resistors you use have a low value of L_S.

Surface-mount resistors are almost always film resistors. These resistors have no leads at all, so L_S is very low. The film is deposited on a ceramic sheet. Because of their extremely small size, surface-mount resistors have very low power ratings — from 1/10 to 1/4 W.

Wire-Wound Resistors

Wire-wound resistors are made just as you might expect. A high resistance wire is wound on an insulating ceramic tube and attached to electrodes at each end. These are made to dissipate a lot of power — from 1 to 100s of watts! Wire-wound resistors are usually intended to be air cooled, but some styles have a metal case that can be attached to a heat sink or metal chassis to get rid of heat. This type of resistor is common in power supplies and other equipment where lots of power is dissipated.

Because the resistive material in these resistors is wound on a form, they have very high L_S. For this reason, wire-wound resistors are not generally used in audio or RF circuits. Be careful when using a resistor from your junk box or a grab bag in such a circuit! Small wire-wound resistors look an awful lot like film or carbon comp resistors. There is usually, but not always, a wide color band on wire-wound resistors. If you're in doubt, test the resistor at the frequencies you expect to encounter.

Resistor Networks

Resistor networks are often used in order to save space on printed circuit boards. These

[1]See **www.vishay.com/docs/20127/sxa.pdf** for more information.

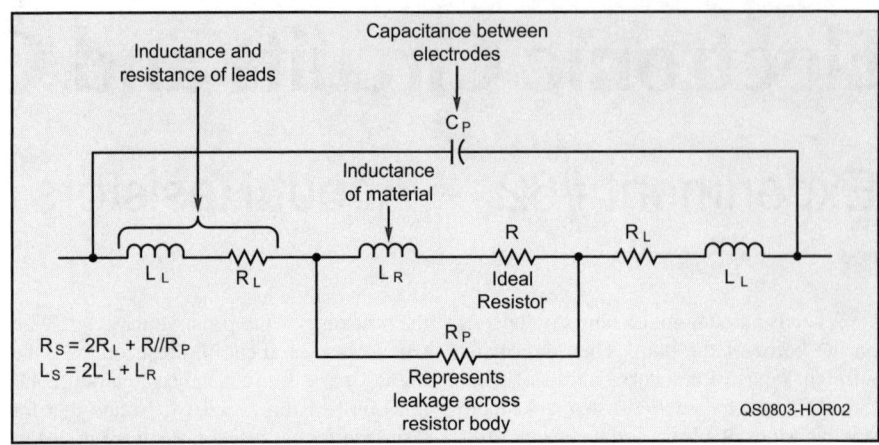

Figure 2 — The equivalent circuit for resistors including the various parasitic inductances and capacitances associated with the resistor's construction.

are miniature printed circuits themselves, placing several resistor films on one substrate. The resistors may be isolated from each other, share a common terminal or be connected in series with taps. There are a number of configurations that can be found in any component supplier's catalog.

Power Dissipation and Voltage Ratings

Power dissipation is the next most important characteristic of a resistor. An overloaded resistor often changes in value over time and can get hot enough to burn itself and surrounding components. Every circuit designer learns the smell of burnt resistor sooner or later!

The common rule of thumb is to calculate how much power the resistor will have to dissipate and then use the next largest size or a dissipation rating higher by a factor of 2, whichever is larger. The power rating is based on unobstructed air circulation around the resistor. For resistors dissipating more than 1 W, arrange nearby components so that air can circulate freely. If possible, mount power resistors horizontally so that convection cools all parts of the resistor equally.

Another important rating is maximum applied voltage. Voltages above this value may cause an arc between the resistor terminals! At high voltages R_P can also become significant, allowing current to leak around the internal resistance. High-voltage resistors must be kept clean. Fingerprints, oil, dirt and dust all create unwanted current paths, lowering R_P and increasing leakage or even causing an arc. This is why resistors for use in high-voltage circuits are long and thin with their terminals far apart — to minimize leakage and withstand high voltage.

Observing the Effects of L_S and C_P

This is all an academic discussion until you can see the effects for yourself. See if you can borrow that SWR analyzer again and gather a bunch of different types of resistors. Values from 20 to 120 Ω will do. Obtain a coaxial-to-binding post adapter, such as a Pomona 1296, and connect it to the analyzer with a BNC-to-UHF adapter.

Mount a resistor on the binding post using short leads and set the analyzer to its lowest frequency. Note the resistance and reactance settings. Sweep the frequency upward while watching the SWR and X meters. Even a carbon composition resistor will show increasing reactance with frequency. You can estimate L_S by noting the frequency at which X = R and using the formula L = X/2πf. Sweep the analyzer through its complete range — some resistors may even show self-resonances, either parallel (high X) or series (low X)! This can be a good way to learn about the different types of resistor construction. It can even make a great club activity with everyone contributing a resistor or two!

Recommended Reading

Chapter 6 of *The ARRL Handbook* contains in-depth discussions of real-world components and makes for illuminating reading![2] For the serious student of resistance, Cletus Kaiser's *The Resistor Handbook* is a thorough handbook on resistors of nearly any type you can imagine.[3]

Next Month

Continuing in this vein of exploring the different types of components, let's forge ahead into the world of capacitors — a component of many more styles than resistors, with a much greater variation of characteristics.

[2]*The ARRL Handbook for Radio Communications*, 2008 Edition. Available from your ARRL dealer or the ARRL Bookstore, ARRL order no. 1018. Telephone 860-594-0355, or toll-free in the US 888-277-5289; **www.arrl.org/shop/**; **pubsales@arrl.org**.

[3]C. Kaiser, *The Resistor Handbook*, CJ Publishing. Available from **www.amazon.com**.

Experiment #63 — About Capacitors

Last month's column on resistors alerted you to how seemingly minor differences in resistor construction can have a major effect on resistor behavior in different uses. The variations in capacitor construction are even more significant!

Terms to Learn

Dielectric — Insulating material that efficiently stores energy in the form of an electric field.

Relative dielectric constant — Measure of the relative ability of an insulating material to store energy in the presence of an electric field compared to that of a vacuum.

Capacitor Fundamentals

A capacitor is formed from a pair of conducting plates separated by an insulator (*dielectric*) as shown in Figure 1. If a voltage is applied between the plates, electrons are forced onto one and removed from the other. The resulting charge imbalance creates a voltage and an electric field between the plates, storing energy in the dielectric. Creating such an imbalance is called *charging* the capacitor. Larger area, thinner separation between the plates, or a dielectric with a higher *dielectric constant* increases capacitance. All capacitor types are just variations on this general theme.

If a constant voltage is applied to the capacitor, an opposing voltage builds with the growing charge imbalance. As a result, charging current gradually reduces until the opposing voltage reaches the applied voltage. At that point, current flow ceases. Thus, dc cannot flow between the two plates except during the time the capacitor charges or discharges. AC is considered to flow between the plates as the electrons flow onto and off the plates with each half cycle.

Capacitance in farads (F) specifies the amount of charge stored in a capacitor for a given amount of voltage between the plates. Farads are units of coulombs/volt or coulombs^2/joule. The energy stored in a capacitor in joules is $E = C \times V^2 / 2$. The area of the plates, the spacing and the material used for the dielectric determine the amount of capacitance:

$$C = k \times \varepsilon_0 \times A / d$$

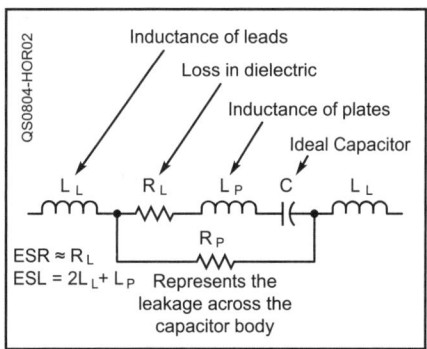

Figure 2 — The equivalent circuit for a capacitor includes parasitic effects. Inductance from the capacitor's shape and connecting leads forms ESL. Loss in the dielectric is shown as ESR. Leakage current across and through the capacitor's body is shown as R_P.

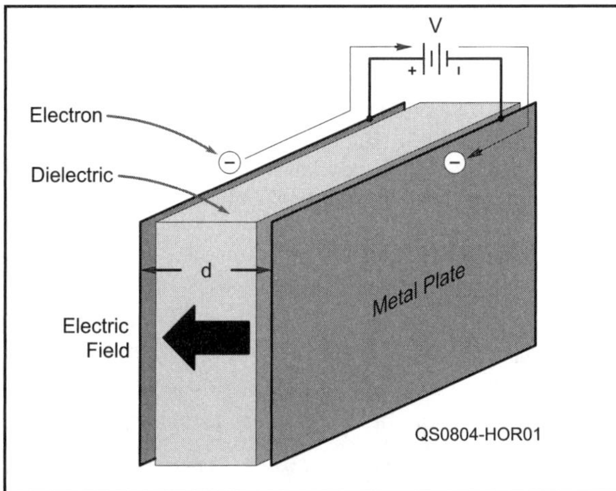

where:
A = the area of the capacitor's plates in square meters.
d = the separation between the plates in meters.
k = the relative dielectric constant (vacuum = 1).
ε_0 = the permittivity of empty space = 8.85 pF/m.

As with resistors, a capacitor's value has a precision or an allowed variation from the labeled or nominal value. Capacitors also have a temperature coefficient because as the materials expand and contract with temperature, the area and separation of the plates also changes.

Figure 2 shows the parasitic effects as parts of the circuit model for a capacitor. The size and shape of the plates and the leads used to connect them to circuits introduce a small amount of inductance called *equivalent series inductance* (ESL). At dc and low frequencies, ESL in the pH to nH range can be ignored, but as the frequency increases, so does its reactance. In fact, the ESL and capacitance form a series circuit (**en.wikipedia.org/wiki/LC_circuit**) with a self-resonant frequency of f_0. Above f_0, the capacitor acts more like a small inductor than a capacitor!

Dielectric materials dissipate a small amount of the stored energy, creating an *equivalent series resistance* (ESR). There is also a little *leakage* current between the plates whenever voltage is present. ESR can be as high as several tens of ohms, but is generally only important when the capacitor current is high, such as in transmitting and power supply circuits.

Leakage resistance, R_P in Figure 2, provides a path for current around or through the dielectric and is typically several megohms. You can ignore leakage resistance except in very low-power and high-impedance circuits.

Capacitor Construction

Figure 3 shows a roll-type capacitor made of two strips of very thin metal foil separated by a dielectric. After leads are attached to the foil strips, the sandwich is rolled up and either placed in a metal can or coated with plastic. *Axial* leads exit from both ends along the roll's axis. Because of the rolled strips, the ESL is high. Electrolytic and many

Figure 1 — A capacitor is formed from two metal plates separated by a dielectric. When a voltage is applied, electrons leave one plate and are forced on to the other, creating an electric field and storing energy in the dielectric.

Electronic Circuits and Components 73

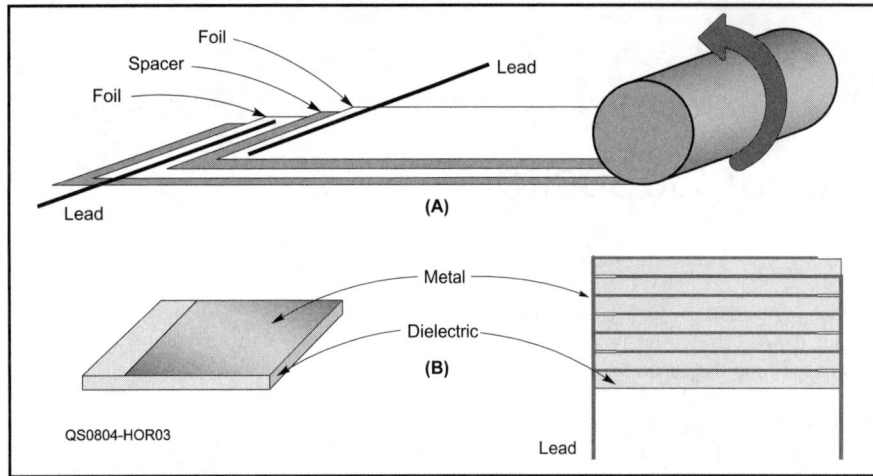

Figure 3 — Two common types of capacitor construction. (A) Roll construction uses two strips of foil separated by a spacer or dielectric strip. (B) Stack construction layers dielectric material (such as ceramic or film), one side coated with metal. Leads are attached and the assembly coated with epoxy resin.

types of film capacitors are made using roll construction.

In the stack capacitor, thin sheets of dielectric are coated on one side with a thin metal layer. A stack of the sheets is placed under pressure and heated to make a single solid unit. Metal side caps with leads attached contact the metal layers. The ESL of stack capacitors is very low, making them useful at high frequencies. *Radial* leads — leads that come out of the capaitor at right angles to its axis — indicate stack-type construction. Ceramic capacitors are the most common stack-style capacitor.

Capacitor Types
Electrolytic
A roll-type capacitor, the dielectric is a thin film of oxide that forms at the surface of the anode where it is contacted by gel that impregnates a fiber sheet that acts as a spacer. Electrolytics have very high capacitance for their volume and cost. They also have high ESL and ESR and are relatively leaky (low R_P). They can be made to withstand substantial voltages. Electrolytics are polarized, meaning that voltage can only be applied in one polarity due to the chemical electrolyte. They generally have very wide tolerances, on the order of ±20%.

A recent advance in electrolytics, *super-* or *ultra-caps* use advanced plate materials to create amazing amounts of capacitance in a small volume. Capacitance of several farads is not uncommon.

Tantalum
A cross between roll and sheet construction, one plate is formed by an extremely porous *slug* of tantalum and an outer metal capsule the other. The dielectric is an oxide coating on the tantalum slug. The slug has a tremendous amount of area and the oxide layer is very thin, so capacitance is high — but ESR is also high. Short leads and small size means that tantalums have low ESL compared to electrolytics. The maximum applied voltage for tantalum capacitors is under 100 V. As with electrolytics, tantalums are polarized and have wide tolerances of ±20%.

Film
Film capacitors have a plastic film dielectric with polyethylene and polycarbonate being the most common material. Most film capacitors are of roll construction, so ESL is moderate although stack types are available. Film capacitors are non-polarized. R_P is high and ESR is low. Special types of film are used for highly stable capacitance values or extremely low leakage. See **www.filmcapacitors.com/specsum.htm** for a good table summarizing the different types of film capacitors.

Ceramic
Ceramic capacitors are widely used in high-frequency applications. Stack construction keeps ESL extremely low so they are useful to hundreds of MHz. They offer low loss and have good leakage specifications. Ceramic capacitors are very rugged and pack a lot of capacitance into a small package. Ceramics are non-polarized and have a wide range of tolerances and temperature coefficients.

Mica and Glass
Silvered mica and glass capacitors are used in RF and transmitting circuits due to extremely low ESR and ESL. A stack-style capacitor, mica and glass form the dielectric layers. Because these capacitors are used at higher voltages than ceramic capacitors, the thicker dielectric layers limit available capacitance to 100 nF or less. Both types typically have a 5% tolerance.

Trimmer or Variable
Air variable capacitors, with plates separated by air, have very low loss and ESL, working well at RF. Because the plates have a wide separation, the larger capacitors can withstand the high voltages present in high power RF amplifier output stages and impedance matching circuits. Air variable capacitors have a voltage rating of 30,000 V/cm of plate separation (at sea level). Vacuum variable capacitors are available with even higher voltage ratings.

An adjustable variation of the mica capacitor in which the stack is compressed by a screw is called a *compression trimmer*. Ceramic and plastic variables are also available with values of up to several hundred pF.

Comparing Self-Resonant Frequencies
It's an illuminating exercise to compare the self-resonant frequency of several different types of capacitors, all of which have the same nominal value of capacitance. This is where the SWR analyzer can be put to good use once again.

Obtain several 0.01 µF film capacitors of different construction styles. Radial leads almost always indicate roll-type construction and axial leads a stack-type construction. A ceramic capacitor can serve as the reference.

Connect a 47 or 51 Ω resistor across the output of the analyzer as in the previous experiment. Sweep the frequency through the analyzer's range and note where reactance begins to increase beyond a few ohms. This will be the upper end of the comparison range.

Now connect each capacitor in series with the resistor and repeat the sweep, first with long leads and then with short leads. Self-resonance should be observed as a sharp return to minimum reactance after a slow increase. Above f_0, reactance will again increase. Compare the self-resonant frequencies of the different film styles with that of the ceramic capacitor. Which would be better at RF?

If you have access to some small tantalum or electrolytics of less than 1 µF, sweep them, too, and compare their performance.

Recommended Reading
Review the capacitor section of Chapter 6 of *The ARRL Handbook* or browse **www.faradnet.com**. Cletus Kaiser's *The Capacitor Handbook*, 2nd Edition (CJ Publishing) covers all subjects capacitive, as well.[1] Capacitor manufacturers often have excellent comparisons of the different styles available on their Web sites.

Next Month
We're going to make some circuit measurements using a new type of oscilloscope that plugs directly into your computer's USB port!

[1]*The ARRL Handbook for Radio Communications*, 2008 Edition. Available from your ARRL dealer or the ARRL Bookstore, ARRL order no. 1018. Telephone 860-594-0355, or toll-free in the US 888-277-5289; **www.arrl.org/shop/**; **pubsales@arrl.org**.

Experiment #78 — Bridge Circuits

Bridges are never far away in radio. There are SWR bridges, impedance bridges, noise bridges and other bridge circuits inside many pieces of equipment. Just as with their full sized cousins, electronic bridges require balance to be useful and that's the subject of this column.

Half-Bridges

The simplest way to begin is to discuss a bridge circuit with which you're already familiar, even though you might not know it. The half-bridge is really just a fancy name for a *voltage divider* as shown in Figure 1. For this discussion, we'll use the general case of the input to the divider an ac signal and both components in the divider being impedances; Z_1 and Z_X, where Z_1 is known and Z_X is unknown. The usual formula for the output voltage from a two-impedance divider is $V_{out} = (V_{gen} \times Z_X) / (Z_X + Z_1)$.

The utility of the bridge is not in developing a particular output voltage, but rather in using the output voltage to measure the value of an unknown impedance, Z_X. To do that, we have to rearrange the equation a little:

$$Z_X = \frac{Z_1}{\frac{V_{gen}}{V_{out}} - 1}$$

This makes sense: If $V_{out} = 0$, then $Z_X = 0$. If $V_{out} = V_{gen}$ then Z_X is infinite. If $V_{out} = V_{gen}/2$ then $Z_X = Z_1$. The only problem with this method of measurement is that it requires a very precise knowledge of three things — the values of V_{gen}, V_{out} and Z_1 — to make a precise measurement of Z_X. Since Z_1 is a fixed value, it can be measured once and would not change much. Voltages are more difficult to generate and maintain at a precise amplitude. Wouldn't it be nice if we could make those constraints go away somehow?

A Sense of Balance

The invention of the full bridge in Figure 2 by Christie in 1833 and its popularization by Wheatstone 10 years later was a major advance in electrical measurement. In those early days, precision instrumentation just didn't exist. It was possible to laboriously calibrate resistances and voltages against standards, but high precision bench-top voltmeters were decades in the future. The major advance of the full bridge circuit was to

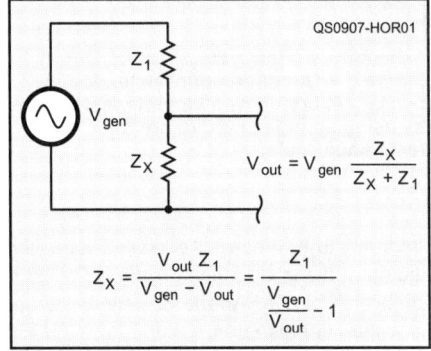

Figure 1 — The basic half-bridge is nothing more than a voltage divider. The unknown impedance can be determined from the input and output voltages and the value of the known impedance.

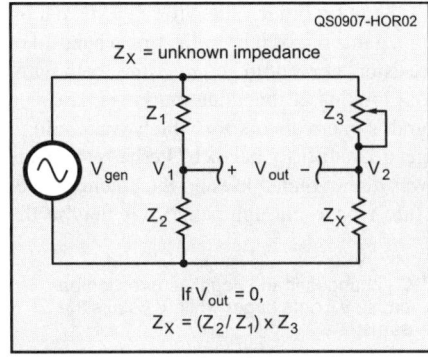

Figure 2 — The Wheatstone bridge consists of two parallel voltage dividers. Output is taken from the midpoints of the dividers. A calibrated impedance, Z3, is adjusted until the bridge is balanced ($V_{out} = 0$), at which point the value of the unknown impedance is equal to that of the calibrated impedance.

eliminate two of the precision measurements — those of the voltages — by converting the measurement from one of value to one of *balance* instead.

If instead of one voltage divider, there are two identical dividers — side by side and driven by the same voltage — the voltage at the midpoint of the dividers V_1 and V_2 will be exactly the same. That condition won't change, no matter what the value of the *excitation voltage*, V_{gen}! For sure, the absolute value of the midpoint voltage will change, but those two values will be the same and the output of the bridge, $V_{out} = V_1 - V_2$, will be exactly 0 V. With a means of detecting when the output voltage is 0 V, both voltage measurements have been eliminated in a stroke!

This made very precise measurements possible to the early experimenters because it is much easier to make an instrument that detects the presence and polarity of small amounts of voltage than it is to make an instrument that can tell you that value of voltage. If such an instrument (called a *galvanometer*) is connected between V_1 and V_2, it is easy to tell if the bridge is balanced ($V_{out} = 0$) with a very high precision. Even in the mid 1800s, it was possible to balance a Wheatstone bridge to less than 1 mV of error.

Furthermore, the two dividers don't even have to be identical for the bridge to be balanced. It is only necessary that the *ratios* of the impedances in each divider be the same. For example, in Figure 2, if $Z_1 = Z_2 = 10$ kΩ and $Z_3 = Z_X = 1$ kΩ, V_{out} will still be zero because the voltage at both divider midpoints will be $V_{gen}/2$ and V_{out} will still be zero.

If $V_{out} = 0$ the bridge is balanced and the value of the unknown impedance is related only to the other impedances: $Z_X = (Z_2 / Z_1) \times Z_3$. Z_1 and Z_2 are fixed and so their ratio is known and becomes a constant. Thus, the procedure for measuring the unknown impedance, Z_X, requires only that the bridge be brought into balance and the value of Z_3 measured. Measurement bridges use a calibration scale for Z_3 so its value can be read directly, without having to be measured separately.

Bridge-Building

Let's make a simple bridge and find out how that works. We'll build the circuit of Figure 2 with $Z_1 = Z_2 = 2.2$ kΩ and a 5 kΩ potentiometer as Z_3. (The potentiometer can have a linear or tapered resistance characteristic — the type of variation of resistance with shaft rotation.) Use a "panel mount pot" with a shaft and if you have one and a knob to fit the shaft that has a position indicator as shown in Figure 3.

Before assembling the circuit, take an index card, punch a hole for the pot's bushing in the center of the card, and attach it to the pot with the mounting nut and washer. Attach the knob to the shaft so that the position indicator is as close as possible to the index card. Short two terminals of the pot as shown in Figure 2 so that the pot becomes a variable resistance.

Now create a calibration scale for the pot. Attach a digital ohmmeter and set the pot to one end of the scale. Make a mark on the index card at that point and label it with the resistance shown on the ohmmeter. Set the pot to the other end of the scale and label that point, too. Starting at minimum resistance, adjust the pot in steps of 500 Ω, labeling each step on the index card, until you reach maximum resistance. You'll have something that looks like Figure 3, although if you have a tapered resistance pot, the calibration marks won't be equally spaced around the knob. If you like, go back and add more calibration marks.

Connect the calibrated pot as Z_3 in the circuit, set it to 2.5 kΩ, and attach a voltmeter to measure V_{out}. Set the meter to its most sensitive voltage scale, perhaps 300 mV full scale. Dig into your parts box and pull out any resistor between 2 and 3 kΩ, such as a 2.7 kΩ unit. Connect that as your unknown, Z_X, and apply 5 V dc to the bridge as V_{gen}. Adjust the calibrated pot until the bridge is balanced — at a meter reading as close to 0.000 V as you can set it. You will find the adjustment to be surprisingly sensitive. Once the bridge is balanced, the calibrated scale will show the value of the resistor attached as Z_X.

Verify that the balance point of the bridge is independent of the value of V_{gen}. Adjust your power supply output voltage with the bridge balanced. It shouldn't change the balance at all, except perhaps for any thermal effects in the resistors. (I told you it was sensitive.) Return V_{gen} to 5 V and make sure the bridge is still balanced. Fire up your

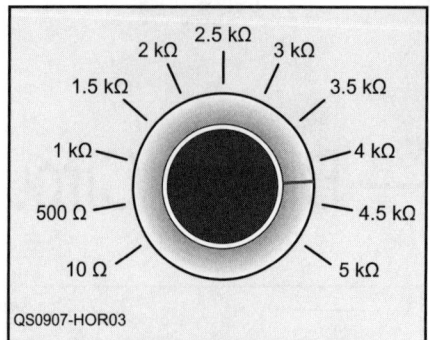

Figure 3 — A panel-mount potentiometer can be calibrated as a variable resistance by using an index card to create a resistance scale.

soldering iron. When it is good and hot, touch the tip of the iron to the unfortunate resistor connected as Z_X. As the resistor heats up, its value will also go up. Pop quiz — if the value of Z_X goes up and Z_3 doesn't change, does V_{out} change in the positive ($V_1 > V_2$) or negative direction?[1] If you have a can of compressed air or contact cleaner or freeze-spray, chill Z_X and watch the balance tip in the other direction.

Now try other values for Z_X. Make a bit of a game out of it by collecting a handful of resistors at random and measure them without looking at the value markings. Did you grab some resistors for which you couldn't get a satisfactory balance? Perhaps the value was higher than 5 kΩ and the potentiometer didn't have enough range. Or maybe the value was very low (say below 100 Ω) and it was difficult to adjust the pot for a precise balance.

There is a solution. Remember that the dividers don't have to be identical for the bridge to balance, only the ratios of the two impedances must be the same. You can expand the range of the bridge by changing Z_3 to a larger or smaller value of potentiometer. Calibrate the pot as before and see if those high or low value resistors are any easier to balance.

Parts List

- Two each 2.2 kΩ, ¼ W resistors.
- Panel mount potentiometer, 5 kΩ and index card.
- Selection of resistors as test samples.

Recommended Reading

Pretty amazing how much fun you can have with four resistors, isn't it? There are many variations on the basic Wheatstone bridge circuit. Begin by reading the Wikipedia entry on the Wheatstone bridge (**en.wikipedia.org/wiki/Wheatstone_bridge**) and then following the links to other bridges. Also read the entries about the Wien Bridge Oscillator, a different bridge application.

Next Month

We've studied and used L networks in past experiments.[2] They are the basis for the more general Pi and T networks more commonly found in amateur equipment. We'll hook up to those circuits next month.

[1]V_{out} changes in the negative direction because V2 gets larger while V1 stays the same.

[2]Previous Hands-On Radio columns and a complete parts list for all experiments are available to ARRL members at **www.arrl.org/hands-on-radio**.

Experiment 115
All About Tapers

What's the difference between a variable resistor, a trimmer or trimpot, a rheostat and a potentiometer or pot for short? Nothing! They are all the same basic device — a resistive structure (the element) made of wire or some conductive material that can be contacted anywhere along its length by a sliding contact or wiper. There are usually three terminals — one at each end of the element and one for the wiper (see Figure 1). The name used depends on the application.

The element can be straight, requiring the wiper to move back and forth in a straight line — those pots are often called *sliders*. Pots used for rotary controls on a front panel (or *panel* pots) have a circular element and a shaft moves the wiper as it turns. *Trimmers* or *trimpots* don't have a control shaft or knob and are adjusted with a screwdriver or tuning wand. They are very small and inexpensive — intended for tuning or bias adjustments that are made very rarely.

A *rheostat* needs only two terminals — one at the end of the element and one for the wiper — and it is used as a variable resistance in series with a circuit, usually a power circuit of some sort. Rheostats are generally made to dissipate large amounts of power (several watts and up) compared to the small pots used in most electronic circuits.

Tracking the Wild Pot Taper

This month we'll take a pot shot at one of those ubiquitous but rarely discussed features all pots have whether the manufacturer says so or not — the taper. Not the pig-like animal from South America (spelled tapir), this type of taper tells you how the resistance of the pot changes with the setting of the wiper. Does it matter? Like the name, it depends on your application.

You might think that equal movements of the wiper would result in equal changes in resistance. Most of the time you'd be right — that's referred to as a *linear taper*. Unless the manufacturer says differently, it's a pretty good bet that a pot has a linear taper. Most of the time that works out just fine — for adjusting a voltage regulator or setting a bias point. Other applications are more sensitive to the relationship between the control position and the resistance from the wiper to each end of the element.

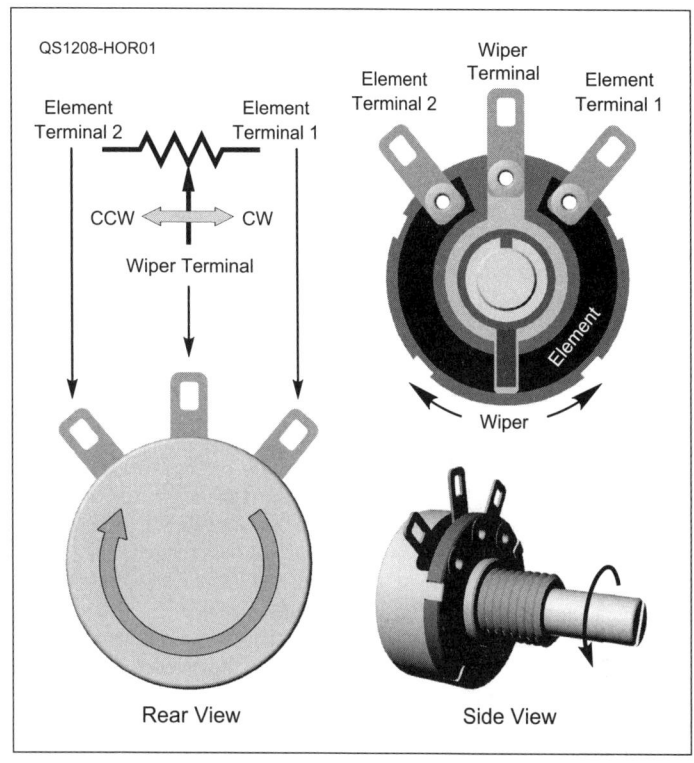

Figure 1 — Construction of a typical panel pot in which turning a shaft moves the sliding wiper contact along the resistive element.

Audio Taper

If you look through the component catalogs you'll see something called *audio taper* or *log taper*. There are also reverse audio taper and reverse log taper pots. What are these? It all goes back to the human ear.

Named for Alexander Graham Bell in 1928, the decibel was a new name for what were previously called transmission units (TU). The TU was a replacement for the earlier MSC (Mile of Standard Cable) unit that represented the loss of one mile of standard telephone cable at a frequency of 795.8 Hz (5000 radians/s) such that 1.056 TU = 1 MSC.[1] The MSC represented the approximate minimum value of attenuation that could be detected by the average listener. (Now you know!)

Early telephone engineers quickly determined that the human ear responded roughly logarithmically to sound level. Thus, if the usual linear taper pot was used to control volume, all of the effective volume adjustment occurred at one end of the control. Listeners expected the perceived volume to

> **Watt's in a Name?**
>
> The word *potentiometer* comes from *potential*. A potentiometer is basically a variable voltage divider that is used to control the voltage or potential in a circuit, thus the name. There is also a potential or voltage measuring instrument with the same name.

[1]en.wikipedia.org/wiki/Decibel

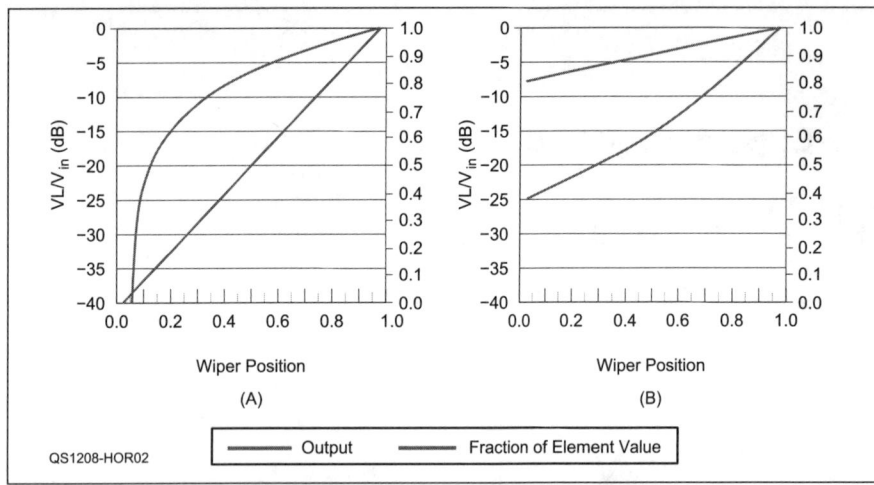

Figure 2 — The output from a linear taper pot (A) changes too rapidly at the bottom end of its range for a linear change in perceived output volume. The exponential change in resistance of a log taper pot (B) produces a linear perceived change in volume. The blue line represents the fraction of resistance between the wiper and common terminal of the pot element.

steadily increase throughout the control's adjustment range roughly linearly. For example, if the control was labeled 0 to 10, listeners expected a setting of 5 to be about one-half as loud as 10. (As fans of the band Spinal Tap know, a setting of 11 is even louder!)

The solution was to make the resistance of the element change logarithmically so that the voltage from the wiper would change in such a way as to create a linear perceived change in sound volume. Figure 2 illustrates the difference between a linear and a logarithmic taper along with the change in decibels of the output signal (red) from the wiper with the input signal connected across the pot element. A wiper position of 1.0 corresponds to a setting of 10 on a control calibrated from 0 to 10.

As you can see from the graph of the linear taper (Figure 2A) the output changes nearly linearly as the position of the control is moved away from "full on." The output is down 3 dB (half volume) at a setting of 0.7 and down 6 dB (quarter volume) at the half-way mark. As the position approaches 0.3, however, the change in output accelerates dramatically, rapidly dropping to what the listener would perceive as *off*. About a third of the control's adjustment range is thus wasted.

Figure 2B, on the other hand, shows what happens if the resistance of the element changes more slowly at the bottom end of the adjustment range. This is an example of an audio taper that produces a linear change of nearly 10 dB across the control's range. Depending on the exact construction of the pot, the listener would find the setting of 1 produces about 1/10 the volume of a setting of 10.

This taper's function of resistance versus wiper position is actually exponential — the inverse function of logarithmic. The taper is called logarithmic because the exponential change in resistance produces a linear change of output volume in logarithmic terms.

There are also "reverse taper" pots that are used in applications such as BALANCE and BASS or TREBLE controls so that a common shaft causes the output signal from one pot to increase while the other decreases. Again, the changes are logarithmic so the listener perceives an even change in volume.

Custom Tapers

Of course, hams like to experiment and there are many instances in which the commercially available tapers aren't right for the job. In that case, you need to "roll your own" taper by combining a pot with some external resistors in series and parallel as in Figure 3. In that circuit, the pot is represented by the series combination of R_{PHI} and R_{PLO}. The wiper contact is at the junction of R_{PHI} and R_{PLO}. The sum of $R_{PHI} + R_{PLO} = R_{POT}$ is constant and represents the value of the pot's element.

Input voltage, V_{IN}, is applied to the series resistor, R_{SER}. The resistor R_{PAR} is connected in parallel across the pot's element. Output voltage, V_L, is taken across the load resistor, R_L, which is connected between the wiper and circuit common.

By varying the resistor values, the experimenter can create an amazing variety of tapers suitable for just about any special need. (Dare I say, a whole herd of tapers...) The equations to do so, however, are an impenetrable symbolic thicket so I've provided a spreadsheet to let you solve the problem empirically, which is a fancy way of saying "cut and try." The *Excel* spreadsheet and the *LTSpice* schematic file, both available on the Hands-On Radio website, will allow you to enter any value for the resistors and pot.[2] You can also create custom functions for the element resistance versus wiper position.

As provided on the website, the spreadsheet is pre-loaded with one of the commonly available "two-piece approximation" functions for wiper resistance that simulates a log response. You'd think the abrupt change in output level would sound awful to the listener but it seems to be acceptable. Two piece tapers are common with several different varieties sold.

After loading the spreadsheet, enter values for R_{SER}, R_{PAR}, R_{POT} and R_L — those cells are highlighted in red. Select one of the functions for wiper setting versus resistance from the provided set (highlighted in green) by selecting the column of values, then using COPY followed by PASTE SPECIAL > VALUES (not PASTE) in the WIPER FUNCTION column of the spreadsheet. All other values and the chart will automatically update.

You can also create your own WIPER FUNCTION by using formulas based on the WIPER POSITION column or by just entering a set of 20 values by hand. Feel free to experiment and let your tapers run wild!

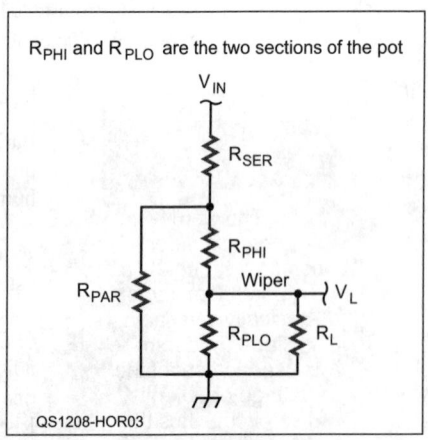

Figure 3 — A circuit that allows an experimenter to create a custom taper by varying the resistor values external to the pot. R_{PHI} and R_{PLO} represent the pot with the wiper connected to their junction.

[2] All previous Hands-On Radio experiments are available to ARRL members at **www.arrl.org/hands-on-radio**. *LTSpice* was the subject of columns 83 through 86.

Experiment 101
Rotary Encoders

Look behind the front panel of most commercial radios and you won't find a fat harness of wires running to multiple variable resistors. You're far more likely to encounter the subject of this month's column — rotary encoders.[1] With nearly every adjustment now performed under the control of a microprocessor it's logical (so to speak) that the controls should be digital, too. And since the encoders are what you put your hands on, what better topic for this column?

Encoder Types

The word *encoder* is used for more than one purpose — you may be more familiar with a *tone encoder* that generates DTMF or CTCSS audio tones for controlling repeater systems. The rotary or shaft encoder we'll discuss here translates rotary mechanical motion into a digital code representing the motion of the shaft.

Mechanical encoders use a disc with a circular pattern of metal strips that make contact with sliding contacts. The disc is mounted to a rotating shaft so that as the shaft turns, circuits are opened and closed to form the digital bits. Mechanical cams are also used to open and close switches. *Optical encoders* substitute phototransistors for sliding contacts (see Figure 1). A plastic *interruptor disc* replaces the strips and contacts with patterns of opaque and clear areas that turn the phototransistors on and off. In both cases, the encoder uses an array of two or more switches to create the on-off pattern representing the shaft's motion or position.

Regardless of whether the switches are optical or mechanical there are two families of encoders. The on-off configuration of the switches in an *absolute encoder* represents the exact position of the shaft as it rotates, whether it is moving or not. In an *incremental encoder* the switches only indicate whether the shaft is moving and if so, in what direction. For both types of encoders, it is up to the interface circuit to keep track of the shaft's motion and translate that into a useful control input.

You Say You Want Some Resolution

Encoders also vary in their resolution — the number of digital bits available to represent shaft position or motion. Absolute encoders are similar to analog-to-digital converters in that they translate an analog input variable (position) into a digital value called a *data word*. The number of bits in the data word determines how many different values of position can be generated by the encoder. An n bit data word can represent 2^n different positions as values from 0 to 2^n-1; 8 bits = 256 different positions (0 to 255) and each bit represents $360/256 = 1.4°$.

The resolution of incremental encoders depends on how many pulses or switch on/off changes occur during a complete revolution. A 12 pulses/revolution encoder generates one pulse for every $360/12 = 30°$ of movement. From the frequency of the pulses, the speed at which the shaft is rotating can be determined as pulses/s × degrees/pulse = degrees/s. This simple frequency to angular rate conversion is the basis of a *tachometer*.

Incremental encoders are also available with an *index* output that closes a switch once per rotation. This provides a means of sensing absolute position by resetting a counter or otherwise synchronizing a circuit or software routine.

Knowing the Codes

The simplest type of digital code used by absolute encoders is binary in which the data word counts from 0 through 2^n-1. If we had a three-bit absolute encoder as shown in Table 1, the data word would count through eight values from 000 through 111 in a complete revolution and then change back to 000 and begin again. Binary code has a problem, however, at the transitions from one value of the data word to the next. For example, take a look at the change between data word values 001 and 010. Two bits are changing at the same time — bit 1 from 0 to 1 and bit 0 from 1 to 0. If both bits always changed at exactly the same instant this wouldn't be a problem. But let's say that due

[1]en.wikipedia.org/wiki/Rotary_encoder

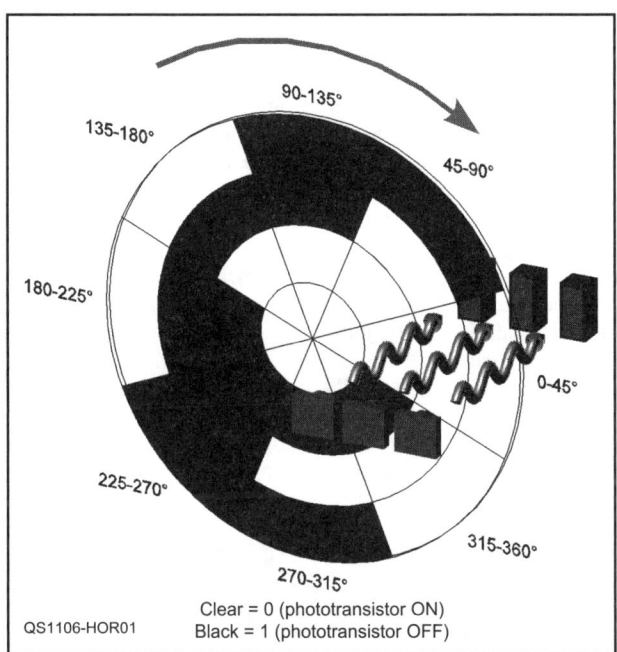

QS1106-HOR01
Clear = 0 (phototransistor ON)
Black = 1 (phototransistor OFF)

Figure 1 — An optical encoder uses phototransistors instead of mechanical switches. In this figure, a three bit Gray code interruptor disc blocks or passes light from an LED to a phototransistor. The combination of which phototransistors are ON and OFF forms a three bit data word representing the position of the encoder shaft with bit 0 on the outer ring.

Table 1
Three Bit Binary and Gray Codes (see Figure 1)

Angular Position	Binary			Gray Code		
	Bit 2	Bit 1	Bit 0	Bit 2	Bit 1	Bit 0
0 to 45°	0	0	0	0	0	0
45 to 90°	0	0	1	0	0	1
90 to 135°	0	1	0	0	1	1
135 to 180°	0	1	1	0	1	0
180 to 225°	1	0	0	1	1	0
225 to 270°	1	0	1	1	1	1
270 to 315°	1	1	0	1	0	1
315 to 360°	1	1	1	1	0	0

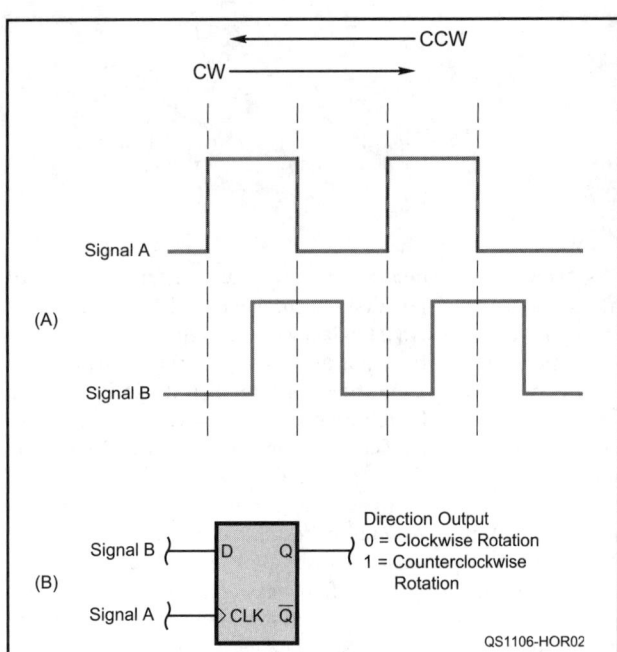

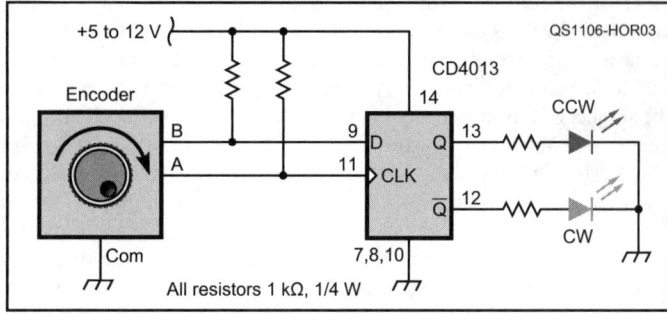

Figure 2 — The quadrature encoder's output contains information about both speed of rotation (pulse frequency) and direction (phase relationship). By comparing signals at a specific signal transition, the direction of shaft rotation is determined. A D-type flip-flop can be used as a direction of rotation sensing circuit.

Figure 3 — A CD4013 D-type flip-flop's output drives LEDs to show the direction of rotation visually. Pull-up resistors (1 kΩ) convert the encoder switch closures to electrical signals.

to minute manufacturing misalignments, bit 1 changes just before bit 0 changes. That means the digital circuit reading the switches would see the data word change from a value of 001 to 011 for a short time before returning to 010. This can cause a lot of problems and it occurs at half of all position changes in a binary code counting in either direction!

For this reason, *Gray code* (named for its inventor, Frank Gray) is used. Look carefully at the Gray code portion of Table 1. You will see that only one bit changes between any two data word values. That means minor misalignments between the switches don't result in large ambiguities in position. The encoder disc in Figure 1 illustrates 3 bit Gray code. Note that only one track changes between opaque and clear at any time.

Encoders are also available that output data as *octal* — binary bits in groups of three that have values from 0 to 7 — and *binary-coded decimal (BCD)* in which four-bit groups form individual digits with values from 0 to 9.

Most encoders used in radios are incremental encoders because it's not usually necessary to know exactly where a knob is pointing or where a dial is set — there are no dials! Absolute encoders are more likely to be found in electromechanical control systems and are quite a bit more expensive than incremental controllers. The rest of this column will focus on incremental encoders.

Using an Incremental Encoder

To be useful in controlling a radio's power output or filter bandwidth it's also necessary to know the direction of the rotation. The radio's controlling microprocessor reads the digital signals from the encoder and determines which way the shaft is rotating, how fast, and how far.

In incremental encoders, the bits form a *quadrature* output in which two digital square waves with a 90° phase difference create a two-bit Gray code as shown in Figure 2A. The digital signals A and B combine in four sequential A-B combinations from left to right: 0-0, 1-0, 1-1 and 0-1. It's easy to tell how fast the encoder's shaft is being rotated simply by counting pulses of either A or B.

Telling the direction of rotation is a bit more subtle. Look carefully at each low-to-high transition of signal A. If the encoder shaft is rotating clockwise (CW), the transition occurs when signal B is low. If the rotation is counterclockwise (CCW), the transition occurs when signal B is high. A circuit that senses the low-to-high transition of signal A can evaluate the state of signal B and determine the direction of rotation.

The D-type flip-flop shown in Figure 2B can accomplish this function by using signal A as its clock input. At each low-to-high transition of signal A, the value of signal B is transferred to the Q output. If the shaft is rotating CCW, the Q output is high and vice versa if the rotation is CW. (This function is easy to perform in software, too!)

Figure 3 shows a circuit you can build to see this for yourself. You'll need a mechanical quadrature encoder, such as one of the Bourns PEC16-series encoders available from **www.digikey.com** for a bit more than 50 cents each. (The PEC16-4120F-N0012-ND part number is a convenient-to-use panel-mount encoder with PC mounting pins and 12 pulses per revolution. A complete data sheet can be downloaded from Digi-Key.)

These and other similar encoders are very simple to wire, being made of just the two switches (A and B) and a common connection (COM). Use 1 kΩ resistors to "pull up" each switch so that when the switch closes, the output is low. (If you use an optical encoder, the phototransistor equivalent is an *open-collector* output and you will have to supply dc power to the LEDs.) Connect each input to the appropriate pin of a CD4013 D-type flip-flop as shown in the schematic. (Be sure to connect the flip-flop's pins for SET (8) and RESET (10) to ground as shown or the flip-flop won't change states.) Any small LED will do for the indicators. When you first apply power, either LED may light but as soon as you begin twirling the encoder the LEDs should light according to the direction. If the LEDs seem reversed, you probably have the A and B signals reversed.

To convert the incremental encoder into an absolute encoder experiment with the programmable up-down counters described in Experiment #36.[2] The index pulse feature of a suitably equipped encoder could be converted into a reset pulse for the counter although you may have to experiment with some spare logic gates to convert the index signal to a reset pulse that appears at the right time.

Parts List

4 — 1 kΩ, ¼ W resistors
2 — Miniature LEDs
1 — CD4013 D-type flip-flop
1 — Incremental mechanical encoder

Recommended Reading

To get an idea of how many different types of encoders are available, enter ROTARY ENCODER in the PART SEARCH window of the Digi-Key home page, then click ENCODERS on the next page.

[2]All previous Hands-On Radio experiments are available to ARRL members at **www.arrl.org/hands-on-radio**.

Experiment #76 — Diode Junctions

The Basic PN Junction

Pure silicon (or any semiconductor material) conducts poorly, compared to good conductors such as copper and aluminum. It's only when impurities are added that things get interesting. In pure silicon, each atom shares an electron with each of its neighbors, creating bonds that hold the lattice structure together. These electrons are held firmly in the bonds and are not free to move if an electric field is applied, such as from a voltage across the material, so conductivity is low.

When an atom with a *donor* impurity — such as antimony, arsenic or phosphorus that has an extra electron available to form bonds — replaces one of the silicon atoms, the unused bonding electron is free to move

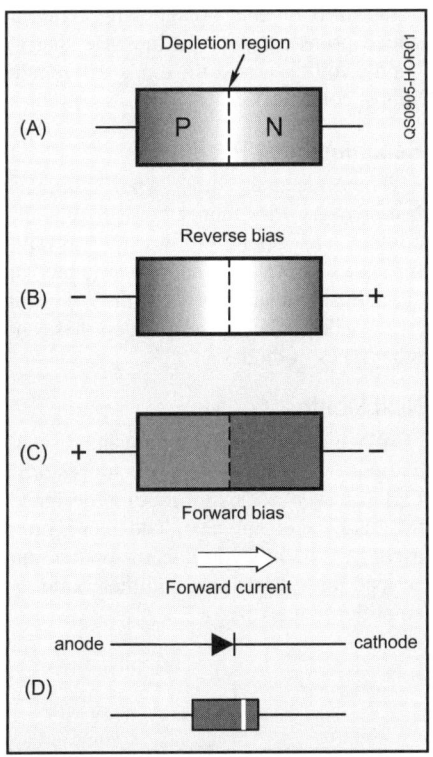

Figure 1 — At (A), a PN junction forms a depletion region at the interface between the two types of material. At (B), reverse-bias widens the depletion region, while forward-bias (C) causes the majority carriers to recombine at the junction. The symbol and physical appearance for a diode are shown at (D).

about the silicon lattice, especially if there is a voltage across the material. (As long as these *free electrons* do not leave the lattice, the material remains electrically neutral.) This is called *N-type* material because of the free negative electrons. An *acceptor* impurity atom — boron, gallium, or indium — has one fewer bonding electron than silicon so that one of the bonds is not made. This is *P-type* material because the unmade bonds from the missing electron act as positive charges.

Missing bonds are referred to as "holes" because unbound electrons give up energy when they are captured to form the bond. Unbound electrons can come from adjacent bonds in the lattice or they can be electrons released from silicon atoms by thermal or other processes.

But wait, how can a hole move? It's really just a missing bond between two fixed atoms. When an electron in an adjacent bond feels enough force due to an applied voltage, it can move to a missing bond in the direction of the voltage, "filling" the hole and leaving a new hole behind. The net effect is the same as if the hole moved in the opposite direction.

Putting Them Both Together

By themselves, N-type and P-type material can't do much, but when you bring them together, something very interesting happens. As shown in Figure 1A, at the interface between N-type and P-type material, free electrons can move across the interface and "fall into the holes", a process known as *recombination*. This only happens for electrons and holes very, very close to the junction between the materials, creating a region on either side in which the majority carriers have recombined, leaving the material depleted of majority carriers. This is the *depletion region*. It has poor conductivity because there are no carriers to flow in response to voltage. The combination of P-type and N-type material with a depletion region between them forms a *PN junction*.

The depletion region is a very thin (approximately 500 nm) insulating layer, so if a small voltage is applied in either direction across the junction, very little current will flow, even though conductivity of the N-type and P-type material is quite good.

If a *reverse-bias* voltage across the PN-junction is applied as in Figure 1B, the majority carriers attempt to flow *away* from the junction as the electrons move toward the positive voltage and holes toward the negative. This widens the depletion region and reduces conductivity even more, so that only a very small *reverse leakage* current, I_s, flows across the junction.

If, however, a *forward-bias* voltage is applied, as in Figure 1C the majority carriers attempt to flow *across* the junction. As they approach the depletion region, the carriers become so close together that they recombine. They are replaced by new electrons injected into the N-type material from the negative voltage source and removed from the P-type material by the positive source. (Remember, this is the same as if the positive conductor were injecting holes into the P-type material.) Thus, *forward current* appears to flow across the PN-junction, although the holes and electrons are actually just recombining furiously in what was previously the depletion region.

The Fundamental Diode Equation

The *fundamental diode equation* describes the relationship between forward

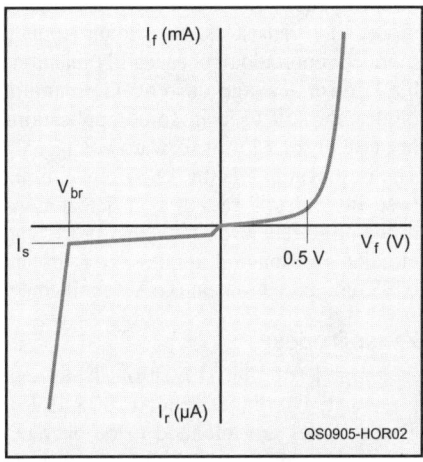

Figure 2 — A generic I-V characteristic curve for a silicon diode showing the forward and reverse conduction regions. Reverse breakdown occurs when the applied voltage forces carriers across the depletion region in the reverse direction.

Electronic Circuits and Components

voltage across the junction, V_f, and current through it, I. (A link to math tutorials on exponential equations can be found at **www.arrl.org/gclm**.)

$$I = I_s \left(e^{\frac{V_f}{\eta V_t}} \right) \text{ or } V_f = \eta V_t \ln(I / I_s)$$

where e is the base for natural logarithms (approx 2.72) and V_t represents a voltage created by the thermal energy present in all materials, about 25 mV at room temperature. V_f is measured from the P-type material (the anode) to the N-type material (the cathode) and is the voltage required for the majority carriers to come together in the depletion region and recombine for a given level of current. A generic diode I-V characteristic curve is shown in Figure 2.

The two remaining parameters depend on the material and method used to make the diode. I_s is the *reverse-bias saturation current* that flows when reverse bias is applied with typical values of pA to µA. η is the *emission coefficient* that depends on material and current level. It is near 2 for silicon at mA current levels and drops close to 1 at large currents. By choosing materials and manufacturing processes, different types of diodes are created.

The depletion region will break down if enough reverse bias is applied. When the reverse breakdown voltage, V_{br}, is reached, current flows backwards across the junction and the diode is in reverse breakdown. Limiting the applied reverse voltage to less than the diode's peak inverse voltage (PIV) rating prevents reverse breakdown in which currents can be large enough to destroy the diode.

Silicon versus Germanium Diodes

Germanium was used extensively in early diodes and transistors, but silicon is much more common today because PN junctions have lower leakage current. Germanium diodes are still used in some applications because of their lower forward voltage, V_f. The best example is the 1N34 diode, often used in "crystal" radios as a detector. At milliamp current levels for signal-processing circuits, a germanium diode's V_f is around 0.3 V, compared to near 0.6 V for silicon.

Zener Diodes

Named for their inventor, American physicist, Clarence Zener (1905-1993), Zener diodes are intended to be operated in reverse breakdown mode with a precisely controlled voltage at which reverse breakdown occurs. Once reverse breakdown is reached, the voltage across the Zener diode is relatively constant over a wide range of current, making it useful as a voltage reference or regulator.

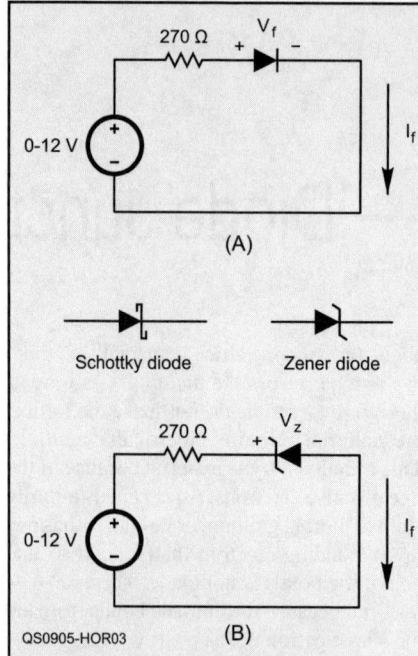

Figure 3 — Measure a diode's I-V characteristics by applying voltage through a current-limiting resistor. The symbols for Schottky and Zener diodes are shown. (A) shows diodes connected with forward bias. (B) shows a Zener diode connected for reverse-bias to act as a voltage regulator.

Schottky Barrier Diodes

A diode's forward voltage causes power to be dissipated, just as in a resistor — P = $V_f \times I$. At the high current levels in power supplies, this can be many watts. The amount of time required for the diode to respond to a sudden reversal of applied voltage and stop conducting is also important, particularly in high-frequency switchmode supplies. Longer response times allow energy to be lost as current flows backwards during the short *recovery period*.

The *Schottky barrier diode* is formed by applying a metal such as gold or platinum directly to N-type material, forming a depletion region at the metal-semiconductor interface. Schottky diodes have a lower V_f for the same amount of current and dissipate less power. They also respond to changes in applied voltage more quickly than PN junction diodes.

Measuring Diode Performance

Figure 3 shows a circuit for measuring a diode's *I-V characteristic curve*. A variable voltage source, such as a workbench power supply, and a current limiting resistor are connected in series with the diode. Use a pair of digital voltmeters (DVMs) to measure current and voltage and plot it on graph paper. (A three-terminal regulator can be used to make a variable voltage supply as in Hands-On Experiment #70.[1])

Start with a 1N4001 or 1N4148 silicon diode. The graph's vertical axis should span 0 to 50 mA and the horizontal axis 0 to 1 V. Measure current directly with your DVM, or measure voltage across the current limiting resistor and use Ohm's law to calculate current.

Vary the power supply voltage from 0 to 12 V in steps of 0.25 V, plotting V_f and I at each step. You should see a curve that starts at the origin and stays close to the horizontal axis until V_f approaches 0.5 V, at which point the curve will bend upwards as the diode begins to conduct.

Substitute a 1N34 or 1N34A germanium diode and repeat. You will see that the diode's V_f is lower than that of the silicon diode. Repeat with a Schottky barrier diode, such as a 1N5817, to get a third I-V curve. Repeat once again with a low voltage Zener diode, such as the 4.7 V 1N4732 or 1N4732A. It should produce a curve almost identical to that of the first silicon diode.

Turn the Zener diode around so that the cathode is connected to the current-limiting resistor. Make a new graph with the same vertical scale, but with the horizontal axis spanning 0 to 12 V. Measure current and Zener voltage, V_Z, while adjusting the power supply voltage in 1 V steps from 0 to 12 V. There will be very little current through the diode until reverse breakdown occurs, at which point current will increase sharply but the voltage across the diode will remain almost constant.

Recommended Reading

There are many other types of diodes: fast-recovery, PIN, tunnel, hot-carrier, point-contact, etc. Spend some time browsing through *The ARRL Handbook* chapter on "Electrical Fundamentals" and the Wikipedia (**www.wikipedia.com**) entries on these and other types of diodes.[2]

Next Month

When an amplifier circuit using discrete components, such as a bipolar transistor or FET (or even a vacuum tube) is designed, the graphical technique of load lines is often used. Next month, you'll learn about dc and ac load lines, draw one, and then build the amplifier it describes.

---

[1]Previous Hands-On Radio columns are available to ARRL members at **www.arrl.org/hands-on-radio**.
[2]*The ARRL Handbook for Radio Communications*, 2009 Edition. Available from your ARRL dealer or the ARRL Bookstore, ARRL order no. 0261 (Hardcover 0292). Telephone 860-594-0355, or toll-free in the US 888-277-5289; **www.arrl.org/shop/**; **pubsales@arrl.org**.

Experiment 99
Cascode Amplifier

If you begin digging into RF amplifier circuits you'll soon encounter a strange looking beast referred to as the *cascode* amplifier. Looking a little like a totem pole, the circuit combines two single transistor amplifiers into a useful combination that's especially useful in amplifying the weak signals in receivers.

Cascade or Cascode?

At first glance, it's easy to think the name of the circuit is *cascade* which refers to stages connected with the output of one feeding the input of the next, as shown in Figure 1. Amplifier stages are often cascaded to increase total gain. That's not the function of the cascode circuit shown in Figure 2. True, the circuit has two stages and the output of the first is connected to the input of the second but there is a special requirement to qualify as a cascode.

The name hails from the days of vacuum tubes and is derived from cascade. In fact, it means "cascade to cathode", coined in 1939 to describe a circuit in which a pair of triode tubes replaced a single pentode.[1] The circuit's input stage is a common emitter or common source amplifier as was described in Hands-On Radio experiments 1 and 2.[2] The output of the first stage is then connected to the input of a common base or common gate amplifier, both described in experiment 28. (The circuits are also covered in Chapter 3 of *The ARRL Handbook*.[3]) This corresponds to the original common cathode input amplifier and common grid (or grounded grid) output amplifier. Figure 2 shows two different forms of the cascode amplifier — one is made from bipolar transistors and the other from FETs.

What Problems Make the Cascode Attractive?

Let's back up and discuss what the de-

[1]**en.wikipedia.org/wiki/Cascode**
[2]All previous Hands-On Radio experiments are available to ARRL members at **www.arrl.org/hands-on-radio**
[3]*The ARRL Handbook for Radio Communications*, 2011 Edition. Available from your ARRL dealer or the ARRL Bookstore, ARRL order no. 0953 (Hardcover 0960). Telephone 860-594-0355, or toll free in the US 888-277-5289; **www.arrl.org/shop; pubsales@arrl.org**.

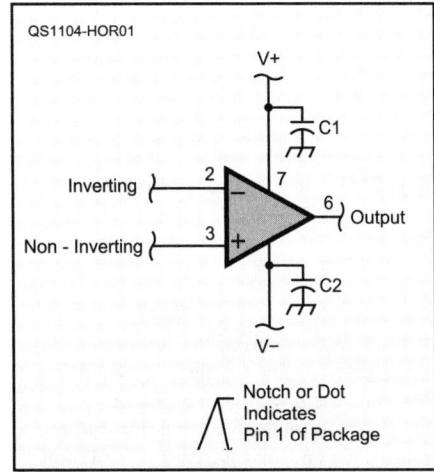

Figure 1 — Example of a two stage cascade amplifier circuit.

signers of cascade amplifiers are trying to accomplish. There are five basic needs (1-5). In many receiver circuits signal amplitude is very small, so high gain (1) is needed at high frequencies (2). The tuned circuits used in receivers are very sensitive to loading, so the circuits connected to them should have high input and output impedances (3 and 4) to avoid altering the tuned circuit's response. (You can find more information about the effects of loading on tuned circuits in Chapter 2 of *The ARRL Handbook*.) Finally, the circuit should be as stable as possible (5) to prevent oscillation or other spurious responses. That is a tall order for a circuit with only one transistor. Here's how the cascode circuit does its job — we'll use the FET version with the understanding that the bipolar junction transistor (BJT) version operates similarly.

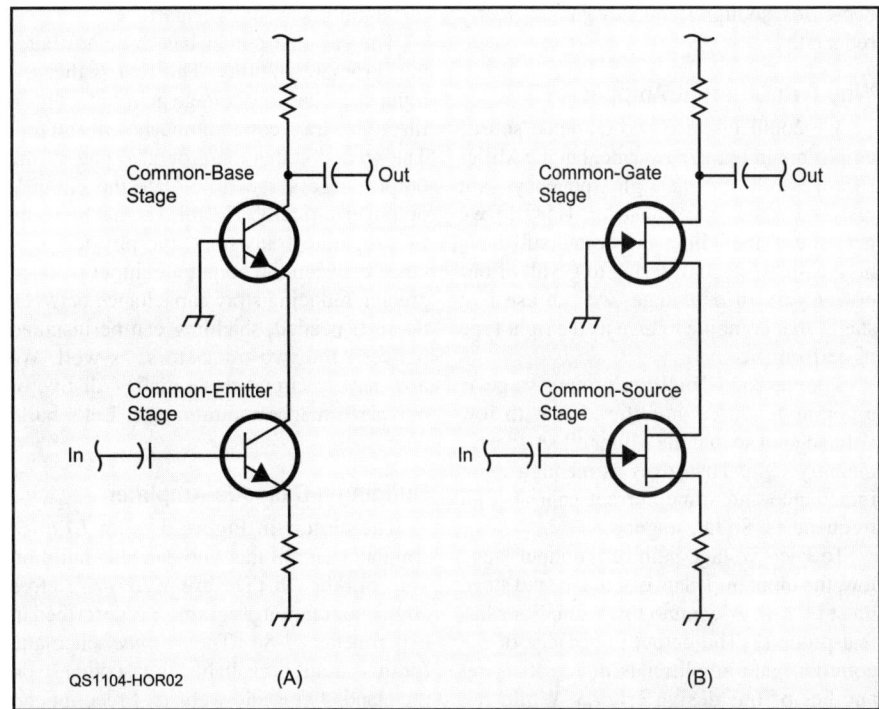

Figure 2 — Cascode amplifier using bipolar transistors (A) and FETs (B). The cascode configuration consists of an input common emitter (or common source) amplifier driving a common base (or common gate) amplifier. The combination results in a stable amplifier with wide bandwidth and high input and output impedances.

If you go back and study the common source amplifier, you'll find that it has medium-to-high input impedance — that's a good thing. Its output impedance, though, is approximately equal to the impedance connected to the drain and the voltage gain of the circuit is proportional to that impedance. That means the requirement for low output impedance and high gain are in conflict. In addition, the input and output terminals of any single-transistor amplifier are close together, creating stability problems at very high frequencies from stray capacitance.

The Miller Effect

There is another problem with the common source amplifier caused by the *Miller effect*. In a nutshell, the Miller effect describes what happens in a high gain common source (or common emitter) amplifier because of the internal capacitance between the drain (the output) and the input (the base), C_{DG}. (Replace C_{DG} with C_{CB} — collector to base capacitance — for a bipolar transistor.) If you measure C_{DG} with a capacitance meter it is very small — in the pF range — so you might not expect it to affect circuit performance very much. In a high gain common source amplifier, however, the output voltage at the drain is much larger and out-of-phase with the input voltage. That effectively multiplies C_{DG} by the circuit's voltage gain, A_V, creating negative feedback to the input. Feedback due to the Miller effect increases with frequency as the reactance of C_{DG} decreases, so high frequency gain is greatly reduced.

Why Two Different Amplifiers?

We could just use two common source amplifiers in regular cascade, but the Miller Effect would eat up high frequency gain just as for a single transistor. How do we get around the Miller effect and still have high gain? Instead of trying to get all of our power gain in one stage, we can use two stages that combine to give us the right type of performance.

The cascode amplifier's input stage is a common source amplifier, but with low voltage gain so that the Miller effect doesn't magnify C_{DG}. That gives us the high input impedance and some current gain at high frequencies. So far, so good!

To keep voltage gain of the input stage low, the input impedance of the second stage must be low. A common gate amplifier has that property. The output impedance of the common gate amplifier is high, satisfying another of our design criteria. While the common gate configuration's current gain is just under unity, it has voltage gain over a wide bandwidth. Thus, the current gain of the input stage multiplied by the voltage gain of the second stage results in high power gain

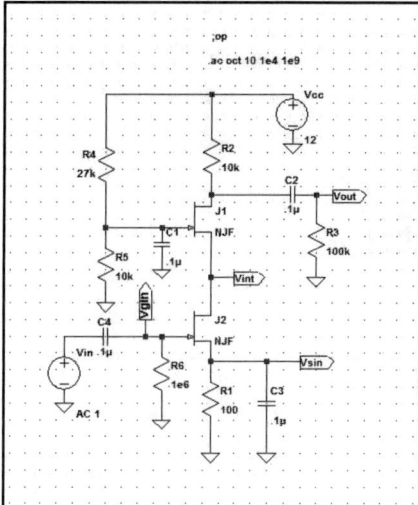

Figure 3 — Screen shot of cascode amplifier schematic using the *LTspice* circuit simulator.

overall. We've solved the input and output impedance problems, the bandwidth problem, and the power gain problem.

There is one more property the circuit must have and that is stability. Solving the first four problems in a single-transistor circuit — particularly high gain at high frequency — makes it hard to prevent feedback from causing oscillation. Shielding the output from the input is also difficult when the transistor's input and output terminals are just a millimeter or two apart!

The cascode circuit has three attributes that promote stability. The first is that the input stage has low voltage gain so feedback through stray capacitance is minimized. The second is that the grounded gate of the output stage serves to isolate the circuit's output electrically. The third is that by using two separate transistors, the physical distance between the output and input is much greater, reducing stray capacitance between them. If needed, shielding can be installed between the two transistors, as well. We now have a circuit that satisfies all five of our performance requirements. Let's build one, shall we?

Building a Cascode Amplifier

The circuit in Figure 3 is an *LTspice* simulator circuit that you can also build on your workbench. (The use of *LTspice*, a free circuit simulation program, was described in experiments 83-86.) The *LTspice* schematic for this circuit is available as cascode.asc on the Hands-On Radio website. I recommend that for practice you construct the schematic yourself using the *LTspice* schematic editor. Use the "njf" component from the schematic library of models for the transistors. To build the circuit prototype, almost any N-channel JFET transistors will give good results for experimental purposes. The J310 and MPF102 are inexpensive, widely available and popular with homebrewers.

The simulation of the circuit illustrates why the cascode is a popular wideband amplifier. The input and output impedance are high while at the same time, the circuit's gain is nearly 12 dB from below 100 kHz up to many MHz. Start with the DC OPERATING POINT simulation (see the earlier experiments for instructions) and be sure that bias voltages and quiescent currents are reasonable.

Set up the AC SWEEP simulation by first making sure that the voltage source, V_{in}, is configured as an ac source with an amplitude of 1. Then select EDIT SIMULATION COMMAND in the SIMULATION menu and select the AC SWEEP tab. Enter the number of points to evaluate per octave and the minimum and maximum frequencies for the sweep. Click the RUN symbol, then use the cursor to select voltages V_{in} and V_{out}. You should see around 11 dB of gain over a wide range. Add a trace to show V_{int}, the voltage between the first and second stages. You'll see 0 dB of gain, confirming the technique of minimizing the Miller effect.

To experiment with the cascode circuit, change the quiescent currents by changing the ratio of R4 to R5 to change the gate bias for J1. Changing R1 will change the drain current for both FETs. Vary the ratio of R2 and R3 to see the effect of changing load impedance. You can also change the transistors to an NPN type, such as a 2N3904, although you'll have to add biasing for the common emitter amplifier at the bottom and adjust biasing for the common base amplifier at the top.

Parts List

Capacitors, 4 each 0.1 µF ceramic.
FET, 2 each J310 or MPF102.
Resistors, all ¼ W; 100 Ω, 2 each 10 kΩ, 27 kΩ, 100 kΩ, 1 MΩ.

Additional Reading

The cascode circuit is discussed in many circuit design textbooks and in *The Art of Electronics*.[4] You can find many practical working designs on line and in the December 2007 *QST* article "The Hybrid Cascode — A General Purpose AGC IF Amplifier" by W7ZOI and WA7MLH. There are Feedback items by the authors in the January and February 2008 issues, as well.

---

[4]Horowitz and Hill, *The Art of Electronics*, 2nd edition, Cambridge University Press. See **frank.harvard.edu/aoe/** for more information.

Experiment #73 — Choosing an Op Amp

Lots of circuits use operational amplifiers (op amps), those handy little packages of gain drawn as simple triangles on schematics. Lots of different op amps. For example, searching for op amps in Jameco's online catalog (**www.jameco.com**) returned over 18,000 part numbers! How is a circuit designer supposed to pick the right one? Whittling the problem down to a manageable size depends on understanding what characteristics of the op amp are most important to your circuit.

Basic Properties

Let's review the basic properties of op amps. An introduction to op amps is provided in Hands-On Radio experiment #3.[1] The perfect op amp has infinite gain, infinite bandwidth, infinite input impedance, and zero output impedance. In demanding applications, however, the actual characteristics of real op amps, in which infinite turns into "high" and zero into "low", become important in how the circuit functions.

The specifications section of a typical op amp data sheet will list at least a couple of dozen parameters. Which of those parameters are important depends on what the circuit is supposed to do. This experiment will demonstrate some of the parameters that most effect common ham radio op amp circuits. Start by downloading the data sheet for a 741 op amp from **www.national.com/ds/LM/LM741.pdf**. On page two you'll find its *Table of Electrical Specifications* — use the columns for the 741A version. We're going to use this venerable part, in production for more than 40 years now, because its limitations are easy to observe. You can extend what you learn to its more modern descendants.

Gain-Bandwidth Product

Voltage gain, A_V, is the ratio of output voltage to input voltage. Any real-world amplifier, whether we're referring to an op amp or a big power amplifier, only delivers gain over a finite range of frequencies. The higher the frequency, the lower the gain. Above a few Hz, as frequency rises and gain falls, the op amp behaves as if the product of *open-loop gain*, A_{VD} ("Large Signal Voltage Gain") and bandwidth is a constant. This is the op amp's *gain-bandwidth product* or *GBW*. For example, if an op-amp has an A_{VD} of 10,000 at 100 Hz, then GBW = 10,000 × 100 = 1 MHz. Knowing GBW, you can predict what the op amp's open-loop gain will be at different frequencies. At 1 kHz, A_{VD} = GBW / f = 1 MHz / 1 kHz = 1000. GBW is specified as a minimum guaranteed value with a typical and maximum value often specified, as well. (The 741 data sheet shows GBW as "Bandwidth" in MHz.)

Figure 1 shows a general frequency response typical of all op amps. The circuit in the figure is a non-inverting amplifier with A_V = 1 + R_f/R = 11 = 20.8 dB. Build the circuit using a bipolar ±12 V power supply. You can use any version of the 741.

Input a 100 mV$_{P-P}$ sine wave from a function generator and verify that the circuit gain is 10 at 100 Hz. (Use an oscilloscope, not a voltmeter, to avoid coloring the results with the voltmeter's frequency response.) Now increase the frequency of the input signal, keeping its amplitude constant, until

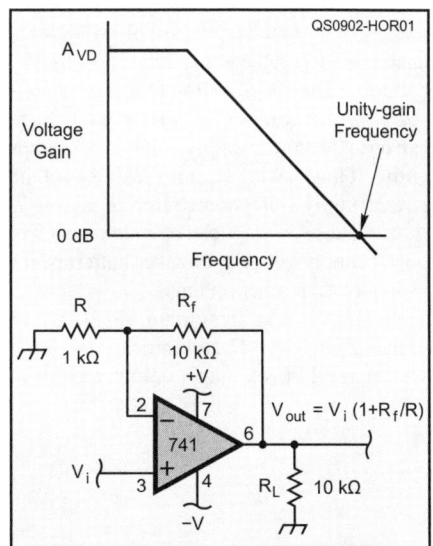

Figure 1 — An op amp's gain decreases linearly (in dB) with frequency, reaching a gain of 0 dB at the unity-gain frequency. Gain-Bandwidth Product (GBW) is the product of gain and bandwidth in this region. The non-inverting amplifier circuit shown is used to evaluate GBW and other op amp parameters.

the output voltage drops to 0.71 of the value at 100 Hz (–3 dB). This is the bandwidth of the amplifier, its decreasing gain acting as a low-pass filter. Calculate GBW. It should be higher than the minimum value for GBW in the data sheet. Find the unity-gain frequency by increasing the input signal frequency until gain drops to unity or 0 dB. Plot gain versus frequency at several values between 10 Hz and the unity-gain frequency using the frequency response spreadsheet developed for experiment #18. You should see a curve that looks a lot like the one in Figure 1.

GBW is important primarily in high-gain ac circuits, such as narrow audio filters in which insufficient gain results in poor filter performance. For example, in a Sallen-Key active filter, the op amp must have $A_{VD} \geq 90 \times Q^2$. For a sharp filter with a Q = 100, the A_{VD} of the op amp must be at least 900,000 or 119 dB.[2] Why would our 741A not be usable in such a circuit?

Slew Rate

Regardless of gain, the op amp's output circuitry can only supply a limited amount of current. This limits how fast the op amp's output voltage can change for a given load. This maximum rate of output voltage change is called *slew rate, SR*, and it is measured in units of voltage over time. In the data sheet, you'll find slew rate specified as a guaranteed minimum value of V/μs.

Input a 0.5 V$_{P-P}$, 1 kHz square wave to the amplifier circuit and verify that the output is a 5 V$_{P-P}$ square wave. Trigger the 'scope on the output signal's rising edge and display both the input and output signals. Increase the sweep speed until you can clearly see the rising edge of the output signal. It will appear somewhat like the drawing in Figure 2.

Assuming your function generator's output has a higher slew rate than that of the 741, the output signal's rising edge will be more slanted than that of the input signal. Measure the op amp's slew rate by measuring the total voltage swing divided by the amount of time it takes the signal to make that swing, as shown

---

[1]Previous Hands-On Radio columns are available to ARRL members at **www.arrl.org/hands-on-radio**.

[2]Maxim Semiconductor Application Note 1762, *A Beginner's Guide to Filter Topologies*, 2002; **www.maxim-ic.com/appnotes.cfm/an_pk/1762**.

Table 1
Popular Op Amp Specifications

Part Number	Application	A_{VD} (V/mV)	GBW (MHz)	Slew Rate (V/µs)	Output Voltage Swing (2 kΩ load)
741A	General purpose	50	0.437	0.3	Within 5 V of power supply
LM324A	Low power	25	1	Not specified	Within 1.5 V of power supply
LF356	Wide bandwidth	15	5	7.5	Within 5 V of power supply
TL071	Low noise	15	3	5	Within 5 V of power supply
TLC2772	Rail-to-rail	15	2.18	1.7	Within 750 mV of power supply

in the figure. This value should be greater than the minimum specified slew rate.

Now increase the input signal frequency. You'll see that the slew rate doesn't change because it's not a function of the op amp's gain. Lower the input voltage and you'll see that doesn't affect slew rate, either.

Increase the input signal frequency until you can see both the rising and falling edges of the input square wave. Keep increasing frequency until you reach a point at which the output waveform is unable to reach a maximum before the input signal's falling edge occurs. The op amp has turned a square wave into a triangle wave.

If you intend to use an op amp circuit for signals with fast rising and falling edges, such as for digital data or video, you'll need to consider slew rate. For example, in a 9600 baud system, each data bit is about 10 µs long. If the signal's amplitude is 5 V and the rise or fall time are to be less than 5% of the total bit period, the op amp must have a slew rate of no less than 5 V / 0.5 µs = 10 V/µs. The specs indicate that the 741A op amp is not suitable for this application.

Output Impedance

Another important parameter for many circuits is the op amp's ability to drive a load. Lower values of output impedance allow the op amp to drive lower impedance loads. Output impedance is often specified indirectly as a minimum value or graph of output voltage swing for a given range of load values. You'll find output impedance to be important when using op amps to drive headphones or as drivers of relays or power transistors.

Input a 0.25 V_{P-P}, 1 kHz sine wave and remove the circuit's load resistor, R_L. Mea-

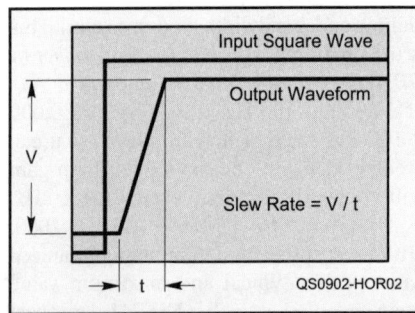

Figure 2 — An op amp's slew rate (SR) can be measured with a square wave input signal and triggering a dual trace 'scope on the input signal's rising edge. The slope of the output signal in V/µs is the op amp's slew rate.

sure the circuit's *open circuit output voltage*, V_{OOC}. Replace the load resistor with a 500 Ω potentiometer and lower its resistance until the output voltage is reduced by half. At this point, the resistance of the pot is equal to Z_0.

Voltage Limits

An op amp's ability to produce an output voltage close to the power supply voltages is an important parameter in circuits powered by batteries. The specifications for *output voltage swing* show how close the output voltage can be to the power supply voltages. Op amps with a voltage swing within 1 V or less of the power supply voltages are referred to as *rail-to-rail*. These op amps can operate with lower power supply voltages (weaker batteries, for example) than other op amps.

Return the amplifier circuit's load resistor to the original 10 kΩ and increase the input voltage until the op amp's output waveform flattens at both the positive and negative peaks. Measure the power supply voltage and the voltage of the flattened peaks. They should be well within ±5 V of the power supply voltages.

Popular Op Amps

Even knowing that only a few of these parameters are important, there are still too many different parts for the hobbyist to worry about. How do you actually find one that isn't an exotic or expensive part? Table 1 shows the properties of several popular op amps suitable for Amateur Radio electronics. These op amps are in stock with all major distributors and at least one will be satisfactory for most purposes. Keeping a few of these op amps on hand (including the versions with two or four op amps per package) will let you build a lot of useful circuits. The major semiconductor manufacturers feature part selection guides and online tools on their Web sites, as well.

Recommended Reading

There are many more subtle characteristics of op amps, one of the most widely used analog components. There are plenty of good online sources of information with one of the best being the Texas Instruments reference *Handbook of Operational Amplifier Applications*, downloadable at **focus.ti.com/lit/an/ sboa092a/sboa092a.pdf**.

Next Month

Let's tackle another basic subject next month — the resonant circuit and Q. These deceptively simple circuits virtually define "radio" so it's a good idea to understand them reasonably well.

Experiment #70 — Three-Terminal Regulators

Although you may have been expecting an article on construction techniques this month, I'm writing about an application of technology from an earlier column, instead. It's hurricane season as I write (betwixt Gustav and Hanna) and soon to be windstorm season in my home, the Pacific Northwest. This prompted me to review and update my *go-kit* of portable radios, packed in a plastic storage container. As I plugged in multiple wall-wart supplies to charge their batteries, I thought, "There has to be a better way!" And soon, there was.

But first, it was necessary to understand the go-kit's requirements for power. In my collection of portable radios, I have a VX-5 and two VX-1 Yaesu FM handhelds as well as a Bearcat BC246T scanner. I also have a four cell NiMH/NiCad battery charger. The VX-1s and the BC246T each require 6 V dc to charge or operate while the VX-5 and charger need 12 V dc. I want to charge every piece of equipment while it is inside the go-kit from any convenient 12 V dc source, such as a vehicle or power supply. I wanted all non-proprietary power connectors to be Anderson Powerpole type, the ARES® standard.[1] The go-kit should have a single external power connection. Visible indicators should show the status of 12 V and 6 V power. This set of requirements resulted in the system diagram of Figure 1. (Starting with determination of the requirements is crucial to creating a satisfactory design.)

This column describes the 12 V to 6 V converter. I decided to build my own converter, based on a three-terminal voltage regulator of the type you first met in Hands-On Experiment #8 on linear voltage regulators.[2] I chose the LM317T,[3] which is widely available and easy to use. By using a pair of programming resistors, output voltages can be selected from 1.2 V up to about 3 V below the input voltage, which can be as high as 40 V.

---
[1]Anderson Powerpole connectors (**www.andersonpower.com**) are rapidly being adopted as the standard power connector for ARES® teams.
[2]Hands-On Radio experiments are available online to ARRL members at **www.arrl.org/hands-on-radio**. The first 61 experiments are also available as *ARRL's Hands-On Radio Experiments* from the ARRL at **www.arrl.org/shop**, order number 1255.
[3]**www.national.com/pf/LM/LM317.html**.

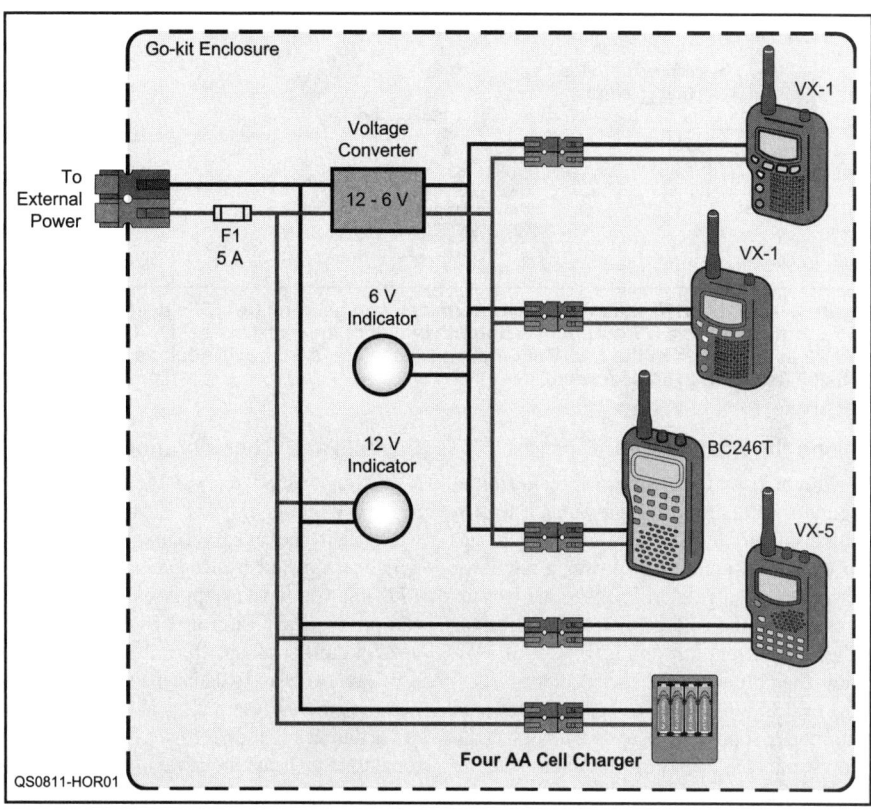

Figure 1 — A 12 V power distribution system for a go-kit. A 6 V regulator provides power for smaller radios. The indicator shows the status of each supply. Powerpole connectors are used throughout.

With an adequate heat sink, the LM317 can supply up to 2.2 A of current.

How the LM317 Works

As described in Experiment #8, a linear voltage regulator is basically a smart resistor, changing its value to maintain a constant voltage at the output, despite changing input voltage and load current. The LM317 attempts to maintain a constant 1.2 V between the ADJUST and OUTPUT pins by supplying more or less current to the load. From the LM317 data sheet, I selected the schematic shown in Figure 2.

R1 and R2 form a voltage divider across the output voltage, V_{OUT}. The voltage at the junction of R1 and R2 (the ADJUST pin) is $V_{ADJ} = V_{OUT} \times R2/(R1+R2)$. The voltage across R1 is then $V_{R1} = V_{OUT} - V_{ADJ} = V_{OUT} \times (1 - R2/(R1+R2))$. If the LM317 maintains $V_{R1} = 1.2$ V, then $V_{OUT} = 1.2 \times (1 + R2/R1)$.

For $V_{OUT} = 6$ V, the ratio R2/R1 must be $(6/1.2) - 1 = 4$. Any combination of R2 and R1 can work with the following constraints: R1+R2 should not consume excessive current and R2 has to be small enough that the voltage error from the small current flowing out of the ADJUST pin (up to 100 µA) does not affect V_{OUT} by raising the voltage across R2. (This is described in detail in the Application Hints section of the LM317 data sheet.)

The data sheet recommends a value of 240 Ω for R1, so I picked the more common value of 270 Ω. That choice means R2 = 270 × (6/1.2 – 1) = 1080 Ω, so I chose 1100 Ω. The exact value of V_{OUT} should then be ½ × (1 + 1100/270) = 6.1 V, just fine for my purposes. As an exercise, pick some other common output voltage, such as 5 V or 9 V, and do the calculations of R1 and R2 using standard values. See how close you can get to the desired output voltage by altering the resistor values.

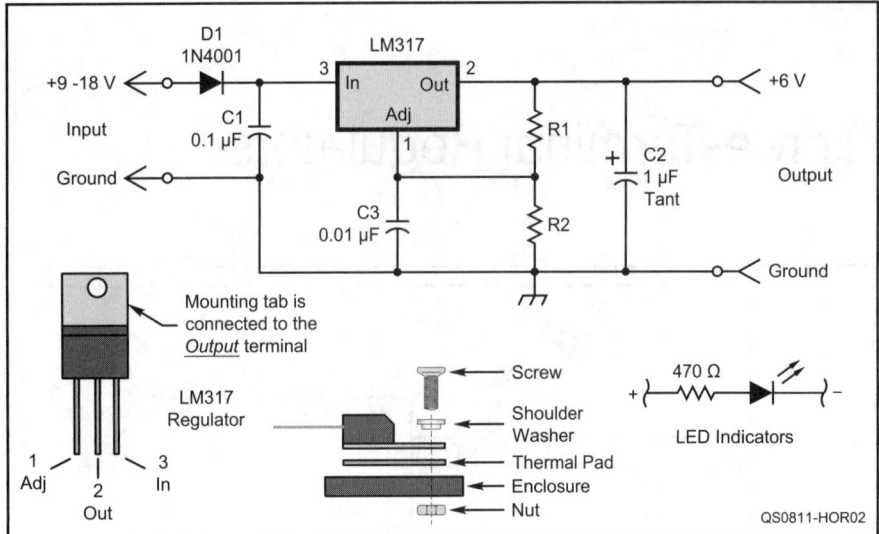

Figure 2 — An LM317T adjustable three-terminal voltage regulator IC is used to convert 12 V input power to a 6 V output. The output voltage of the circuit is set by the ratio of R1 and R2 as explained in the text. Both indicators use standard LEDs such as RadioShack 276-209 (red) and 276-304 (green).

Taking the Heat

The next part of using a three-terminal regulator is determining how much heat it will dissipate and figuring out how to get rid of it. The amount of power a regulator dissipates is calculated in the same way as a resistor — the voltage across the regulator times the current through it ($P = V \times I$). We know the voltage across the regulator; $V_{IN} - V_{OUT} = 13.8 - 6.1 = 7.7$ V. (I used the nominal automotive supply voltage as an input voltage the circuit was likely to encounter during long periods of use.)

What is the current it will have to supply? The highest load current (and the highest power dissipation) occurs when all three radios are charging. From their specifications, the VX-1s draw 150 mA each (max) and the BC246T 500 mA (max) for a total of 800 mA, neglecting the 10 mA or so of current that a visual indicator would draw. (This is unlikely to be sustained, but could last long enough to generate a fair amount of heat, so it is a conservative estimate.)

The heat to be dissipated is then 7.7 V × 0.8 A = 6.2 W, enough that a heat sink will be required. (The process of selecting a heat sink is described in Hands-On Radio Experiment #24 in *QST* for January 2005.) Since the circuit will be installed in the closed up go-kit, I decided to make the heat sink the circuit's enclosure so that the maximum amount of air could circulate around it.

If the LM317 gets too hot, it will enter *thermal shutdown mode*. Internal protective circuitry senses the temperature and cuts output current to reduce the temperature to a safe level. This often leads to the regulator turning current off and on up to several times a second. Because the indicator will dim from the lower average voltage, I'll be alerted to the over-current condition.

Caveats and Considerations

Why not use a fixed-voltage regulator such as the 6 V output 7806? That would certainly work, but it's not as easy to find as the LM317, which is stocked by RadioShack (p/n 276-1778). If you have a supply of LM317s, you can pretty much whip up any regulator you need in short order and the only extra expense is two resistors and a disc capacitor. Be aware that the *pinouts* (the pin by pin connections) are different for the 7800 series and LM317 regulators and that the LM317 mounting tab is *hot* — connected to the output pin.

Three-terminal regulators also require that you pay attention to their *stability* requirements. Almost all of these ICs specify that a bypass capacitor be placed at the INPUT pin of the regulator if the dc source filter capacitors are more than a few inches from the IC. The long leads would enable noise or RF signals at the input to disturb the regulator's operation.

The ADJUST pin is also bypassed to ground to prevent noise signals from affecting the regulator maintaining 1.2 V across R1. Since it's very likely that the go-kit will be used in the presence of RF, a bypass capacitor is cheap insurance against interference to the regulator.

A 1 µF tantalum capacitor is used at the output to prevent the regulator's output from ringing in response to *transient* (short duration) changes in load current. Tantalum capacitors are used because of their excellent performance above 100 kHz, at which point the inductance present in all electrolytic capacitors would affect performance significantly. (See Experiment #63 for a discussion of capacitor types and performance.) If an electrolytic capacitor is used, a value of 25 µF or greater is recommended.

If you read the Application Hints section of the LM317 data sheet, you will also find paragraphs describing how to prevent currents flowing back through the regulator under certain circumstances. I felt these were unlikely to be encountered and so only added reverse-polarity protection diode D1 in the input circuit. This 1N4001 is rated at 1 A, enough for the expected maximum current.

The Finished Product

The enclosure is an electrical wiring junction box with the regulator mounted on it. It's heavy enough to absorb the heat load at full charge and has enough surface area to dump the heat inside the go-kit container. The regulator circuit is built on a piece of PC board scrap "air circuit" style. The junction box is attached to the inside of the container with mounting screws. (Photos of the completed converter are on the Hands-On Radio Web site.)

The indicators are regular LEDs with a series 470 Ω resistor. The 6 V indicator draws (6 – 1.5) / 470 = 10 mA and the 12 V indicator about 22 mA. Both are bright enough to be seen through the plastic wall of the container. You can use incandescent bulbs if you prefer — or none!

Input voltage is supplied through a pair of Powerpole connectors attached to the outside of the go-kit using an aluminum mounting plate. (Power Werx p/n 1462G1 — www.powerwerx.com) An in-line, 5 A fuse is included for safety. I couldn't be more pleased with the results — instead of unpacking all of the radios, untangling their charging cords and trying to find outlets for the wall-warts, I now just plug the entire kit into a 12 V power supply or if I'm in the car, to the cigarette lighter. Sweet! I'll bet you have a batch of radios and accessories that could use the same treatment! Now that you know how to design and build your own dc regulators, what's stopping you?

Parts List (regulator only)

- Capacitors — 0.1 µF and 0.01 µF disc ceramic; 1 µF, 16 V tantalum capacitor.
- Fuse — 5 A in-line.
- Integrated circuit — LM317T voltage regulator with thermal pad, shoulder washer, mounting screw, lockwasher and nut.
- Diode — 1N4001 rectifier.
- Resistors 270 Ω and 1100 Ω.
- PC board or perfboard scrap.

Recommended Reading

The NEC application note on using three-terminal regulators (tinyurl.com/6rmafe) is full of good information. A presentation on 12 V wiring for radios (www.arrl.org/FandES/ead/materials/12-VDC-Distribution.ppt) covers issues associated with power distribution for radios.

Next Month

I promise we'll do the construction techniques experiment mentioned last time!

Experiment 98
Linear Supply Design

Way back in Experiment #8, "Linear Regulators," we learned about using transistors, op amps and three terminal regulators to tame dc power and create nicely controlled power sources for electronic circuits.[1] Many Hands-On Radio experiments require just such a supply and even one capable of supplying both positive and negative voltages. While kits and surplus lab supplies are available, designing and building your own simple linear power supply is a great way to get started with electronics. And if you save a little money along the way, so much the better.

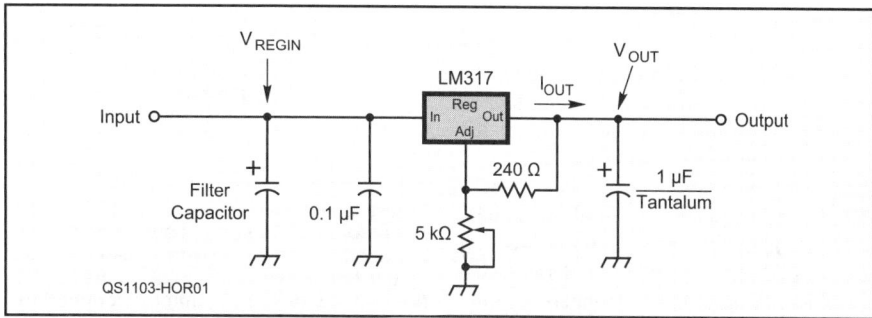

Figure 1 — The basic regulator circuit including the filter capacitor. The 0.1 µF and 1 µF capacitors are for stability and should be mounted close to the LM317.

AC to DC

The usual source of power for power supplies is the 120 V ac line. A suitable transformer is selected and the secondary windings are connected to a full wave rectifier — either center tapped or bridge. (Full wave rectifiers are explained in Experiment #6.) The output of the rectifier is filtered with a large capacitor, and a regulator circuit takes it from there. That's your basic linear or analog power supply in a nutshell.

Step one, however, is to determine how much voltage you need the transformer to supply. As with many design problems, you have to work backwards from the desired end result. Let's say you want a supply that can deliver ±15 V. That is the required output from the regulator circuit, assuming no other components are in series with the output current to create voltage drops.

The regulator IC in Figure 1 will have some voltage drop across it in order to increase and decrease output current and maintain the output voltage at a steady level. In this sense, a linear regulator acts like a *smart resistor* that changes its value to maintain a constant output voltage by varying its I × R voltage drop.

Assuming you are using a three terminal regulator such as the LM317, the next step

[1]All previous "Hands-On Radio" experiments are available to ARRL members at **www.arrl.org/hands-on-radio**.

is to determine the *minimum* voltage drop across the IC, called the *dropout voltage*. If there isn't enough voltage across the regulator from input to output, the IC can no longer regulate the output voltage and is said to *drop out* of regulation. If you download an LM317 datasheet you can find a graph of dropout voltage versus temperature, with different curves at different current levels.

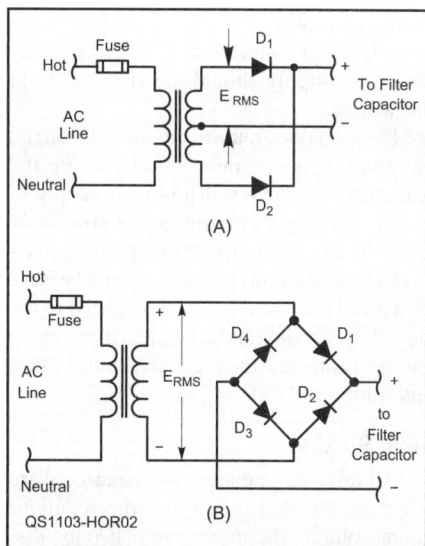

Figure 2 — Full wave rectifier circuits. A full wave center tapped circuit is shown at (A), while (B) is a full wave bridge circuit.

The higher the output current, the higher the dropout voltage. The rectifier circuit's output voltage must be at least higher than the desired output voltage plus the dropout voltage, plus some margin for other factors, such as ac line voltage variation.

Let's say our maximum output current will be 1 A. From the datasheet, the dropout voltage is a bit less than 2 V across the normal workbench temperature range. A commercial product designer would have to accommodate a range of input voltages, component tolerances, transformer efficiency and all sorts of other variations to minimize the cost of the supply, while guaranteeing proper performance under all specified conditions. We will simply use a rule-of-thumb to double the maximum dropout voltage, add that to the power supply output voltage and call that the minimum regulator input voltage.

$$V_{REGIN} = V_{OUT} + 2 \times V_{DROPOUT}$$
$$= 15 + 2 \times 2 = 19 \text{ V}$$

Let's keep working backward. Since we are going to operate from the 60 Hz ac line and are using a full-wave rectifier circuit, the filter capacitor will be recharged with current pulses at 120 Hz. Between pulses, the rectified ac voltage will be below the capacitor voltage and the capacitor will discharge into the regulator, creating *ripple* in the regulator input voltage. To determine the minimum capacitance, we need to specify the maximum amount of ripple. The smaller the amount

Electronic Circuits and Components   **89**

of ripple we allow, the larger the capacitor must be, according to the following equation:

$$C = (I_{OUT} \times t) / E_{RIPPLE}$$

where t is the time between current pulses from the rectifier; in this case 1/120 Hz = 8.33 ms. Our output current is 1 A and if we allow 5% ripple (19 × 0.05 = 0.95 V), the capacitor must be at least 1 × 0.00833 / 0.95 = 0.0087 F = 8700 µF. The nearest larger standard size is 10,000 µF (0.01 F). What size capacitor is needed if a half wave rectifier is used instead? (A 60 Hz recharge rate requires 17,400 µF.) Try changing the amount of ripple to see the effect on capacitor size.

We are almost ready to pick our transformer, but we need to decide whether the rectifier will be a full wave center tapped (Figure 2A) or full wave bridge (Figure 2B) circuit because each diode in the current path adds voltage drop. Center tapped rectifiers have the advantage of only one diode in the current path and the disadvantage of requiring a center tapped secondary. Bridge rectifiers can use a single secondary winding but have two diodes in the current path. We'll go with a full wave bridge circuit because of the simpler (and possibly less expensive) transformer and widely available full wave rectifier bridge ICs. If the forward voltage drop, V_F, of each diode is 0.7 V, the peak output voltage of the transformer secondary at our 1 A current load must be:

$$V_{PK} = V_{REGIN} + 2 \times V_F = 19 + 2 \times 0.7 = 20.4 \text{ V}$$

You'll remember from last month that converting peak voltage to RMS requires multiplying V_{PK} by 0.707, so V_{RMS} = 20.4 × 0.707 = 14.4 V. A transformer with a 15 V_{RMS} secondary would do nicely and give us a little more voltage headroom. If we are designing a bipolar supply, then we need a transformer with a 120 V primary and two 15 V secondary windings. Each winding has to supply 15 V at 1 A, so it must be rated at 15 VA and the whole transformer 30 VA at a minimum. It would be wise to overrate the transformer by at least 25% to keep it cool, so a 40 VA transformer will do nicely. Again, we add design margin by going up to the next higher rating.

A Supply Afloat

What if a transformer with a single 30 V, center tapped secondary were available — could we use that? Yes, but only as long as the positive and negative regulator circuits could share the common connection to the center tap. If you want independent or *floating* outputs, the transformer windings will have to be independent, too. Figure 3 shows

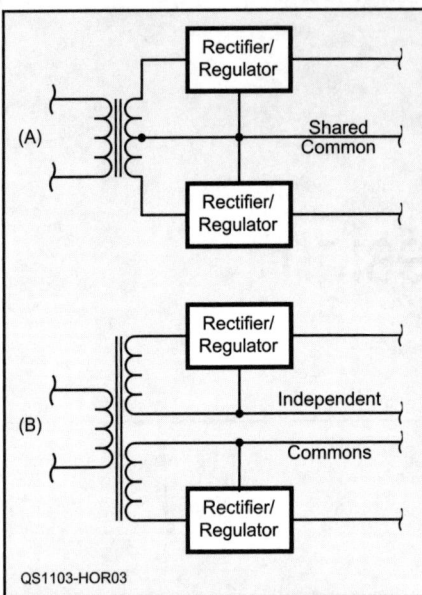

Figure 3 — At (A) the two output regulator circuits share a single common connection to the transformer secondary center tap. At (B) the regulators are floating and have no common connection.

Dwelling on the Negative

What about generating negative output voltage? The LM317's sibling, the LM337, makes a great companion regulator with the same simple circuit. Watch out for the slightly different pin assignments!

the difference. For a supply to be floating from some other circuit, it must not share any electrical connection to the circuit. Our dc power supply should clearly float from the ac line.

Floating does require attention to safety, however. It is common practice to tie the common connection of a bipolar supply to the ac safety ground unless you absolutely need the supplies to float from ground. Even if floating, the power supply should be built in a metal enclosure that is connected to the ac safety ground and operated from an ac circuit protected by a ground fault circuit interrupter (GFCI).

Heat Sinking

All this design margin and headroom are great except that the higher the regulator input voltage, the more power it must dissipate. The power to be dissipated is easily calculated:

$$P_{DISS} = (V_{REGIN} - V_{OUT}) \times I_{OUT}$$

Notice that as the output voltage decreases, the regulator must dissipate *more* power. If the lowest output voltage we allow is 3 V, the regulator will be dissipating (19 − 3) × 1 = 16 W. That's a fairly hefty heat load. Read Experiment #24 for more information on heat sinking. From the datasheet, you might consider using the TO-3 package LM317K since it can handle more heat.

Alternate Power Sources

My assumption has been that the power supply will operate from ac line voltage and you will be selecting a transformer and rectifiers. An alternative is to repurpose another power supply.

Make a visit to your local PC recycling facility or used computer store and ask them if they sell used laptop power supplies. Take along a pocket DVM, plug the supplies into a wall outlet and verify they operate (some require special control signals from the computer — avoid those). Confirm that there is an open circuit from the ac hot and neutral to the output power connections. A low resistance connection between a three conductor ac line cord ground and the output connector's common terminal is okay, although you'll have to remove this connection if you want a floating power supply.

Another inexpensive way to make a power supply is to use *wall wart* transformers and power supplies. The regulation of these supplies is not bench top quality, but as the source of the regulator input power, they are fine.

You can also use gel cells or a pack of D cell batteries to supply the regulator circuit. A benefit of using batteries is that they create a floating supply by definition, provide ultra clean dc power and make a good portable power supply.

Recommended Reading

Power supplies are a great way to get started with analog circuit design — and you get a good workbench tool, too. A book that tackles learning electronics from this perspective is the *Tab Electronics Guide to Understanding Electricity and Electronics* by G. Randy Slone. If you'd like to review a complete project, the 13.8 V, 5 A power supply in the 2010 and 2011 *ARRL Handbook* is a good example.[2]

[2]*The ARRL Handbook for Radio Communications*, 2011 Edition. Available from your ARRL dealer or the ARRL Bookstore, ARRL order no. 0953 (Hardcover 0960). Telephone 860-594-0355, or toll-free in the US 888-277-5289;

Experiment 97
Programmable Frequency Reference

A recent article in the NZART *Break-In* magazine reminded me that a frequency reference was once part of every ham's shack, using a stable oscillator based on a 100 kHz (or 1 MHz) crystal.[1] The oscillator's fundamental and harmonics were used to identify the various band and segment edges. Although modern commercial gear may not need a frequency reference, what of the homebrew rig? With that in mind, this month's experiment uses a multistage counter to create a programmable reference for calibration or alignment or for generating a digital clock signal.

The Ripple Counter

Ripple counters were introduced in Hands-On Radio Experiment #36 — The Up-Down Counter.[2] Because they are asynchronous, the ripple counter isn't a very good choice for keeping a consistent count of input events because the change of state propagates (or ripples) through the chain of flip-flops. As a result it takes some time for the counter to stabilize after each count.

In our application, asynchronous operation is not an issue. It's only important that the counter has enough stages to divide the input signal by a large enough number. (There is one caveat we'll discuss at the end.)

We're going to use the 74HC4040 12 stage ripple counter. Enter 74HC4040 DATA SHEET into an Internet search engine and download a copy for reference. The internal circuit of the IC consists of *toggle flip-flops* that change the state of their Q and $\overline{Q}$ outputs whenever the T input changes from low to high. Figure 1 shows the logic diagram. The *truth table* for the counter is shown in Table 1.

The data sheet should also provide a *timing diagram* of the relationship between the control, input and output signals. The 74HC4040 has one control input — the master reset at pin 11 connected to the R_D

[1]A. Woodfield, ZL2PD, "Programmable CMOS Clock Generator," *Break-In*, New Zealand Association of Radio Transmitters, Sep/Oct 2009, pp 6-7.
[2]All previous Hands-On Radio experiments are available to ARRL members at **www.arrl.org/hands-on-radio**.

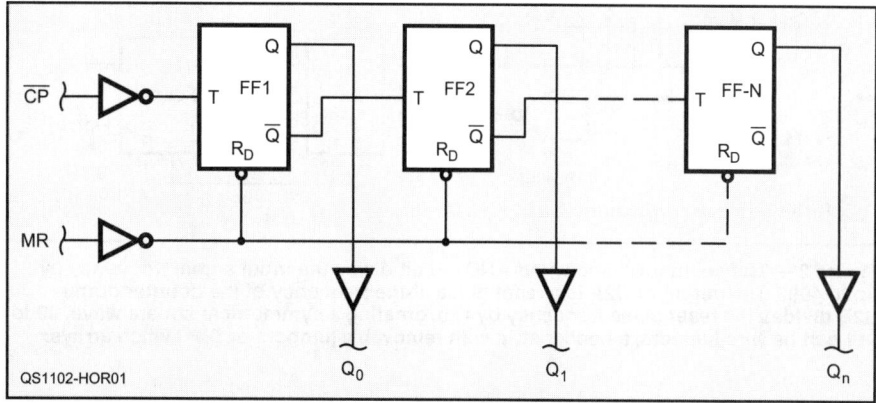

Figure 1 — The internal logic diagram of the 74HC4040 counter IC.

Table 1
Function Table of 74HC4040 Counter IC

Inputs		Outputs
CP	MR	Q_n
⇑	L	No Change
⇓	L	Count
×	H	L

Notes:
H = High voltage level.
L = Low voltage level.
× = Don't care.
⇑ = Low to high voltage transition.
⇓ = High to low voltage transition.

input of each flip-flop — and one signal input — the clock pulse at pin 10. (MR and CP on the Philips data sheet, respectively.) If MR is high, all of the flip-flops are forced to the state in which Q is low and $\overline{Q}$ is high, regardless of what $\overline{CP}$ is doing. You can see this in the timing diagram because no flip-flop changes state until MR is low.

Once MR is *released*, the next high to low transition of the $\overline{CP}$ signal causes the first flip-flop's Q output to go high and the $\overline{Q}$ output to go low. Each successive high to low $\overline{CP}$ transition causes the first flip-flop's Q and $\overline{Q}$ outputs to change state. No action occurs if the $\overline{CP}$ input changes from low to high. Since the $\overline{Q}$ output of one stage is the input to the following stage, each flip-flop changes state at ½ the rate of the preceding flip-flop.

The counter's outputs Q_0 to Q_{11} produce a square wave with a frequency of the input divided by powers of 2 from $2^1 = 2$ to $2^{12} = 4096$, respectively. The counter advances from 0 (all Q outputs low) to 4095 (all Q outputs high) and then returns to 0 for a total of 4096 states.

If our objective were to divide the input signal by any integer power of 2, our work would be done — use the output corresponding to that ratio. Generally we want some other divisor than a power of two and that's where the fun begins.

Getting Wired

The function of the row of diodes and the JK flip-flop labeled U2A in Figure 2 is to detect that the counter has reached a specific count (N) and reset the counter to zero. Imagine that the shorting jumpers labeled J0 to J11 are all installed. Thus, all of the diode anodes share a common connection to the 4.7 kΩ pull-up resistor. Since each cathode is connected to a Q output, if *any* Q output is low the current through that diode will pull the anode connection low. For the anode connection to be high, *all* of the Q outputs must also be high. This is a wired-AND connection — all of the inputs to the wired-AND (the counter outputs) must be high for the output (the anode connection) to be high. (If the diodes were turned around, exchanging

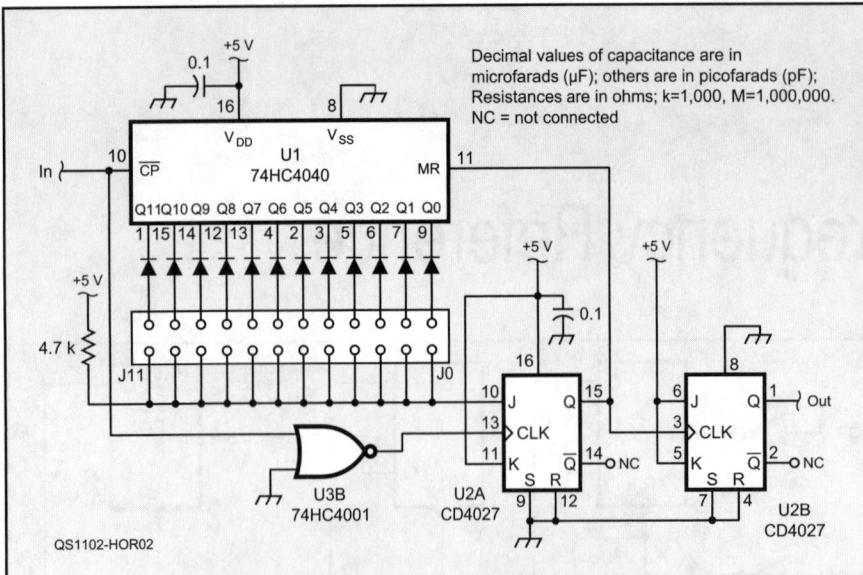

Figure 2 — The counter U1 and wired-AND circuit divide the input signal frequency by up to 4096. The output of U2A is a reset pulse at the frequency of the counter output. U2B divides the reset pulse frequency by two, creating a symmetrical square wave. J0 to J11 can be wire jumpers, a header strip with removable jumpers or DIP switch arrays.

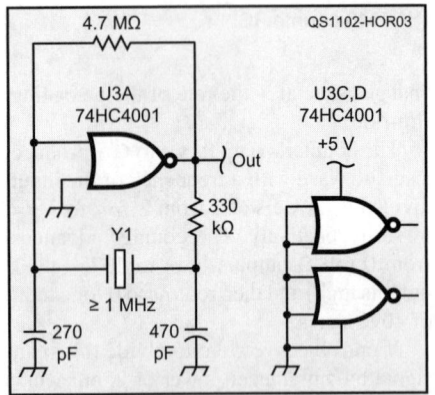

Figure 3 — A crystal oscillator circuit for high frequency and tuning fork crystals. For tuning fork crystals, replace 270 pF with 10 pF and 470 pF with 47 pF.

the anode and cathode, that would be a wired-OR connection for which the output would be high if *any* of the inputs were high.)

The wired-AND connection forms our *count detector* used to reset the counter and control the ratio of frequency division. We can make the wired-AND output go high after the desired count is reached by using the jumpers to select which counter outputs are connected to the wired-AND. Here's how it works: Say that we want U1 to divide the counter's input clock by 35. First, convert 35 to a 12 bit binary number with one digit for each counter output: 000000100011B (the right-most B denotes a binary number) with Q_{11} corresponding to the *most significant bit* (MSB) on the left and Q_0 to the *least significant bit* (LSB) on the right. This describes the counter outputs after 35 low to high transitions of the input signal: Q_0, Q_1 and Q_5 are high and all the rest are low.

If these three jumpers (J0, J1 and J5) are connected and the other jumpers disconnected, the wired-AND will go high as the count of 35 is reached. The other jumpers are not connected because we don't want the wired-AND to go high at any other count but 35. When 35 is reached that is the first time the combination of counter outputs will cause the wired-AND to go high. It doesn't matter if any higher count causes Q_0, Q_1 and Q_5 to be high (such as 37, 39, etc) because the counter will not reach those values.

The wired-AND output is the J input to the flip-flop U2A, so the Q output of U2A will go high at the following low to high transition of the input signal at U2A's clock input. (The input signal is inverted by U3B, a NOR gate wired as an inverter, so that it acts at the same time as the ripple counter's clock.) U2A's Q output is also connected to U1's reset input, causing all of the flip-flop Q outputs, the wired-AND output, and U2A's J input to go low. At the next input signal low to high transition, one input cycle later, U2A's Q output returns to low, creating a pulse one input signal cycle long. The pulse occurs every 35 counts, so the input frequency has been divided by 35!

Squaring the Cycle

We could stop here, but if we are going to use the circuit as a frequency reference, it would be preferable to have a symmetrical output. That is the function of U2B, the other half of the 74HC4027. This flip-flop divides its input signal frequency by two, producing a square wave. So, if the jumpers are set to divide the input frequency, f_{in}, by N, the circuit's output frequency will be $f_{in}/2N$. The sharp edges of the 74HC series logic signals are rich in harmonics, making an especially good *marker generator* for a homebrew receiver. If you do intend to use the circuit as a receiver calibrator use a metal enclosure, filter the input power well, and connect only the output signal to a short antenna.

The caveat regarding the ripple counter's asynchronous operation involves propagation delay through the counter as compared to the shortest input cycle period. For the circuit of U2A to function properly, its J input must be high *before* its clock input signal's low to high transition. Since the J input won't be high until the most significant ripple counter output connected to the wired-AND goes high, the total propagation delay through the ripple counter must be *less* than input signal period or a reset pulse won't be generated. Using typical values for propagation delay from the CLOCK input to Q_n at power supply voltages of 4.5 V, it takes $17 + 11 \times 10 = 127$ ns from the input for Q_{11} to change. That puts an upper limit of $1/127$ ns = 7.87 MHz on the input signal if the Q_{11} output is to be used, although the limit is higher with lower values of N.

Figure 3 shows an oscillator circuit that can be used for crystals above a few hundred kHz and for tuning fork crystals that operate at 32 kHz or lower. You can use any input signal as long as it meets the 74HC4040 requirements for logic high and low levels. Crystal manufacturers usually publish application notes showing how to make an oscillator at any frequency.

Parts List

74HC4001 — quad NOR gate IC.
CD4027 IC — dual J-K flip-flop IC.
74HC4040 — 12 stage ripple counter IC.
12 — 1N4148 signal diodes.
2 — 100 nF capacitor.
270 pF and 470 pF capacitors (assuming high-frequency crystal).
4.7 kΩ, 330 kΩ and 4.7 MΩ resistors.

Recommended Reading

The ARRL Handbook's updated "Digital Basics" chapter has a good discussion of counters and other digital concepts.[3] Your library may have the terrific *CMOS Cookbook* on the shelves, as well, with many other great counter circuits.[4]

[3]*The ARRL Handbook for Radio Communications*, 2011 Edition. Available from your ARRL dealer or the ARRL Bookstore, ARRL order no. 0953 (Hardcover 0960). Telephone 860-594-0355, or toll-free in the US 888-277-5289; www.arrl.org/shop; pubsales@arrl.org.

[4]D. Lancaster, rev by H. Berlin, *CMOS Cookbook*, Howard W. Sams & Company, 1988.

RF Techniques
Experiment 91 — Common Mode Choke

The *common mode choke* is an RFI fighting friend, although most hams know it as a dandy balun. Construction can be as simple as a coiled coax feed line, but most often the choke is made of a few turns on a toroid core, the subject of this month's column.

Differential versus Common Mode

Before the choke's long and winding story begins, it's important to understand *differential mode* and *common mode* voltages and currents. Applied to a voltage, *differential* means that the voltage is measured between two arbitrary points — a voltage difference — and not with respect to some absolute reference, such as the Earth. Differential mode current is really a pair of currents with the same amplitudes but flowing in parallel and opposite directions, completely independent of ground. So a differential mode signal exists as the voltage between two conductors, V_{DM}, and a pair of equal and opposite currents, I_{DM}, flowing on them. Neither conductor need be connected to ground in any way.

A special case of differential mode signals occurs if the conductors are very close together with respect to the signal's wavelength and either parallel to each other or concentric. In this case, the conductors form a transmission line in which the differential mode signal exists as an electromagnetic field traveling in the space between the conductors. Coaxial cables and open wire line are examples of *tightly coupled* conductors forming a transmission line for differential mode signals.

Common and differential mode voltages and current may coexist on the same conductors, but are measured differently. On our two conductor transmission line, a common mode voltage, V_{CM}, has the same absolute value on both conductors with respect to the circuit's or system's reference voltage. Similarly, common mode currents, I_{CM}, would flow on all conductors of the line in the same direction with the same value. We'll only discuss two wire transmission lines for this experiment, but remember that they are a special case of a multiwire cable, such as an antenna rotator control cable or a ribbon cable for a parallel data interface.

Here's an example — imagine that a voltmeter shows one wire of an open wire feed line at 6 V with respect to ground and the other wire at 4 V with respect to ground. The differential voltage is the difference of $6 - 4 = 2$ V. Common mode voltage is the average of the two voltages: $(6 + 4) / 2 = 5$ V. Figure 1 illustrates the differential and common mode voltages in this case.

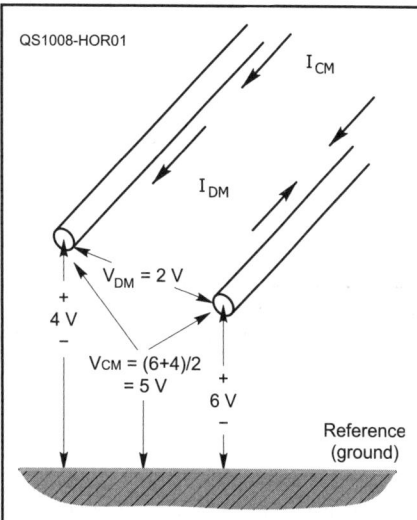

Figures 1 — Differential mode and common mode voltages are illustrated as V_{DM} and V_{CM}. Differential mode currents, I_{DM}, flow in opposite directions, while common mode currents, I_{CM}, flow in the same direction on both conductors.

Common Mode Chokes

There is a special case of common mode current for coaxial cable. Due to the skin effect, the outer conductor's inside and outside surfaces are effectively separate conductors at RF. A single current flowing on the outside of a shield can be treated as a common mode signal — a common situation. Hams know that the outside of a coaxial cable shield can pick up signals of all sorts, just as any isolated wire acts as an antenna. This common mode current can cause problems by reradiating the signal and distorting an antenna's radiation pattern.

If conducted into the shack, the current can flow between pieces of equipment and cause RF feedback or disrupt the equipment functions. If conducted into a neighbor's cable TV converter or home entertainment system, the resulting RFI can lead to disruption of a different kind — that of good neighborly relationships. In general, it is undesirable to have common mode currents flowing on the outer surface of antenna feed lines or of audio or control cables because of the unpredictable and usually unwanted effects.

While it's impossible to shield the shield to remove it as a conductor from the circuit completely, it is quite possible to block or dissipate the current. In this way, the unwanted currents are prevented from flowing — they are *choked* off. The component that performs this trick is called a *common mode choke*. (Other terms include *common mode RF choke* and *common mode ferrite choke*, describing more precisely the function or construction of the choke.)

Use in Antenna Systems

A popular use for a common mode choke is as a *current balun*. The current balun is used at the interface between an unbalanced feed line (such as a coaxial cable) and a balanced load (such as a dipole). No impedance transformation is performed by this balun because its sole function is to block common mode current that might flow back down the feed line's outer surface. Similarly, a *feed line isolator* is just a common mode choke installed on a feed line somewhere between the source and the antenna.

The choke can be made by winding coaxial feed line into a coil, by winding it around a toroidal ferrite core, or by placing ferrite beads (which are just one-turn toroid cores) over the outside of the cable. In all three cases, the differential mode signals inside the cable are completely unaffected. Only currents flowing on the outside surface of the coax's shield experience the choke's effects. Several examples of all three types of common mode chokes can be found in the *ARRL Handbook* or the *ARRL Antenna Book*.[1,2]

---

[1]*The ARRL Handbook for Radio Communications,* 2010 Edition, p 27.12. Available from your ARRL dealer or the ARRL Bookstore, ARRL order no. 1448 (Hardcover 1462). Telephone 860-594-0355, or toll-free in the US 888-277-5289; **www.arrl.org/shop**; **pubsales@arrl.org**.

[2]R. D. Straw, Editor, *The ARRL Antenna Book,* 21st Edition, pp 25-21ff. Available from your ARRL dealer or the ARRL Bookstore, ARRL order no. 9876. Telephone 860-594-0355, or toll-free in the US 888-277-5289; **www.arrl.org/shop**; **pubsales@arrl.org**.

Figure 2 — A common mode choke made of coaxial cable wound on a ferrite toroid core. Note that each turn passes through the core in the same direction.

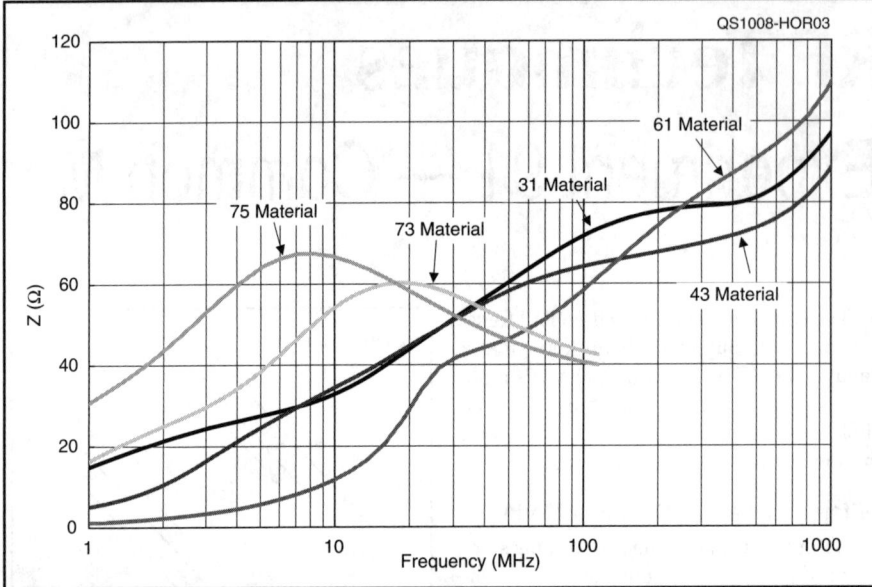

Figures 3 — These curves show the impedance versus frequency of a single ferrite bead made of different types of ferrite material. Each bead is 3.50 mm × 1.30 mm × 6.00 mm. (Information courtesy of Fair-Rite Corporation)

Use for RFI

Applying common mode chokes to suppress RFI can also be beneficial. It is here that the simple choke (we'll drop the common mode for now) really shows its value. With modern transmitters having very low spurious emission levels, such as harmonics and parasitics, most cases of RFI from amateur transmissions are caused by common mode currents picked up from strong local signals.[3]

The signal is picked up by RF or audio cables, speaker leads, power cords — any conductor more than a small fraction of a wavelength long at the signal's frequency — and appears as common mode current at and in the affected device. Even though the components required to filter out such a signal cost very little, most consumer equipment (and a surprising amount of industrial and professional equipment) does not include them. That means the current must be dissipated or blocked before it can enter the equipment. The choke rides to the rescue.

Ferrite Core Chokes

The small size of chokes wound on ferrite cores or beads slipped over cables fits the tight confines of both the ham shack and stereo system. Creating an effective choke requires that a ferrite material with an impedance that is largely resistive be used. The resistive and reactive components of ferrite's impedance changes with frequency. (The characteristics of ferrite are explored in detail in the *ARRL Handbook* and in the For Further Reading references.) For general purpose HF and lower VHF use, #31 material is the best choice. At VHF and above, #43 material performs best.

These materials are especially made for RFI suppression. They are designed to present a high impedance to common mode currents. Ferrite materials designed for inductive uses (in which energy is stored and not dissipated) are less effective than suppression materials. It is important to use the correct type of material to create an effective choke.

To make a choke from a toroid core, the process is fairly simple — wind as many turns of the cable through the toroid as practical, as shown in Figure 2. T-140 (1.4 inch OD) and T-240 (2.4 inch OD) cores are the most practical for use with audio and RF cables and ac power cords. A dozen turns or so is usually the practical limit, particularly if the connector is still installed on the cable. If you can remove and reinstall the connector, more turns may be possible. The goal is to create a choke with an impedance of at least several hundred ohms at the frequency of interest. All turns *must* pass through the core in the same direction or the magnetic flux created by each turn will cancel with that from an oppositely oriented turn, negating the effect of the core.

What about ferrite beads? The effects of a bead, in essence a single-turn core, are less than for a multiturn toroid. (A clamp-on *split core* is a special type of bead sawn in half with a plastic case that holds it together.) More than one bead can be added to a cable, however, acting in series to add their effect together. Figure 3 shows the impedance from one bead of different materials at different frequencies. A general rule is that beads are suitable for suppressing RFI at VHF, but it takes quite a few of them at HF to have the same effect as a few turns of cable through a toroid.

Testing a Choke

You can test the effect of a common mode choke by making some RFI and then getting rid of it. If you have a handheld VHF transceiver, transmit with the antenna close to a cable on your home stereo, TV system or other device. Try the RF, audio and power cables until you find one that enables the device to respond to your transmission. Obtain a T-140 or T-240 size toroid core of #31 or #43 material. (Mouser Electronics, **www.mouser.com**, is one source.) Wind turns of the responding cable onto the core one at a time, repeating the transmission and observing the device. You'll likely see small reductions in RFI until at some *threshold*, the RFI disappears. This is typical of the effects of a common mode choke, your best RFI toolkit friend.

For Further Reading

The "RF Techniques" chapter of the *ARRL Handbook, 2010 Edition* covers ferrite materials in detail. The downloadable RFI tutorial by K9YC **www.audiosystemsgroup.com/ RFI-Ham.pdf** is excellent reading and more technical details are available in W1HIS's online paper, **www.yccc.org/Articles/W1HIS/ CommonModeChokesW1HIS2006Apr06. pdf**. Be sure to ferret them out!

---

[3] ARRL RFI literature uses *fundamental overload* to refer to any disruption of a device's function by a strong local signal, whether the device is a radio receiver or not. Most fundamental overload is caused by common-mode RF current.

Experiment 112

RFI Hunt

This month's story begins with a *clank* — a rather loud sound emanating from somewhere inside our gas cooking stove. I'd just put up a 105 foot dipole about 30 feet above the kitchen. Operating on 20 and 15 meters was no problem but when I dropped to 30 or 40 meters — *clank*. The oven door latch solenoid was activating whenever a dot or dash was sent, holding for a few seconds after the RF stopped and — *clank* releasing, only to — *clank* activate again on the next transmission. Ellen is tolerant but not so tolerant that I could ignore it, and so the story begins.

General Approach to RFI

Let's back up a little bit. Solving RFI problems requires a somewhat organized approach — without it you'll chase your tail forever or until you give up and go off the air entirely. We can start by dividing RFI into two categories: RFI *caused by* your ham radio equipment and RFI *to* your ham radio equipment. This was clearly in the former category.

In the cases for which your transmissions are causing RFI, there are a series of cases to evaluate. First, is the *victim* device experiencing the RFI designed to receive a radio signal as part of its normal operation — such as a TV or cordless phone? If so, the first thing you must do — even according to the FCC rules — is to determine or ensure that your station is not generating a spurious signal on the frequencies intended to be received by the victim. For example, in the case of RFI to a weather radio, your station should not be generating any signals on the frequencies of the weather alert transmissions. If it is, go no further and either fix the transmitter or add the necessary transmit filters.

Assuming that the device is a receiver of some sort and your transmissions are *clean* (free of spurious emissions), determine whether or not it is simply the strength of your signal causing the problem. For example, a commercial broadcast FM receiver's front end circuitry could be overwhelmed by a strong 2 meter FM signal from your nearby mobile rig. This is called *fundamental overload* and is a symptom of a receiver being overwhelmed by a strong signal via its normal receive path. The usual fix here is a filter in the receive signal path to the receiver (such as in the antenna feed line). Obviously, the stove was not designed to receive RF signals so it was on to the final case.

This final case is the most common form of RFI from ham transmissions now that most TV reception is not via over-the-air analog signals. In this case, the signal from your station is being picked up as *common-mode current* by external cables or wires attached to the victim device. (See Hands-On Radio experiment #91, "The Common-Mode Choke," for a discussion of common and differential mode signals.)[1] Those currents are then conducted into the device where they disturb its normal operation in some way. If you can block those signals from getting into the device — usually by using some kind of common-mode choke or shielding — you can solve the problem. That's what I expected would solve my problem with the stove.

There is one additional frequent RFI case — *direct pickup* — in which the wiring

[1]All previous Hands-On Radio experiments are available to ARRL members at www.arrl.org/hands-on-radio.

inside the device picks up the signal without any external cables or wires. This is often the case for battery powered devices and can be very difficult to solve. Guess which mine turned out to be?

Let's Get Cooking

The stove is made of heavy sheet metal and, as with every other appliance these days, has a microprocessor that controls its functions. The only external wiring is the ac line cord. As Figure 1 shows, the stove sits directly under the antenna. Testing showed that the RFI only occurred at power levels greater than 25 W and only on the bands at and below 10 MHz. With the antenna so close, the stove was clearly in enough V/m of field strength to cause interference.

My first — and only — candidate for picking up common-mode signals was the ac line cord to a socket directly in back of the stove. Opening up my shack notebook and starting a troubleshooting log for recording each step, I grabbed some ferrite split cores (the common variety available at RadioShack) and snapped them on the ac power line where it entered the back of the stove through a hole in the sheet metal. [While the split core type of ferrite bead is

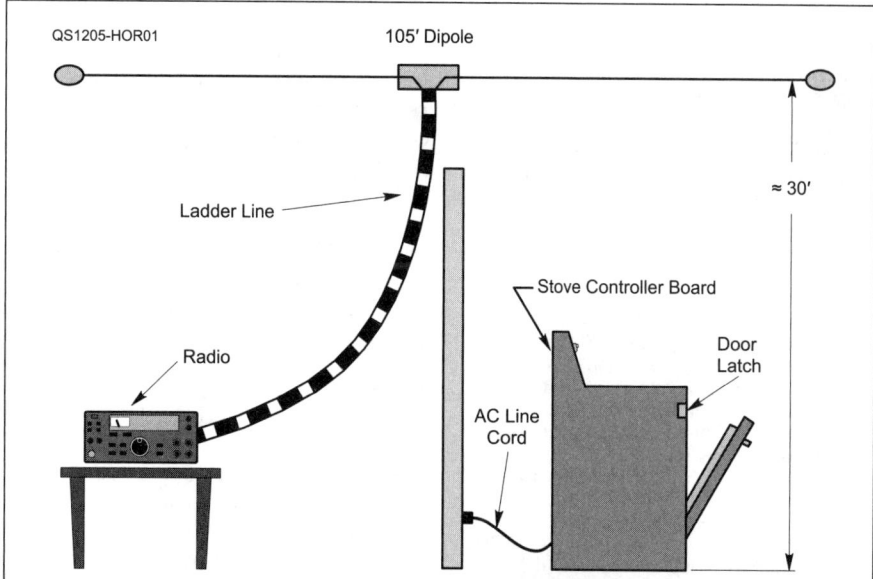

Figure 1 — The gas stove experiencing the RFI was directly underneath the HF dipole less than 30 feet above it. Even though the stove was made of sheet metal and the wiring completely enclosed, there was still enough RF picked up by the latch solenoid wiring to cause a problem.

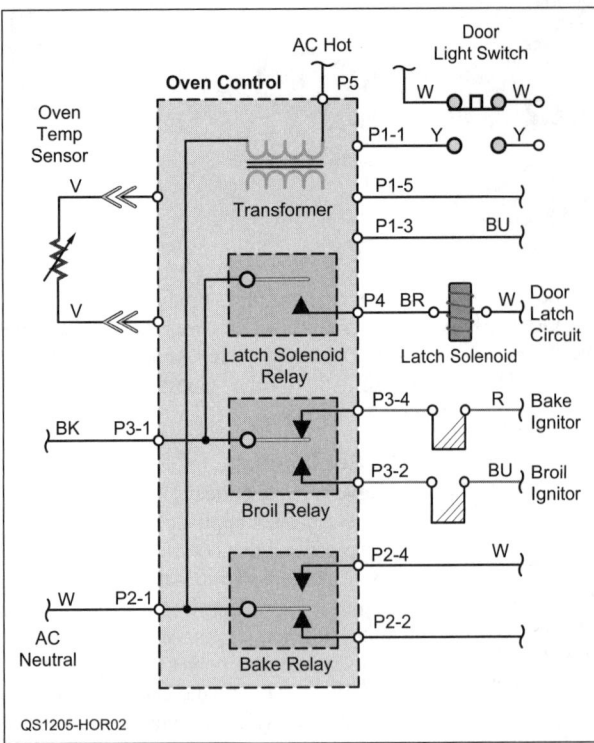

Figure 2 — This excerpt from the stove service manual shows the connection terminal numbering for each input and output to the controller board. This helps identify which terminals to bypass and what voltages may be present.

handy, it generally is more appropriate to suppression of VHF signals. HF RFI generally requires multiple turns around a toroidal core or multiple split cores of appropriate mix. — *Ed*.]

At the lowest power at which the RFI occurred, the cores had no effect on the RFI. I have since obtained cores made of #31 material, optimized for RFI suppression below 10 MHz. See K9YC's online tutorial about RFI and ferrite, the RF Interference chapter of *The ARRL Handbook* or *The ARRL RFI Book*.[2-4]

The next step was to apply sterner filtering. I purchased a Delta 10DEEG3B ac filter line with a 10 A rating,[5] attached short pigtails with a plug and receptacle (a metal junction box could also be used), and inserted the filter in the line. Again, no effect on the RFI, even with both the ferrite cores and the line filter. I was going to have to open up the patient.

Finding Resources
The owner's manual for the stove was available but it didn't provide a schematic nor was there a schematic on the inside of the stove's rear cover. By entering the stove's model number (Whirlpool SF385PEE) into an Internet search engine I was able to locate a service manual with the internal wiring diagram showing the wiring for the sensors, power and switches, summarized by Figure 2. I was getting warmer, so to speak.

Safety Check
Before we go any further, it's important to remember some basic safety guidelines. When dealing with an ac line powered appliance, any component that you attach to the ac line *must* be rated for ac line use. Capacitors should be rated for at least 6 kV ac, and both capacitors and inductors should use flame retardant insulating materials. This is not the place for components intended for use in low voltage electronics. If you insert any components in series with the ac line, be sure the connections are well insulated with sleeve or heat-shrink tubing. Use cable clamps or small enclosures so that the components and connections can't touch the body of the appliance and aren't subjected to vibration. Appliances often have elevated internal temperatures (particularly stoves) so the components and materials must be adequately rated for high temperature use, as well. And never work on energized equipment.

I'm Going In!
The control board was covered by a single piece of sheet metal and mounted behind the display panel at the top of the range. Along with the hot and neutral ac line wires (safety ground was connected where the line cord entered the stove) there was a bundle of individual wires connected to the control board that carried sensor, switch and control signals. With the external ac filter and cores still on the external line cord, I added two more cores to the ac line and control bundle — no effect on the RFI. Note that I left all of the attempted fixes in place while I continued to work since the fix can be cumulative — there might not be a "magic bullet" that solves the problem by itself.

The next step was to bypass individual connections to the controller board using capacitors. I'd purchased some line rated 0.0068 μF capacitors and connected one from the hot and neutral to a nearby chassis sheet metal screw. This had the effect of raising the power level required to trigger the solenoid, so I was on the right track. I then added a bypass capacitor across each switch input or control output, one at a time. Bypassing the door latch control output eliminated the clank at all power levels and bands (I can run up to 100 W). Bypassing the other inputs or outputs had no effect.

To verify I'd really found and fixed the problem, it was important to remove the bypass capacitors to see if the problem came back. It turned out that I needed bypass capacitors both on the ac line inputs and across the door latch control output — either alone didn't solve the problem completely. (That there is more than one path for the RF to cause a problem is fairly common.) I also verified that removing the ferrite cores and the ac line filter had no effect so the external components were removed. I left the core on the ac line inside the stove for good measure as shown in Figure 3 in the *QST* Digital Edition and the QST-in-Depth website.[6] In retrospect, the door latch solenoid has the longest wiring run in the stove except for the ac line cord so it was logical to be the weak link.

All this was duly noted in my shack notebook — you *do* keep a shack notebook, don't you? I also put this information out where other Internet search engines could find it by sending an e-mail to the RFI Reflector containing the specifics of the symptom, the affected device, and the solution (available at **lists.contesting.com/_rfi/2010-12/msg00044.html**). Posting this wouldn't be a bad idea, if you are able to thoroughly solve a similar problem at your location.

Summary
The point of this RFI travelogue is not to teach you how to get rid of RFI in a specific type of gas stove. The description of the journey is its own reward in that it is an example of how to whittle the problem down one step at a time and then solve it in an orderly manner. A deliberate, documented approach may seem like plodding overkill but in my experience, this type of tortoise beats the rabbit more often than not!

---

[2]J. Brown, K9YC, "A Ham's Guide to RFI, Ferrites, Baluns, and Audio Interfacing," **audiosystemsgroup.com/RFI-Ham.pdf**.

[3]*The ARRL Handbook for Radio Communications*, 2012 Edition. Available from your ARRL dealer or the ARRL Bookstore, ARRL order no. 6672 (Hardcover 6634). Telephone 860-594-0355, or toll-free in the US 888-277-5289; **www.arrl.org/shop**; **pubsales@arrl.org**.

[4]*The ARRL RFI Book*. Available from your ARRL dealer or the ARRL Bookstore, ARRL order no. 6834. Telephone 860-594-0355, or toll-free in the US 888-277-5289; **www.arrl.org/shop/**; **pubsales@arrl.org**.

[5]Available from Digi-Key (**www.digikey.com**), part number 603-1135-ND.

[6]**www.arrl.org/qst-in-depth**.

Experiment 114
Recording Signals

In Hands-On Radio Experiment 112, we went on the hunt for a cure to my sufferin' stove for RFI caused by my transmitted signal.[1] As hams know, however, RFI works both ways — a ham can receive interference from external sources, too. Tracking down the source of RFI to your operating can be at least as difficult as solving an RFI problem to an appliance in your home. This month, we'll do a neat science experiment with a free software tool you can use not only for RFI hunting but for other jobs as well.

Noise Signatures
When you are receiving an interfering noise signal, one of the first steps in identifying the source is to determine its *signature* — the signal's combination of amplitude and frequency that is almost always unique to that particular noise source. When you tell your tale of woe to your friends, the first question will almost always be, "What does it sound like?" Is it a buzz, a tone, a wideband rasp… or what? This gives important clues as to what is causing the noise. The frequencies on which you hear the noise are also clues.

Another part of the signature that can be just as important is the signal's behavior in time. For example, does the noise pulse regularly or is it present all the time? Does it appear throughout the day? Does the amplitude vary? All of these are also important clues. For example, HF operators in rural and suburban areas are often quite familiar with the pop… pop…pop… of an electric fence charger.

Sitting at the radio for hours listening to noise is not very exciting, despite what is depicted in the movie *Contact*. Yet it might be very important to learn that your noise comes and goes at regular hours — such as when a local street lamp turns on and off at dusk and dawn. The solution is a data recorder.

By Jove, You've Got It!
I found out about this month's software gadget in the course of helping the St Charles High School Radio Club, (KD0QLW), set up their Radio Jove Project equipment to listen to noise from the planet Jupiter.[2] This experiment involves building a two dipole array for

[1]All previous Hands-On Radio experiments are available to ARRL members at **www.arrl.org/hands-on-radio**.

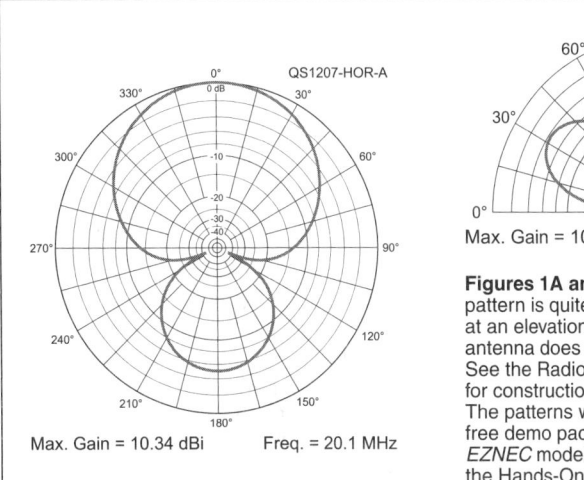

Max. Gain = 10.34 dBi   Freq. = 20.1 MHz

Max. Gain = 10.34 dBi   Freq. = 20.10 MHz

Figures 1A and B — The two-dipole array pattern is quite broad with a beamwidth of 92° at an elevation angle of 42°. As a result, the antenna does not need to be precisely aligned. See the Radio Jove project documentation for construction and installation instructions. The patterns were computed by *EZNEC*'s free demo package (**www.eznec.com**) — an *EZNEC* model for the antenna is available on the Hands-On Radio website.

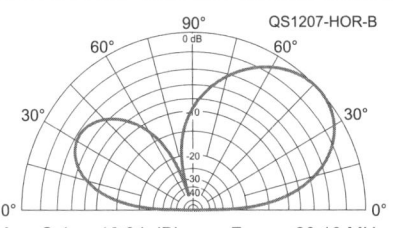

Figure 2 — The radio telescope receiver consists of an ICOM IC-7000 HF transceiver tuned to 20.15 MHz USB and a stereo audio cable connected to the laptop's MIC input. [N0AX photo]

20.1 MHz (see the digital edition for a photo of the antenna and club members), connecting a radio, and listening to the receiver's output noise as Jupiter moves through the beam of the array while the Earth rotates. If you have a directional antenna for 15 meters (or can build the two dipole array) you can hear the noise, too. Figure 1 shows the *EZNEC* azimuth and elevation pattern of the array.

Jupiter is a powerful source of noise at HF and was one of the first radio sources identified in the sky by radio astronomers. The

[2]**radiojove.gsfc.nasa.gov**

noise can be received below 35-40 MHz down to frequencies below 15 MHz, at which the signals are either absorbed or reflected by the ionosphere. Frequencies around 20 MHz are usually recommended — just tune your receiver to a clear frequency around 20 MHz and…then what?

Jovian noise can often manifest as a sound similar to the crackly pops and crashes that you might hear from any thunderstorm. Listening by ear, there is no obvious signature of either amplitude or frequency so you must use time as your meter stick. The Radio Jove project directs the experimenter to a free program called *Radio-SkyPipe II* (**radiosky.com/skypipeishere.html**) that uses a PC sound card to record the output of a receiver, graph it on a stripchart like display, and save the data to a file. Once the data is recorded, you review it to find the slow rise in noise level from Jupiter passing in front of your antenna.

Taking Measurements
Radio-SkyPipe II couldn't be easier to set up and install. Download and run the self-in-

stalling package from the website — you're ready to go. You will be prompted to enter your location (this is an astronomy tool, after all) and select your sampling rate and signal source. The default settings worked fine for me. I purchased the upgrade to the *Pro – Home Use* version so that I could work more easily with the data files. I'm using a slow, old laptop for recording and it seems to handle this simple task just fine.

Connecting the radio is also quite simple — use a plug-to-plug stereo audio cable from the headphone jack to the sound card's MIC input jack. Most readers will have a radio that can receive near 20 MHz or just outside the 15 meter band. Tune to a clear frequency free of local carriers or birdies and plug in the audio cable. Click the START CHART button and the red trace will begin crawling across the display at 10 samples/sec. In the OPTIONS menu, select the STRIP CHART tab and un-check Y AXIS AUTOSCALE. Right-click on the Y axis and click SET YMIN TO 0. This fixes the vertical scale at 0 to 10000, which is convenient for displaying noise measurements.

Adjust the receiver volume so that the background noise level is around 4000. You'll see the trace move up and down as the volume is adjusted so that the system looks somewhat like that in Figure 2. Point your antenna due south and wait for Jupiter to pass slowly by.

The time at which Jupiter crosses the north-south meridian over your location — called the *transit* — can be found on the Naval Observatory's website (**aa.usno.navy.mil/data/docs/mrst.php**). If you have a medium-to-low local noise level and your antenna is not completely blocked to the south, you'll see a gradual rise and fall in the noise level. After Jupiter passes by, click STOP CHART and save the file to your hard drive. After the file is saved, you'll see the entire session on the display. Maximize the window and use the control buttons at the left to move around. There are extensive on-line Help files available to get you started viewing data. Congratulations — you're a radio astronomer just like Grote Reber, W9GRZ back in the 1930's![3]

Figure 3 shows a portion of the strip chart I obtained on April 11, using the student's dipole array. It covers approximately 2 hours centered on the observed transit time. The 2 hour time period corresponds to 30° of the Earth's daily rotation. (360° × 2 hours / 24 hours = 30°) The Naval Observa-tory transit times (see Table 1) put the actual transit time at 14:33 (2:33 PM), which is 1933 UTC at my location in St Charles, MO. The slight rise in noise level was observable for a little more than an hour (≈16° of rotation).

Experienced radio astronomers suggested the rise in noise was far more likely reception of noise from terrestrial sources or the Sun than from Jupiter. The high school will try again from a quieter location

The peak of the noise was later than predicted, however. I took data for several days — stopping and starting the chart twice a day — and saw a similar increase in noise

[3]**en.wikipedia.org/wiki/Grote_Reber**

Table 1
Jupiter Transit

ST. LOUIS, MISSOURI
Location: W 90°15'00.0", N38°37'48.0", 100m
(Longitude referred to Greenwich meridian)

Time Zone: 5h 00m west of Greenwich

Date	Rise Az	Transit Alt.	Set Az.
2012 Apr 10 (Tue)	07:43 69	14:36 67S	21:31 291
2012 Apr 11 (Wed)	07:39 69	14:33 67S	21:28 291
2012 Apr 12 (Thu)	07:36 69	14:30 67S	21:25 291
2012 Apr 13 (Fri)	07:33 69	14:27 67S	21:22 291
2012 Apr 14 (Sat)	07:29 69	14:24 67S	21:19 291
2012 Apr 15 (Sun)	07:26 69	14:21 67S	21:17 291
2012 Apr 16 (Mon)	07:23 69	14:18 67S	21:14 291

Each column contains time and azimuth angle for rise and set or elevation angle and S for South for transit.

level about the same time, shifting a little bit earlier each day, so I am confident that the noise was from Jupiter. I attribute the timing skew to partial blockage of the antenna beam by the house so that the signals were received best after the planet's transit. An open site with no obstructions and lower background noise would allow better observations but this was fine for a first run.

RFI Sleuthing

As you sort through the charts you collect, you'll no doubt notice other interesting phenomena. You'll see large static crashes and periods of high noise levels. You may see bursts of noise and on/off patterns, as well. In my data, it's clear that noise in my neighborhood builds quite a bit in the late afternoon and early evening as folks drive their cars home from work and begin to run their appliances and gadgets. A neighbor's motorcycle leaves a distinctive peak on the trace from ignition noise, for example.

I've been chasing a loud noise source that seems to appear only in the evenings and nearly wipes out 40 meters. My mobile rig is on my workbench listening and recording so that I can find out when it appears and when it goes away. Maybe that will help me determine what is generating it.

The *Radio-SkyPipe II* software can be used to record any audio signal your sound card can accept as input. You can adjust the sampling parameters to record faster or slower, set up automatic sampling, and so forth. The *Pro* version has a squelch-like option to log only when the input signal is above a specified level. Once you get this software running, I am confident that you'll find it a valuable addition to your electronics and radio toolbox!

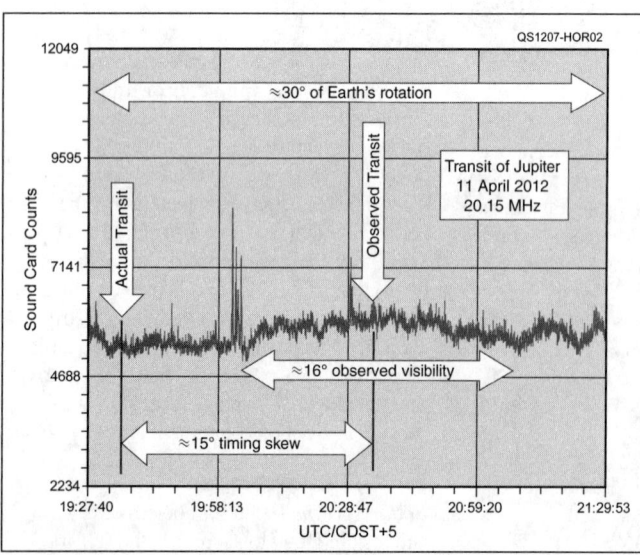

Figure 3 — This portion of the 11 April strip chart clearly shows the gentle rise in background noise as Jupiter passes by in the early afternoon. Blockage from buildings to the south and southeast results in the shift of the received peak later (westerly) in the day.

Experiment 93 — An LED AM Modulator

A recent VHF-UHF Advisory Committee (VUAC) recommendation allows LED light sources as well as the traditional lasers to be used for VHF+ contest contacts. While this might be somewhat obscure to most hams, making contacts using light waves instead of the longer radio waves sounds like fun to me. High efficiency LEDs abound at all wavelengths from infrared to blue-violet and photodetectors are quite sensitive — so why not? After all, light at terahertz (THz) frequencies is to microwaves at GHz frequencies as microwaves are to 160 meter signals. All are a part of the ever increasing scope of ham radio!

This led me (so to speak) to the topic of this month's column — an AM modulator for an LED light source. Certainly, one could simply hook up an LED to an audio amplifier output, but where's the fun in that for an experimenter? Thus, we embark on a two-for-one experiment that not only demonstrates light wave AM, but also a method of pulse modulation.

Basics of Pulse Modulation

Pulse modulation occurs if some characteristic of a *pulse train* (a regular sequence of pulses) is changed so as to represent the information of an input or modulating signal. Three characteristics of the pulse train are commonly employed for pulse modulation: amplitude, position and width. The first varies the maximum value of the pulse while the second varies the position of each pulse with respect to the other pulses, moving the pulses back and forth in time. The third — *pulse-width modulation* or *PWM* — varies the duration or width of each pulse. All three types of pulse modulation are described in the Modulation chapter of *The ARRL Handbook*.[1]

Figure 1 shows the basic concept of PWM. As the voltage of the modulating signal increases, the pulse width increases and vice versa. (Inverting the relationship also inverts the modulated signal's envelope.) The amplitude of each pulse stays the same and the pulses occur at the same *repetition rate, PRR,* or *repetition frequency, PRF.* Assuming that each pulse is perfectly rectangular, the energy in joules (J) contained in each pulse, E_p, is equal to the *instantaneous power, P_i,* of each pulse times its duration, t_p, so — $E_p = P_i \times t_p$. (Instantaneous pulse power at any instant is the pulse voltage, V_p, times the pulse current, I_p.) The average energy, E_{avg}, during each pulse period is the energy in the pulse divided by the entire period allocated to one pulse — that is, the reciprocal of PRF — so that $E_{avg} = P_i\, t_p \times PRF$. Changing P_i creates *pulse amplitude modulation (PAM)* and changing PRF creates a form of *pulse position modulation (PPM)* similar to FM. PWM changes t_p.

By itself, one pulse doesn't convey much information but over many pulses, the average power of the pulse train follows the variations of the modulating signal. The usual method for recovering the modulating information from a PWM signal is to pass it through a low-pass filter, whose output power is approximately the average of the PWM signal's power.

The low-pass filter needn't be an electronic circuit. Any device that responds to the average power of an input signal will act as a low-pass filter, recovering the information from a PWM signal if the frequency of the modulation is not too high or too low. For example, a variable speed dc motor is often controlled by PWM of its drive power. This is a very common way of controlling motor speed in appliances, robotics, and heating and cooling systems. (The sharp pulse edges of PWM signals are rich in harmonics, making them potent sources of RFI if not filtered.)

Modulator Construction

As with analog amplitude modulators, creating PWM modulation requires two basic circuits: a carrier or *clock generator* and a modulator. The output of the clock generator is a regulator series of pulses all having uniform amplitude and width. These pulses serve as a trigger for a variable width pulse generator controlled by the modulating signal. In this experiment both the clock generator and the PWM modulator will be based on the 555 timer IC that was the subject of Hands-On Radio experiment #5. Reading that experiment will help you understand the rest of the experiment better.[2] (The dual timer 556 IC can also be used.)

Our PWM modulator is designed to pro-

[1] *The ARRL Handbook for Radio Communications,* 2010 Edition. Available from your ARRL dealer or the ARRL Bookstore, ARRL order no. 1448 (Hardcover 1462). Telephone 860-594-0355, or toll-free in the US 888-277-5289; **www.arrl.org/shop**; **pubsales@arrl.org**.

[2] All previous Hands-On Radio experiments are available to ARRL members as downloadable PDF files at **www.arrl.org/hands-on-radio**.

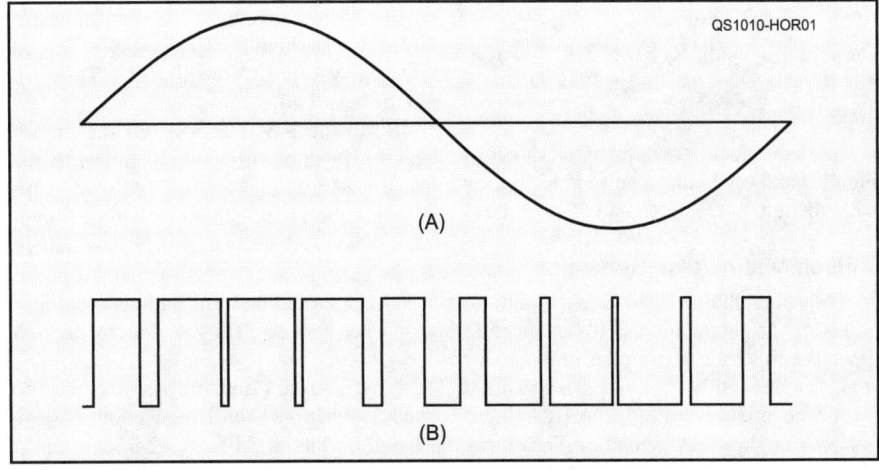

Figure 1 — PWM by a sine wave (top) results in the width of the pulses increasing with increasing sine wave amplitude and vice versa. The resulting variation in average power of the PWM signal creates an AM signal.

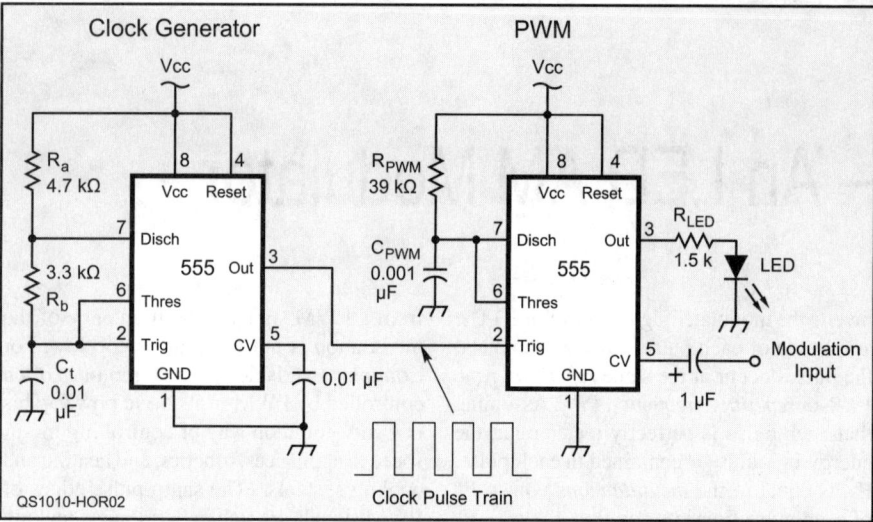

Figure 2 — At left a 555 timer in the astable mode generates clock pulses at 12.7 kHz. At right, a 555 timer in the monostable mode generates a variable-width output pulse at each negative-going edge of a clock pulse. An ac voltage applied to the CONTROL VOLTAGE input of the right-hand timer varies the switching threshold and thus the output pulse width, creating a PWM AM signal.

duce a PWM signal modulated by human speech with a maximum frequency of around 3 kHz. Since each pulse creates a sample of the modulating signal, the PRF of the clock generator must be at least twice the highest modulating frequency or 6 kHz. The selected values for the clock generator place the PRF well above that limit — about 12 kHz.

Clock Generator

The clock generator circuit, on the left in Figure 2, is a 555 timer configured in its astable or *free running* mode. The following two equations describe the output pulse train:

PRF (Hz) = $1.44 / [(R_a + 2R_b) \times C_t]$ = $1.44 / [(4.7\ k\Omega + 2 \times 3.3\ k\Omega) \times (0.01\ \mu F)]$ = 12.7 kHz

and

Output pulse width (s) = $0.693 \times (R_a + R_b) \times C_t$ = $0.693 \times (4.7\ k\Omega + 3.3\ k\Omega) \times 0.01\ \mu F$ = 55.4 µs

Build the clock generator circuit shown in Figure 2 and verify that a pulse train appears with about a 13 kHz PRF and 55 µs pulses.

Variable Width Pulse Generator

The circuit that does the actual modification is at the right in Figure 2. It's another 555 timer but this time configured in its *monostable* mode, with an output pulse triggered by the negative going edges of the input pulses from the clock generator. The output pulse length for this mode of operation is:

Output pulse width (s) = $1.1 \times R_{PWM} \times C_{PWM}$

The modulating signal is applied to the CV (Control Voltage) input. Increasing voltage raises the discharge threshold of ⅔ V_{CC} and lengthens the output pulse. Since the modulating signal (voice) will both increase and decrease about the zero signal value, for maximum range the output pulse with no signal applied should be approximately centered between the maximum (1/PRF) and zero pulse width. With a PRF of 12.7 kHz, ½ × 78 µs = 39 µs. The values shown in the schematic generate pulses of $1.1 \times 39\ k\Omega \times 0.001\ \mu F$ = 43 µs with no signal applied to the CV pin.

The LED current limit resistor value depends on the LED used in your circuit. It's not necessary to use a high efficiency or high current LED. I used a generic red LED out of the junk box that gave plenty of output with 10 mA of forward current. The current-limit resistor value should be $R_{LED} = (V_{CC} - V_{LED}) / I_{LED}$. If V_{CC} = 12 V, a value of 1.5 kΩ will do for most LEDs.

Build the PWM portion of the circuit and verify that with no connection to the clock generator output, the LED stays dark. Connect the clock generator with no input modulation and verify that an approximately 43 µs pulse is output for every input pulse. The LED should be significantly less bright than when connected to V_{CC} directly through R_{LED}.

Now connect an audio source to the modulation input. You'll need about 1 V_{RMS} of audio, which translates to 2.8 V_{P-P}. I used the headphone output of my shop radio and monitored the signal level with a DVM set to measure ac voltage. I could immediately detect flickering in the LED brightness along with the audio — AM optical modulation!

Where is the low-pass filter? It's your eye! The LED can turn on and off very rapidly and reproduces each pulse of current as a pulse of photons. Your eye, however, can't distinguish between the individual pulses of light at that PRF, only the average power of the PWM signal, effectively demodulating it. The averaging response of your eye, plus the brightness versus current characteristics of your LED, are probably not optimized for the combination of pulse rate and width from the PWM modulator. Feel free to experiment with PRF and pulse width to see if you can increase the level of modulation of the LED. I found that substituting a 100 kΩ potentiometer for R_{PWM} allowed for adjustment that made the most difference in the LED modulation.

If you'd like to use a microphone to transmit your voice, the Transmitters chapter of *The 2010 ARRL Handbook* includes a microphone amplifier for dynamic or electret microphones.[3] It has an adjustable gain and will work with a single power supply voltage.

Is PWM actually used for AM communications? Yes, and in fact, it's widely used in the AM broadcast industry as an efficient and effective means of producing an AM signal. For more about generating an RF AM signal using PWM, review "Using two 555s as an AM Transmitter" at **williamson-labs.com/555-circuits.htm**.

Parts List

2 – 555 timer IC (or 1 556 dual-timer).
1.5 kΩ, 3.3 kΩ, 4.7 kΩ and 39 kΩ
 ¼ W resistors.
100 kΩ potentiometer.
LED.
0.01 µF and 0.001 µF ceramic or Mylar
 capacitors.
1 µF, 25 V electrolytic or tantalum
 capacitor.

For Further Reading

The 555 timer IC is described along with many useful circuits in Walter Jung's *IC Timer Cookbook*.[4] An excellent application note from Philips Semiconductor is also available online at **www.555-timer-circuits.com/datasheets.html**. If you're interested in trying to receive and demodulate your PWM light signal, begin with the material at **www.imagineeringezine.com/ttaoc/detector.html**. How long will it be before we hear (or see) "CQ 720 nanometers and standing by"?

[3]See Note 1.
[4]W. Jung, *IC Timer Cookbook*, Second Edition, SAMS Publishing, 1983.

Experiment 102
Detecting RF — *Part 1*

RF is all around us with wavelengths from thousands of miles to a few nanometers and yet the only RF we can sense directly is a narrow slice of spectrum from the infrared (as heat) to the ultraviolet (as a sunburn!). For every other type of RF we have to use some kind of technological device to tell whether RF is present and whether or not it is carrying some kind of information we're interested in. That's the subject of this (and next) month's column — detecting RF.

As all hams know, or find out, a sure fire way of detecting RF is for your neighbor to buy a touch lamp or an inexpensive pair of computer speakers. It's guaranteed that they'll come right over and tell you whenever RF is present — yours or not. In these columns, I'll present a survey of more useful techniques, I'll discuss how they work, and I'll present some simple circuits that you can build and use as shack test instruments.

Detect Versus Demodulate

Before we plunge in, there is a small matter of defining what we mean by *detect*. As we learn from **www.dictionary.com**, the definition of *detect* is "to discover the existence of," as in "I detect mold in my QSL collection." The noun form is *detector*, a device that detects. More specifically, the *CRC Comprehensive Dictionary of Electrical Engineering* defines *detector* as "a device that converts RF input signals to a corresponding dc output signal."[1] Notice that it doesn't say anything about the output signal being representative of anything other than the presence of RF.

A *demodulator*, on the other hand, takes things a step further by performing *demodulation*, "the process by which a modulated signal is recovered back to its original form. It is the general process of extracting the information bearing signal from another signal." So a detector simply tells us whether or not RF is present. If we happen to detect a modulated signal and get the modulating information out of the detector, so much the better. A demodulator, on the other hand, must apply some kind of process to extract the modulating information from the modulated signal. We'll focus on the simpler business of detection.

An Early Detector

The best known detector used in the early days of radio was not the cat's whisker crystal detector. It was the *Branly coherer* shown in Figure 1 — a tube of metal powder or filings with electrodes at each end — invented in 1890 by Édouard Branly. If an electromagnetic wave passed an antenna connected to the coherer, the current through the coherer made the filings stick together a little bit and that changed the resistance between the electrodes. The change in resistance could be used to make clicks in headphones or perform some other action an operator could sense.

After the wave had been detected, it was necessary to tap the coherer to loosen up the filings again and prepare the coherer for more current. The Morse receiver of the day used a coherer to drive a relay that caused the received code to be printed on paper tape and also activated a solenoid that would deliver a sharp tap to the coherer with its armature.

Needless to say, this wasn't a very sensitive or selective receiver but it did the job and allowed radio to get started. Interestingly, technology progressed rapidly and better detectors were soon invented before the *Branly effect* — the reason why the coherer worked — was understood. It was not until a century later (approximately 2005) that the nanoscale physics behind it was unraveled.[2]

Metering Waves

Another type of detector that was used long ago and is still useful today is the *absorption wavemeter*, used to determine the presence and frequency of RF. The wavemeter is not called a "frequency meter" because in the early days, signals were referred to in terms of wavelength. Today, we think in terms of frequency.

As shown in the schematic of Figure 2, this is the simplest of instruments. All that's needed is a means of collecting (absorbing) whatever RF is present (the antenna), a calibrated means of discriminating between RF signals of different frequencies (the tuned circuit), and a means of displaying signal strength to an observer (the meter and diode). The coil is generally fixed in value and the capacitor tuned. In many wavemeters used for circuit troubleshooting the coil and antenna are combined, with the coil placed close to the circuit being tested to pick up signals. Essentially, the wavemeter is a crystal radio with a meter replacing the headphones.

To make your own wavemeter, you'll need to wind a suitable coil and find a suitable tunable capacitor. The example shown here is just one of an infinite number of combinations that can work. An 18 µH inductor

[1] *Comprehensive Dictionary of Electrical Engineering*, IEEE, 1999.

BRUNO HENRY, F1JMM

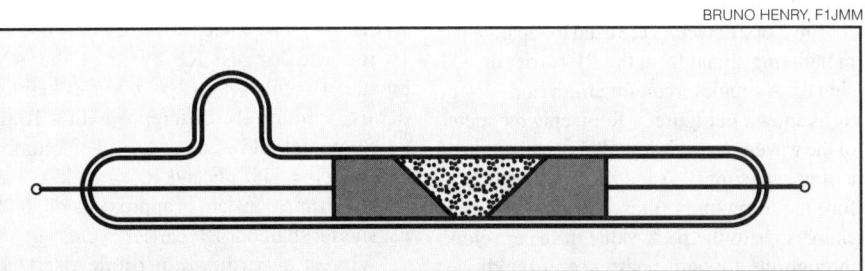

Figure 1 — The structure of the coherer — a sealed glass tube with filings contained between two electrodes.

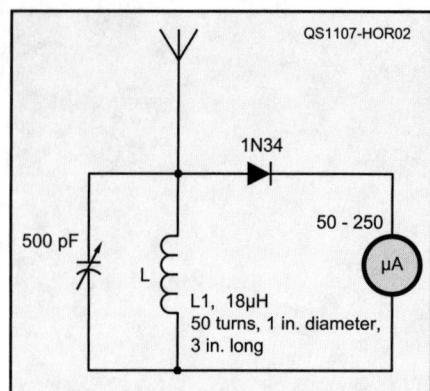

Figure 2 — A simple absorption wavemeter circuit. The tuned circuit acts as a filter to determine signal frequency. Any 50 to 250 µA full-scale meter may be used.

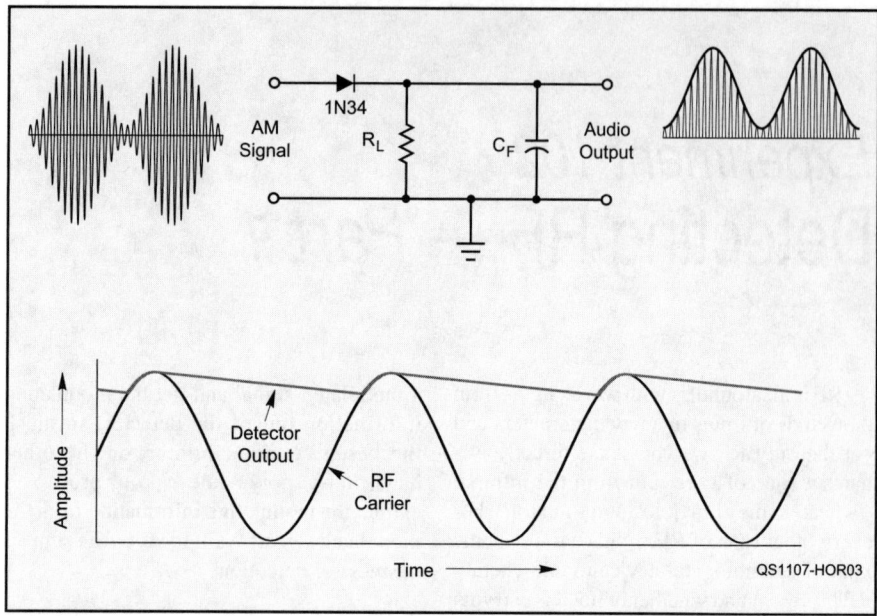

Figure 3 — The simple envelope detector circuit rectifies and filters the incoming modulated RF signal. A low-frequency waveform representing the envelope of the modulated signal is passed to the output.

(50 turns on a 1 inch diameter form, such as a pill bottle, evenly wound over a 3 inch length) in parallel with a 500 pF variable capacitor, can tune to the lowest part of the 160 meter band whose name itself is a nod to our wavelength legacy. A handy online calculator for air wound inductors can be found at **www.crystalradio.net/cal/indcal.shtml** so you can try different inductor values or shapes.

You can find many variations on this simple circuit by searching the Internet for *absorption wavemeter*. A common enhancement to this circuit is adding a diode to form a voltage doubler or using a switch to switch in and out different numbers of turns to change the range of the tuned circuit's resonant frequency. You can calibrate your wavemeter by using a signal generator or your own transmissions.

Envelope Detectors

The rectifier and meter also form a crude *envelope detector* that responds to the peak value of the RF signal. If you tuned the wavemeter to an AM broadcast station or an SSB station transmission, you would see the meter move in sync with the amplitude of the modulating signals. (If you tuned in an FM signal — a *constant-power* signal — the meter would show a steady deflection as long as the transmitter was on.)

Envelope detectors are used to separate the modulating signal from the RF carrier of AM signals. A simple circuit for an envelope detector is shown in Figure 3. Replacing the meter of the wavemeter is the parallel combination of a filter capacitor (C_F) and a load resistor (R_L), forming a low-pass filter. The diode quickly charges C_F to the peak value of the envelope through the forward resistance of the diode, R_d. C_F then discharges slowly through R_L. This difference in charge and discharge time constants allows the output voltage to track the envelope of the AM signal while shunting the RF component to circuit common.

For the envelope detector to work, three conditions must be met. First, in order to be able to reproduce the modulating signal, the cutoff frequency of the low-pass filter formed by R_L and C_F must be higher than the bandwidth of the modulating signal: $1 / (2\pi \times R_L \times C_F) \gg BW$. Second, the filter's cutoff frequency has to be much lower than the RF carrier frequency, f_C, so that it is removed from the output: $1 / (2\pi \times R_L \times C_F) \ll f_C$. Third, the detector's charging time constant ($R_d \times C_F$) must be short enough for the capacitor to charge on RF signal peaks. For small signal diodes, once conduction begins in earnest R_d is around 800 Ω.

To make your own AM broadcast signal envelope detector start by determining the desired modulating signal bandwidth — about 10 kHz — so $R_L \times C_F \gg 1 / (2 \times \pi \times BW) = 16$ µs. The longest RF cycle of the AM broadcast band is $1 / (2 \times \pi \times 550$ kHz$) = 0.3$ µs. Splitting the difference with a 10 µs time constant allows us to use two common values, $C_F = 100$ pF and $R_L = 10$ kΩ. The charging time constant is approximately 0.08 µs, shorter than our RF carrier cycle.

Why is a germanium diode used? If the input signal is very small, such as for a crystal radio, a diode with a very low forward threshold voltage should be used. 1N34 or 1N277 germanium diodes begin to conduct in the 0.2 to 0.3 V range while a silicon diode, such as 1N914 or 1N4148 requires 0.5 to 0.6 V. This is why silicon diodes don't work well in crystal radios.

Try building an envelope detector and use an oscilloscope to view the recovered waveform. You can either try receiving a local AM station (an outdoor ham band dipole will probably pick up plenty of signal if the transmitter isn't too far away) or transmit an AM signal on 160 meters at low power while using a long piece of wire as your receive antenna. You can also add a blocking capacitor (0.1 µF will suffice) in series with the output and route the signal into an audio amplifier.

Experiment by varying the values of the low-pass filter while listening to the signal. What happens to the audio quality as the low-pass filter time constant is increased by raising the value of C_F? How large can C_F be and still recover acceptable speech? (Speech has a bandwidth of about 3 kHz.)

More Detectors

You can put the envelope detector to work on your bench as an RF voltage probe for your voltmeter as shown by Monty, N5ESE, at **www.io.com/~n5fc/rfprobe1.htm**. You may also want to reread Hands-On Radio experiment #53 on "RF Peak Detectors" for another application.[3] Next month, we'll detect RF current and sample an RF waveform.

---

[2]J. Dilhac, "Édouard Branly, the Coherer, and the Branly Effect," *Communications*, IEEE, Sep 2009, pp 20-26.

[3]All previous Hands-On Radio experiments are available to ARRL members at **www.arrl.org/hands-on-radio**.

Experiment 103
Detecting RF — *Part 2*

In last month's column we discussed some simple circuits to detect RF, beginning with that most ancient of detectors, the Branly coherer.[1] This month, we continue our journey of detection, changing emphasis, as we do, from voltage to current.

Field Strength Meter

We suspended operations last month after touching on RF voltage probes and peak detectors as examples of envelope detectors. Another useful example is the *field strength meter* (FSM). In its simplest form, the FSM is just a wideband, untuned AM envelope detector with a meter to show relative field strength. As such, it can be used for *go/no-go* testing and general better or worse evaluation. These are the most common type of FSM, often found at hamfest flea markets for a few dollars — don't pass up that bargain.

If you use the ARRL's *QST* online archives, read the article "Learning to Use Field Strength Meters" by W1FB.[2] It presents an increasingly capable series of passive (powered only by the detected signal) and active (amplified) instruments. The author presents a method for calibrating an FSM and shows how to use it for various useful tasks. We'll present another such task a little further on.

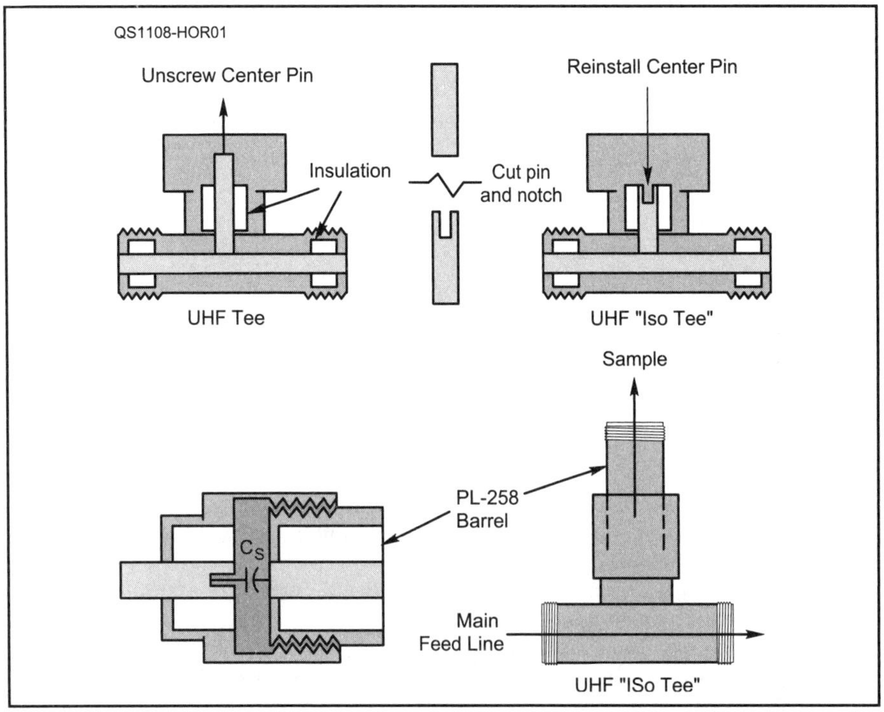

RF Sampler

An RF sampler is not a box full of RF candies, but a method of extracting a little bit of one's transmitted signal for measurement or observation. The goal of an RF sampler is to provide a signal that is an exact replica of the much more powerful signal, but without affecting the transmitter through loading or adding of unwanted reactance.

The voltage divider method places a high value two-resistor series string (R1+R2) across the feed line being sampled with one end of the lower value resistor (R2) connected to the common side of the line. The voltage out of the divider $V_{OUT} = V_{IN} \times R2 / (R1+R2)$. If $R1 = 9 \times R2$, the output

[1]All previous Hands-On Radio experiments are available to ARRL members at www.arrl.org/hands-on-radio.
[2]D. DeMaw, W1FB (SK), "Learning to Use a Field Strength Meter," *QST*, Mar 1985, pp 26-29.

Figure 1 — The "IsoTee" is made by removing and shortening the center pin of a UHF Tee connector. A slot is filed or cut into the pin to allow it to be seated with a small screwdriver. If a PL-258 barrel adapter is inserted, the combination creates a small coupling capacitance, C_S, for sampling a signal in the main feed line.

voltage is $\frac{1}{10}$ the input voltage. Voltage dividers look simple but at high power, the voltages encountered require high voltage resistors that are large enough to limit their effectiveness at high frequencies. For that reason, voltage dividing samplers are generally limited to low power uses.

Capacitive couplers extract a portion of the signal through a very small value of capacitance, usually a few picofarads. This can be done by connecting a leaded capacitor directly to the line or as in the novel *IsoTee* variation shown in Figure 1. The shortened center pin of the IsoTee does not make a connection with the contact of the mating connector, creating instead a very small value of coupling capacitance, C_S, between the end of the shortened pin and the center conductor of the inserted connector.

Because the reactance of C_S decreases with frequency, without some kind of opposing compensation the amplitude of the sample relative to the sampled signal will increase with frequency. For this reason, capacitive couplers are generally only used for relative and not absolute measurements.

Current Transformer

An even better method of coupling involves no contact — less worries about high voltage — and provides a relatively constant coupling over a wide frequency range. A magnetic coupler uses a *current transformer* to sample the current in the main feed line instead of voltage.

We're all used to power transformers and the equation that relates primary and secondary voltages: $V_{SEC} = V_{PRI} \times n$, where n is the secondary to primary turns ratio, n_{SEC} / n_{PRI}. A current transformer has the same structure of primary and secondary windings and the magnetic core, so what's different about it?

In our usual uses of transformers the primary is hooked up to a voltage source with a

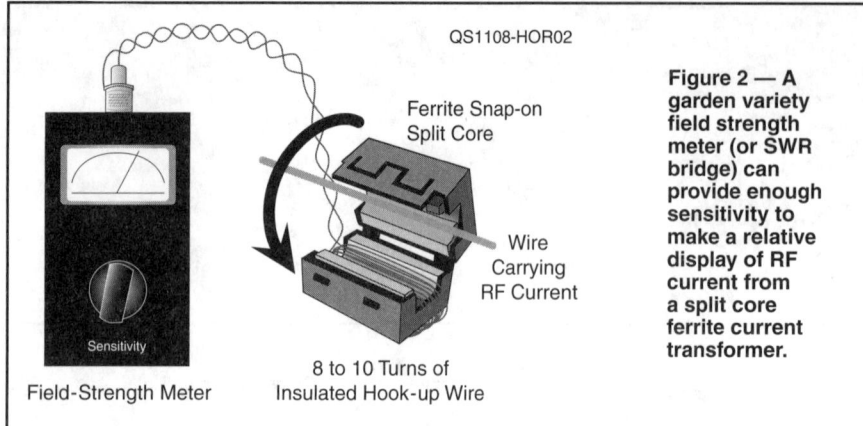

Figure 2 — A garden variety field strength meter (or SWR bridge) can provide enough sensitivity to make a relative display of RF current from a split core ferrite current transformer.

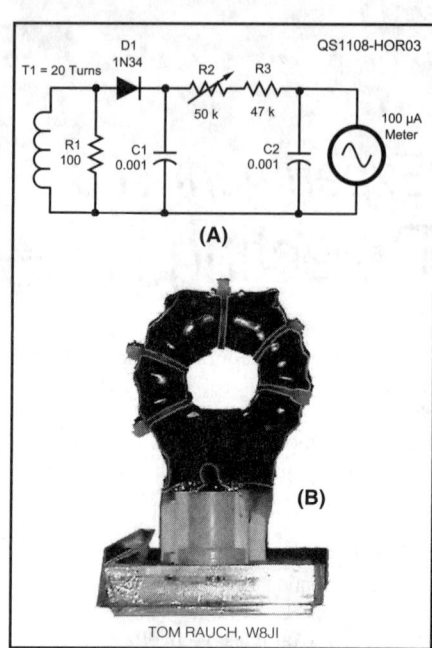

Figure 3 — By using a well designed current transformer, this simple RF current probe provides repeatable, consistent performance over the MF and HF amateur bands. A germanium diode at D1 will provide the most sensitivity. At (A) the schematic, at (B) the transformer.

very low internal impedance, such as the ac power line. The intent is to transfer power from the primary circuit to the secondary circuit and the load presented to the primary by the secondary winding is significant. Under such circumstances, the current in both windings is determined by the secondary load.

A current transformer is used differently in that its primary is connected to a source of current and very little power is transferred from primary to secondary. The primary often consists of a single turn formed by a wire passed through the center of a toroid core on which the secondary is wound.

This results in a fairly high turns ratio and, if moderate to low impedances are connected to the secondary, the impedance presented to the primary winding is low. As a result, there isn't much voltage drop across the primary — in fact the primary circuit is affected very little and the secondary current is then determined almost entirely by the turns ratio: $I_{SEC} = I_{PRI} / n$ and the voltage across the secondary is determined by the secondary load, $V_{SEC} = (I_{PRI} / n) \times R_{LOAD}$.

Because the two windings are completely isolated, current transformers are used in power systems to sense current without having to contact the high voltage conductors. Clamp on meters are common examples of current transformers. This sounds like a good way to sense RF in a high power circuit, doesn't it?

RF Current Probe

If you are just looking for a relative indication of RF current, such as when you are hunting common mode current on cable shields or current on ground wires, you can make your own clamp-on RF current probe using a split core ferrite bead and an inexpensive field strength or SWR meter as shown in Figure 2. Wind the secondary turns through the central hole — do not twist them together. Outside the core, twist the wires together to hold the winding on the core and reduce RF pickup that is not from the primary wire. Slip the core over the current carrying conductor and snap it together.

The exact mix of the core is not important as you are simply trying to convert current in the primary to a signal in the secondary. A core intended for RFI suppression such as a RadioShack 273-105 will work fine, as will the garden variety field strength meter or SWR bridge — just experiment with sensitivity settings until you are able to detect the current. This cheap and easy sensor makes a great RF current sniffer. Don't expect accuracy or repeatability.

If you are willing to put a little more effort into creating a wideband RF current probe, a design by Tom Rauch, W8JI (**www.w8ji.com/building_a_current_meter.htm**) may be just what you need. Figure 3 shows the circuit of the probe and the final assembly.

T1 is the current transformer, consisting of a T-157-2 powdered iron core with a 20 turn secondary, resulting in a current ratio of 20:1 (1 A in the primary results in 50 mA in the secondary). D1 and C1 form the detector with the current meter and calibration resistance making up the low-pass filter's load resistance as discussed in the previous column's section on envelope detectors. C2 provides additional filtering.

Tom notes that for consistent performance over a wide range, it's important to keep leads short and minimize stray capacitance. For this reason, the toroid core and simple circuitry are simply glued to the back of the meter with the calibration resistor's screw adjustment exposed. (Tom's website explains how to calibrate the current probe.)

The toroid must be slipped over the antenna or radial wire to be tested, but the design gives very consistent results over a wide frequency range (1.8 to 30 MHz). If you would like to try your hand at a clamp-on style RF current probe, check out the design by Lyle Koehler, KØLR at **www.nutstreet.net/k0lr/currprob/currprob.htm**.

The current transformer style of RF current probe also makes a good RF sampler as exhibited in the May 2011 *QST* Technical Correspondence item, "A High Power RF Sampler," by Tom Thompson, WØIVJ. The sampler is made so that it can be permanently installed in a coaxial feed line and the output sample is approximately 40 dB below that in the main feed line of a 50 Ω system.

RF Ammeter

The final RF detecting device we will review is the *RF ammeter*. These special meters were once very common but are now mostly used by AM broadcast stations and new ones are quite expensive. Nevertheless, you will find used RF ammeters for sale through surplus dealers and at hamfests and online auction sites. (Be sure you are buying an actual RF ammeter — it's not uncommon for meters labeled "RF Amps" to actually be simple current meters calibrated for use with an external sensing unit.)

The *thermocouple* RF ammeter is really a sensitive voltmeter across a low resistance thermocouple. (The thermocouple must also have low resistance with respect to the antenna or feed-line circuit through which the current is flowing or it may affect the current magnitude.) Current flowing through the thermocouple creates a voltage that is then displayed by the meter. A *hot wire* RF ammeter uses the mechanical expansion of a heated wire to change the deflection of the meter needle. RF current flowing in the wire creates the heat.

Ralph Hartwell, W5JGV, has devised his own style of RF ammeter based on the envelope detecting properties of Schottky diodes. You can learn more about Ralph's approach at **w5jgv.com/rfa-2/rfa-2.htm**.

Experiment #66 — Mixer Basics

The *heterodyne principle*, invented by Fessenden, describes how signals of two frequencies can produce *products*, signals at the sum and difference of the two original signal frequencies. The circuit that makes a superheterodyne (or *superhet*) receiver possible — the most common type of receiver for the past 90 years — is the *mixer*.

You may be more familiar with the microphone mixers used in audio systems — an unfortunate overlap of terms. Those pieces of equipment combine the audio signals, producing an output in which the input signals are present with altered levels. Microphone mixers do not change the frequencies of the input signals, however. Even so, a combiner is a good place to start our study of how mixers work.

Combiners

In Experiment #3, Figure 4 is the schematic of a two channel *summing amplifier*.[1] This circuit adds the two input signals together into a single composite output signal. The ratio of R_F to the input resistor R_1 or R_2 determines the amplitude of the signal contribution from each input in the output signal.

Figure 1 of this experiment shows a very simple *passive resistive combiner*. This circuit is suitable for use with PC sound card inputs and outputs. In this experiment we'll make use of PC based function generator and audio analyzer programs to generate and analyze input and output signals.

Build the passive combiner circuit. The resistor values are not critical — any close value will do. The values of the input resistors should be relatively close if same-value resistors are not available. The value of the load resistor can be any value from 4.7 kΩ to 27 kΩ.

Two sine wave input signals are required: 500 Hz and 800 Hz. If you are using the *Dual Function Generator* program described last month, set both channels to FUNC, SINE, and an output level of 60.[2] Set one channel

[1] All previous Hands-On Radio experiments are available online to ARRL members at **www.arrl.org/hands-on-radio**.
[2] *Dual Function Generator* and *SpectrumView* are part of the *ARRL Software Library for Hams*, Vol 2.0 CD package, available for $19.95 from the ARRL at **www.arrl.org/shop**.

LEFT-RIGHT control to full-left and the other to full-right. (If you are using standalone function generators, output levels of about 0.5 to 1 V_{P-P} will do.)

Connect the input of the PC sound card to the junction of all three resistors as shown and run an audio spectrum analyzer program. If you are using the *SpectrumView* program, the sample rate should be 44.1 ksps, the transform size should be 8192 samples, and the horizontal display axis set to display frequencies from 0 to 1600 Hz.[3] All other settings may remain at their default values. You should see a pair of signals at 500 Hz and 800 Hz, with a value between –20 and –30 dB on the vertical amplitude scale. (If your levels seem low, open your computer's VOLUME CONTROL settings, assuming a *Windows* based system, and be sure that the WAVE and VOLUME levels are set to maximum.)

[3] See Note 2.

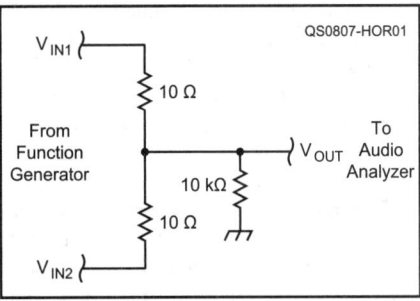

Figure 1 — A passive resistive combiner adds the input signals together into a single composite output. The input signal frequencies are unchanged.

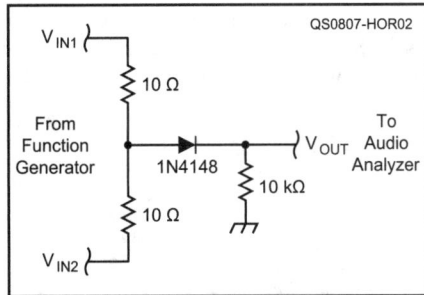

Figure 2 — Adding a diode turns the resistive combiner into a passive diode mixer.

Change the levels and frequencies of the input signals to observe the effect on the displayed output signal. (You may see some signals at low frequencies from hum or signal processing artifacts; these may be ignored for this experiment.) The signal input voltages and frequencies can be changed independently because this is a *linear* circuit. The output signal is a scale replica of the input signals and is composed of the same frequencies as the input signals.

Multiplying Mixers

Instead of adding the signals together as in our combiner, what if the signals were multiplied together instead? Let's assume that each signal is a sine wave; A $\sin(2\pi ft)$, where A is the peak amplitude and f is the signal's frequency. Multiplying two sine waves of different amplitudes and frequencies results in the output signal:

A $\sin (2\pi f_A t) \times$ B $\sin (2\pi f_B t)$ = ½ A × B $[\cos (2\pi[f_A - f_B] t) - \cos (2\pi[f_A + f_B] t)]$

That's a hefty equation, but the important thing to note is what is inside the square brackets of the cosine terms: $[f_A - f_B]$ and $[f_A + f_B]$. These are output signals at the difference and sum of the input frequencies. Note that there are no signals present with the original frequencies, f_A and f_B — they have been converted to *mixing products* with the new frequencies. (If the difference frequency is negative, use its absolute magnitude.)

Let's try an example to see how the math works out. If both input signals have an amplitude of 3 V, signal A's frequency is 1 kHz, and signal B's frequency is 1.5 kHz, what are the mixing products? The output signal amplitudes are ½ the product of the input amplitudes — in this case, ½ × 3 × 3 = 4.5 V. The complex output signal is made up of two components: 4.5 cos (2π 500 t) and 4.5 cos (2π 2500 t). Therefore, the output signal frequencies are 0.5 kHz and 2.5 kHz.

Why not use a multiplier circuit as a mixer? That would certainly work, but a true multiplier circuit suitable for use in a communications receiver is not easy to construct. The alternative is to use circuits whose output signal consists partly of the

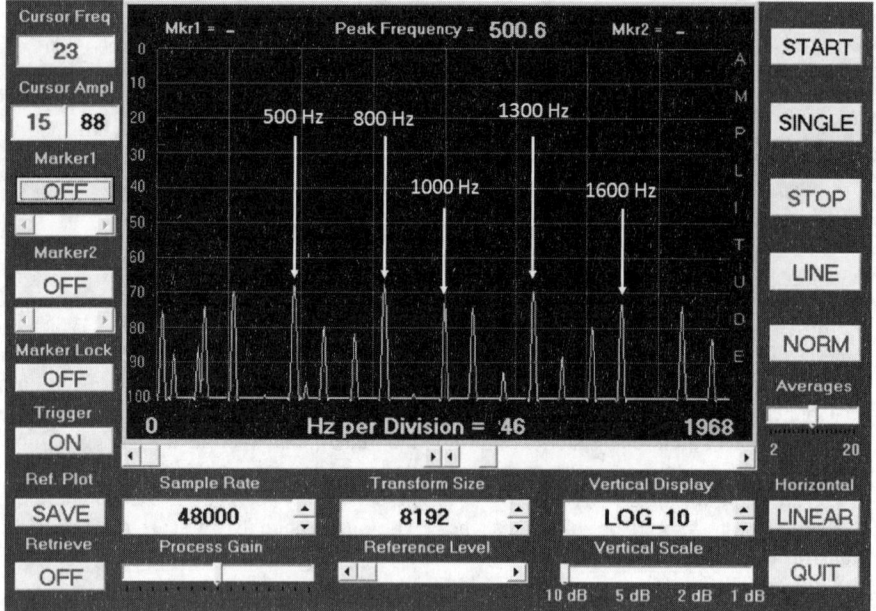

Figure 3 — The output spectrum from the passive diode mixer with input sine waves of 500 and 800 Hz.

input signals multiplied together and partly of other combinations and products.

Surprise — It's A Mixer!

Taking two signals, multiplying them together and getting a completely different pair of signals out might remind you of some other process. What if the input signals were an audio tone (perhaps 2000 Hz) and an RF carrier (perhaps 14,300 kHz)? The output signals would have frequencies of 14,302 and 14,298 kHz. In other words, multiplying created a double sideband, suppressed carrier signal! Remove one of the sidebands with a filter and, *voila*, your SSB signal appears. So a *balanced-modulator* is really a type of multiplying mixer!

Non-Linear Mixers

Another way for two signals to combine in such a way that signals at different frequencies are created is to use a *non-linear* circuit. That is, the output is not a scaled replica of the input — it is distorted. You saw in the previous experiment how changing the makeup of the components of a complex signal changed the waveform's shape. A similar process is at work in a non-linear circuit by which distortion of the input signal creates a complex output signal with components not present in the input signal.

Figure 2 shows a very simple non-linear passive mixer. Adding the diode forces the current from each input to flow through the diode's junction. The relationship between the current through a diode and the voltage across it is an exponential equation called the *diode equation*. Even if the voltages from the input signals combine linearly, the current through the diode will be non-linear. This means that the voltage across the load resistor will also be a non-linear reproduction of the input signals.

The non-linear behavior of the diode not only creates signals at the sum and difference of the input frequencies, but at all of the *linear combinations* of the frequencies. That means there will be signals at $nf_1 \pm mf_2$ where n and m can take on any integer value from 0 to infinity. Any component for which either n or m are 0 (but not both) is the fundamental or a harmonic of one of the input signals.

Don't the unwanted products overwhelm or interfere with the desired products at the sum or difference frequencies? Luckily, the higher the values of n and m (the sum of n and m is the called the product's *order*) the less energy is present in that component, so nearly all of them can be ignored.

Try it! Add the diode to your circuit and take a look at what happens to the circuit's output, keeping the input signal frequencies the same (500 and 800 Hz). As you can see in Figure 3, many components will be visible. (Temporarily short out the diode with a wire jumper to see the original input signals.) Your display may not look exactly like that of Figure 3, but it is the frequencies of the components that are important.

Start by identifying the components corresponding to the original signals. Then find the 300 Hz (800 – 500) and 1300 Hz (500 + 800) components. Can you identify the input signal harmonics? Make a table of the nine linear combinations of the input frequencies for n and m = 1, 2, and 3. See if you can identify the components that correspond to those combinations.

This stew of components is typical of what happens when signals are combined in a non-linear device. The diode is a particularly good example. You might also consider what happens if strong transmitted signals from broadcast or commercial stations are present at non-linear junctions, such as rusty fences or gutters! The many mixing products they generate can also be radiated (inefficiently, thank goodness) by the same fences and gutters, bedeviling the amateur operator!

Parts List
- 2 each 10 Ω, ¼ W resistors.
- 10 kΩ, ¼ W resistor.
- 1N4148 diode.

Recommended Reading

Reginald Fessenden and Edwin Armstrong are two very important names in radio. Reading up on both and the superheterodyne receiver would be a good way to learn the history of the mixer and its use in radio. All are well-represented in Wikipedia at **www.wikipedia.org**.

Next Month

I'll be taking in the Dayton Hamvention before I write the next column. I'll be looking for some interesting and novel new bit of electronics to show you!

Experiment #68 — Phase Locked Loops, the Basics

Phase locked loops are found in many types of radio equipment. They can be used as modulators, demodulators, oscillators, synthesizers, clock signal recovery circuits and the list goes on. Are they mysterious and difficult to understand? Not really, once you get to know each piece and do a little experimentation.

Background

The phase locked loop (PLL) has its roots in receiver design. It was invented in 1932 as a technique for stabilizing an oscillator's frequency.[1] The PLL was then adapted for use in television receivers, synchronizing the vertical and horizontal sweep circuits to the incoming video signal. In the 1960s and '70s, integrated circuit PLL chips became available and the technique soon became even more widespread.

Let's start with the name itself. *Phase* refers to the relative phase difference between an input signal and the loop's internal oscillator. *Locked* means that the oscillator's phase maintains a constant relationship of that of the input signal. This also means the frequencies of the two signals are the same, otherwise the phase difference would change. *Loop* comes from the feedback loop that controls the internal oscillator's frequency to remain in sync with that of the input signal. Thus, a *phase locked loop*.

Feedback is key to the PLL's function. Think back to the description of how an op-amp amplifier circuit works in Hands-On Radio Experiment #3.[2] Amplifying the difference in

[1]www.uoguelph.ca/~antoon/gadgets/pll/pll.html.
[2]Hands-On Radio experiments are available online to ARRL members at www.arrl.org/hands-on-radio. The first 61 experiments are also available as *ARRL Hands-On Radio Experiments* from the ARRL at www.arrl.org/shop.

voltage between its input terminals, the op-amp output voltage changes and the external circuitry is configured to make that change reduce the difference, bringing the circuit back into balance. That kind of feedback loop uses a signal's amplitude (voltage and current) instead of frequency and phase as does the PLL.

Loop Components

The PLL has three basic components, seen in Figure 1 — the phase detector, the loop filter and a voltage-controlled oscillator (VCO). The output from the phase detector (C in Figure 1) is a signal that contains the frequency and phase difference between the input signal and VCO output. The loop filter creates the VCO control voltage based on the difference signal. The VCO changes frequency in response to the

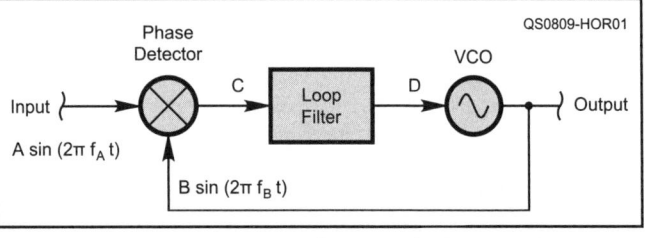

Figure 1 — The basic structure of a phase locked loop. The phase detector acts as a mixer, generating products at the sum and difference frequencies of its inputs. The filter extracts the dc component of the mixer output for the VCO to use as a control voltage.

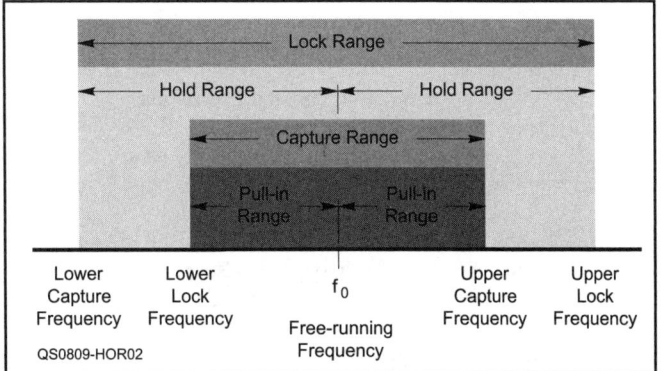

Figure 2 — The four frequency ranges that define a PLL's behavior. Lock range (and hold range) shows how far the PLL frequency can track an input signal. Capture range (and pull-in range) shows how far from the free running frequency the VCO will move to lock onto an input signal.

control voltage until the two frequencies are the same. Simple, no? Maybe we should slow down a little bit and look at each piece.

The VCO is a special type of oscillator that has a frequency controlled by an applied voltage. The frequency of the VCO without any control signal applied is called the *free-running* frequency, f_0. Depending on the circuit design, the VCO may be designed so that f_0 occurs with zero dc voltage input and a bipolar control signal, or at some non-zero dc voltage so the circuit can operate from a single power supply voltage.

Next, you may be wondering why I used a mixer symbol for the phase detector. It's because the phase detector is just that — a type of mixer. Experiment #66 provided the equations describing a mixer's output products, but ignored differences in phase between the input signals. Taking phase into account, the mixing product at the difference of the two input signal frequencies, f_A and f_B, is $\cos(2\pi[f_A - f_B]t + \theta)$, with θ representing the difference in phase between the signals. If the two signals have the same frequency and the phase difference is constant, then $f_A - f_B = 0$, leaving $\cos(\theta)$, a dc voltage that makes a fine VCO control signal.

The high frequency of the sum product at $f_A + f_B$ is not suitable as a VCO control voltage and so must be removed. That is the job of the low-pass loop filter — to remove everything but the phase detector's $f_A - f_B$ product, along with the phase information. Depending on the design of the phase detector and the nature of the signals (sine, square, pulse), the loop filter may also need to convert short bursts of current into a smoothly varying voltage.

PLL Operation

After the PLL is turned on with no input signal, the VCO will oscillate at the free-running frequency, f_0, until an input signal is applied. The phase detector generates sum and difference products, the loop filter removes the sum product, and the VCO output frequency begins to change. Assuming the input and VCO frequencies are not the same, the output of the loop filter (D in Figure 1) will be an increasing or decreasing voltage depending on which signal has the higher frequency.

This changing voltage causes the VCO to respond very quickly, reducing the difference between the VCO and input frequencies. Consequently, the loop filter's output voltage is also reduced, making smaller and smaller changes in the VCO frequency. Within a short time (typically a few milliseconds for RF PLLs) the VCO frequency is equal to that of the input signal and the loop is "locked." Any change in either the PLL input or VCO frequencies is tracked by a change in the loop filter output, keeping the two frequencies the same.

This process of adjust and hold is called *capture*. The minimum and maximum input frequencies to which the loop can move the VCO as it captures an input signal is called the *capture range* as shown in Figure 2. The segments of the capture range above and below f_0 are called the *pull-in range*. The pull-in ranges are not necessarily symmetrical.

If the control signal is proportional to the cosine of the phase difference, it will be zero when the phase difference is 90° (cos 90° = 0). It will be a maximum when the two signals are in phase (cos 0° = 1) or out of phase (cos 180° = –1). This defines the range over which the PLL can keep the input and VCO frequencies locked together. As the input frequency moves farther and farther from f_0, the VCO's free-running frequency, the loop's control action will keep the VCO frequency the same as the input frequency, but with a phase difference that gets closer to 0 or 180°, depending on which direction the input frequency changes.

If the input frequency has moved so far that the phase difference between it and the VCO frequency is either 0 or 180°, any further change will cause the control signal to move back toward its 90° value and the VCO frequency away from the input signal. The loop is no longer locked and the input and VCO frequencies are no longer the same. The range of input frequencies between the value at which the loop is locked with a phase difference of 0° and 180° is called the loop's *lock* range. The lock range above and below f_0 are called the loop's *hold ranges*. The lock range is not always centered on f_0.

Building A PLL

The venerable 565 PLL IC, a fixture in electronics for nearly 40 years, is still widely used. Start by downloading the LM565 datasheet from **www.national.com/JS/search Documentdo?textfield=LM565xcategories-Datasheets**. Familiarize yourself with the pin connections and browse some of the circuit examples.

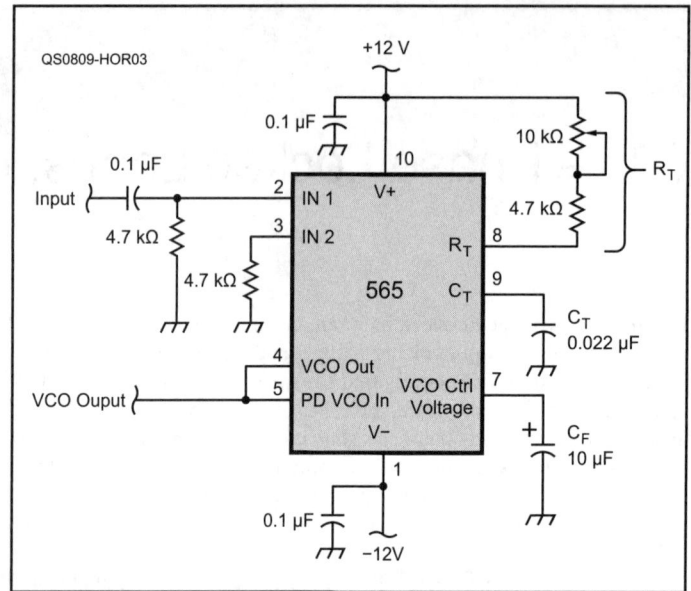

Figure 3 — The 565 integrated circuit PLL contains almost all of the circuitry necessary to build a PLL. Only a few discrete components are needed to set the VCO free-running frequency and loop filter time constant.

Build the circuit shown in Figure 3. You'll need a bipolar power supply to do this experiment. Set the potentiometer to half-range, about 5 kΩ. Without connecting any input signal, apply power and use an oscilloscope or frequency counter to measure the free-running frequency at VCO out. It should be close to $f_0 = 1.2/4R_TC_T \approx 1360$ Hz.

Set your function generator to output a sine wave at the measured value of f_0. 0.5 to 1 V_{P-P} will be sufficient. Apply the sine wave to the PLL's input. Use a dual-channel oscilloscope to monitor both the function generator output and the VCO output. Use the function generator output to trigger the 'scope. The sine waves on both channels should be stable (because they are locked in frequency) but will be somewhat out of phase.

Slowly reduce the generator output frequency until the PLL loses lock — seen as one trace suddenly becoming unstable. That frequency is the lower limit of the PLL's lock range. Return the generator frequency to f_0 and then increase it until the PLL loses lock again at the upper limit of the lock range. Total lock range is the difference between these two frequencies.

Slowly reduce the generator frequency until the PLL suddenly captures the input signal and locks again — both traces will be stable. This frequency, the upper limit of the PLL capture range, will be somewhat lower than the upper lock range limit. Change the generator frequency to something below the lower limit of lock range you measured previously. Slowly increase frequency until the PLL captures the input signal at the lower limit of capture range. Total capture range is the difference between these two frequencies.

Capture range depends on the time constant of the loop filter, determined by C_F and a 3.6 kΩ resistor connected inside the IC. The time constant of the filter equals $R \times C = 3.6$ kΩ $\times$ 10 µF = 36 ms. The larger the time constant, the smaller the capture range because the loop doesn't respond quickly enough. Replace C_F with smaller capacitors, down to 1 nF and see what happens to capture range as the loop reacts more quickly. Leave the circuit assembled for next month's follow-up experiments!

Parts List

- Capacitor — 0.1 µF ceramic, quantity 3.
- Capacitor — 0.022 µF ceramic or film.
- Capacitor — 10 µF, 25 V electrolytic.
- Phase locked loop IC — NE565.
- Potentiometer — 10 kΩ.
- Resistor — 4.7 kΩ, ¼ W, quantity 3.

Parts hint — the end of fishing season is a great time to find bargains on tackle boxes. They make terrific parts and tool organizers!

Recommended Reading

Many electronic experimenters have gotten their start in understanding PLLs by reading the classic tutorial Motorola application note AN535 "Phase Locked Loop Design Fundamentals." It's available at **www.datasheetcatalog.org/datasheet/motorola/AN535.pdf** and would make a good addition to your technical library.

Next Month

This month you manipulated the PLL by hand. Next month, we go live as we use a PLL to demodulate an FM signal.

Experiment #69 — Phase-Locked Loops, Applications

Last month's column introduced the basic functions of a phase-locked loop (PLL). What is the PLL used for besides locking onto a signal? What is that locking mechanism good for? The original PLL generated a signal locked to the sync signals embedded in a video waveform, an example of clock recovery in which the PLL reproduces a timing signal. PLLs can recover such a signal even if it's very noisy, such as from a weak spacecraft signal or in the presence of other signals carrying several data streams.

FM Demodulation

PLLs can also perform FM demodulation. You demonstrated this last month while measuring the loop's capture and lock ranges. By varying the frequency of the input signal, albeit very slowly, you created an FM signal. The PLL locked onto and followed it as long as the input signal frequency remained within its lock range. You also observed the error signal changing along with the input signal frequency. If you had graphed the input signal frequency and the error signal versus time, you would have seen that the error signal represented the variations of the input signal frequency. In other words, the error signal reproduced the modulating signal. Let's demonstrate that ability.

You can make a simple FM generator from a 555 timer IC as shown in Figure 1. See Hands-On Radio Experiment #5 for more information on the 555 timer.[1] The 555 has a CONTROL VOLTAGE input connected directly to the resistive divider at the point that determines the $\frac{2}{3}$ V_{CC} threshold. By changing that threshold with an external voltage, the timing of the circuit is altered. In most 555 circuits, the CONTROL VOLTAGE pin (pin 5 on the DIP package) is bypassed to ground with a small capacitor to prevent noise from causing false triggers or unwanted shifts in the threshold.

[1]Hands-On Radio experiments are available on-line to ARRL members at **www.arrl.org/hands-on-radio**. The first 61 experiments are also available as *ARRL's Hands-On Radio Experiments*. Available from your ARRL dealer or the ARRL Bookstore, ARRL order no. 1255. Telephone 860-594-0355, or toll-free in the US 888-277-5289; **www.arrl.org/shop/**; **pubsales@arrl.org**.

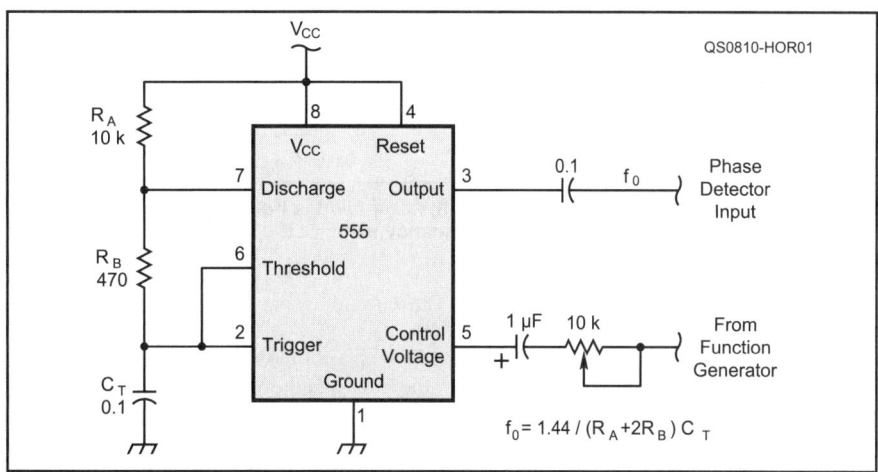

Figure 1 — The 555 timer wired for its astable mode can be made into a simple FM generator by applying the modulating signal to its Control Input pin.

An FM signal can be created by varying the voltage at the $\frac{2}{3}$ V_{CC} threshold. If the threshold voltage is lowered, the timing capacitor will be discharged more frequently and the output frequency will be higher (and vice versa). If a sine or square wave is applied to the CONTROL VOLTAGE input, the output frequency will vary with the amplitude of the input waveform.

■ Start by constructing the circuit shown in Figure 1. A stand-alone function generator with FM capability can also be used.[2] If you use a 1 µF polarized capacitor at the CONTROL INPUT, its positive side should be connected to the IC where a dc voltage of $\frac{2}{3}$ V_{CC} is present. Similarly, the 0.1 µF capacitor at the IC's OUTPUT pin blocks the dc component of the output signal. Don't connect the CONTROL VOLTAGE input (FROM FUNCTION GENERATOR) to anything just yet.

■ Power up the circuit and verify that the output frequency is close to the 565 VCO's free-running frequency, f_0. Adjust the VCO frequency until it's within 100 Hz of the 555 output frequency. If the frequencies are too

[2]Some function generators have a fixed-frequency FM capability built in or have an input labeled FM IN. If the former, when the FM function is activated, a tone will modulate the output. If the latter, any signal connected to the FM IN jack will frequency-modulate the output signal.

far apart to adjust, replace the 555 circuit's 10 kΩ resistor with a 10 kΩ potentiometer in series with a 4.7 kΩ resistor and adjust the 555 output frequency to be closer to the 565's f_0.

■ Now set your function generator to output the modulating signal, a 1 Hz sine wave of about 1 V_{PP} (the exact amplitude isn't critical). Set the 10 kΩ potentiometer connected to the CONTROL INPUT to provide about 5 kΩ of resistance. Connect the modulating signal to the pot and monitor the timer output frequency with an oscilloscope or counter. You should see it increasing and decreasing in step with the modulating signal — higher input voltage will lower the frequency. Adjust the sine wave amplitude so that the output frequency swings up and down about 200 Hz. If you are using a counter, you may have to reduce its frequency resolution to get it to update quickly enough to see the output frequency changing. If you do not see the frequency changing, reduce the 10 kΩ pot's value or increase the function generator output amplitude until you do.

■ Next, connect the 555 output to the PLL's phase detector input. The frequencies of both circuits — the 555 oscillator and the 565's VCO — should be the same. The error signal at the input to the 565's VCO (pin 7) should be going up and down along with the modulating signal. You've just created an FM demodulator!

■ Increase the modulating signal's fre-

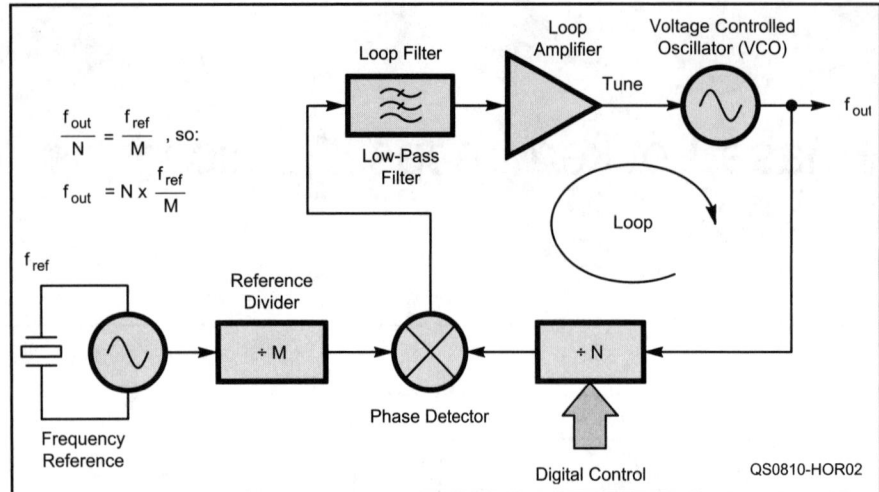

Figure 2 — A PLL acting as a frequency synthesizer divides both the reference and VCO output frequencies to lock at a series of frequency steps of the VCO output.

quency and watch the error signal to see what happens. Checking to be sure that the modulating signal amplitude stays the same (within 5% or so), at what frequency does the amplitude of the error voltage fall to 0.7 times its low-frequency value? That frequency is approximately the loop's bandwidth.

- Change the value of the capacitor connected to pin 7 of the 565 to 1 µF and see if that changes the bandwidth. A smaller value for this capacitor allows the VCO input voltage to change more rapidly so that the PLL can track faster changes in the input frequency and reproduce higher frequency input signals.

- Change the modulating signal to a square wave and vary its frequency. Assuming you are using a 'scope, trigger on an edge of the modulating square wave and zoom in on the error voltage as the input square wave changes voltage. You'll see the loop's transient response as it adjusts the VCO frequency to track the sudden change in input frequency.

Loop Gain and Filtering

What would happen if the error signal — a dc signal — were multiplied by a factor of two before it was input to the VCO? The loop would lock faster. Why? Because the error signal would be twice as large for the same amount of phase error as before, causing the VCO to react that much faster.[3] All of the loop's components — the phase detector, the filter, and the VCO — have some characteristic amount of gain in volts per volt, or in phase or frequency per volt. *Loop gain* is the product of all three gains — phase detector, loop filter, and VCO.

The frequency response of the loop filter is also important. The 565 PLL circuit uses a simple *lead* (low pass) filter formed by an internal resistor between the VCO input and VCC in parallel with the 10 µF capacitor. The capacitance in this low-pass filter can be reduced to allow the loop to change frequency quickly, but the loop's average frequency will be less stable.

For higher-speed uses such as data recovery, the loop may have to track very abrupt changes in input frequency. A common solution is the *lead-lag* filter in which the filter changes from a single-order filter (a single RC combination) to a second-order filter. A lead-lag filter combines a low-pass (lead) and high-pass (lag) response. These filters are used with PLLs recovering FSK RTTY or packet radio data.

The tradeoff between frequency stability and the speed at which a PLL can track an input signal is important. Instability results in more phase noise in the PLL output. The faster a PLL can track changes in the input frequency, the noisier its output is likely to be. It is a difficult design challenge to achieve both a very quiet output and a very fast tracking response.

Frequency Synthesizer

While a full discussion of the use of PLL circuits as frequency synthesizers is beyond the scope of Hands-On Radio experiments, Figure 2 shows the basic structure. Instead of the loop tracking a varying input signal, the input is a very stable reference frequency, usually a low-noise crystal oscillator.

The VCO output frequency is divided by some integer value, N, before being applied to the phase detector. (Assume that the value of divider M = 1.) If the value of N is two, then the loop will be locked when the VCO's output frequency is twice that of the reference signal. By changing the value of N, the VCO frequency in lock will also change. In this way, the VCO output can have very nearly the same low-noise characteristics of the reference signal, but still have a variable frequency.

If we let the value of M change, too, the situation gets a little more complicated. For any value of M and N, the loop will be locked when the frequency of the VCO output divided by N (VCO/N) is equal to the reference frequency divided by M (REF/M). Thus, the VCO frequency will equal REF × N/M. By selecting the right reference frequency and values for M and N, the VCO's output can be made to change in steps, such as the 15 or 20 kHz steps between repeater frequency channels!

Parts List

- Capacitor — 0.1 µF, 25 V ceramic, quantity 2.
- Capacitor — 1 µF, 25 V electrolytic.
- Integrated circuit — 555 timer.
- Resistor — 10 kΩ, ¼ W.
- Resistor — 470 Ω, ¼ W.
- Potentiometer — 10 kΩ linear taper.

Recommended Reading

The URL given last month for "AN535 — PLL Fundamentals" was incorrect. The downloadable PDF can be found at **www.datasheetcatalog.com/datasheets_pdf/A/N/5/3/AN535.shtml**. *The ARRL Handbook* has an extensive discussion on both PLL and frequency synthesizer circuits.[4]

Next Month

Construction techniques can have a major effect on how a circuit performs. Next month, we'll build the same filter three ways, but it won't act like the same filter!

---

[3]Higher loop gain also reduces capture and lock range because it takes less frequency change to reach the maximum phase shift limits.

[4]*The ARRL Handbook for Radio Communications,* 2008 Edition. Available from your ARRL dealer or the ARRL Bookstore, ARRL order no. 1018. Telephone 860-594-0355, or toll-free in the US 888-277-5289; **www.arrl.org/shop/**; **pubsales@arrl.org**.

Experiment #79 — Pi and T Networks

We've experimented with the L network in a previous Hands-On Radio experiment, #21. For any combination of source and load impedance, there is at least one version of the four L networks that will transform one into the other using one L and one C. If that's so, why do we need more complex networks to do the same job?[1]

The first reason is that any single version of the L network can only create an impedance match between half of the possible source and load impedance combinations. If the combination is "wrong," then a different version of the L network is required.

Another reason is that the Q of the L network — which determines the frequency range over which the network creates the match — is equal to the ratio of the two impedances to be matched. The more different the two impedance values, the higher the value of Q and the "sharper" the tuning of the network becomes. We need some flexibility to make the adjustment less sensitive to changes in frequency.

Pi Networks

The Pi network can be thought of as two L networks "back to back" as shown in Figure 1. (We'll treat the impedances as pure resistances to simplify the discussion.) Start with the Pi network at A, divide the inductor in half, and create the two L networks in B. The two circuits at A and B act exactly the same. Splitting the inductor, however, creates an intermediate point in the circuit at which we can imagine the input L network transforming R_{IN} to an image resistance, R_{IMAGE}, and the output L network transforming R_{IMAGE} into R_{OUT}. (R_{IMAGE} is not an actual load, of course; it is just a new ratio of voltage and current.)

Different configurations of inductance and capacitance can form a Pi network. The version shown is by far the most common in Amateur Radio because it acts as a low-pass network. This makes it popular as an output circuit for power amplifiers because it reduces the harmonics that are generated in the amplifier.

This "two-step" transformation allows the designer to choose Q for the Pi network because the value of R_{IMAGE} is variable. If R_{IMAGE} is variable, than the Q of the input and output L networks are variable, too.

The value of R_{IMAGE} has to be smaller than either R_{IN} or R_{OUT}. Why? Take a look at the two L networks. The input L network configuration with the parallel or shunt element at the input requires $R_{IN} > R_{IMAGE}$. Similarly, the configuration of the output L network requires $R_{IMAGE} < R_{OUT}$.

While the particular application will affect your choice of Q, one method is to specify the bandwidth (BW) over which the network is to efficiently transfer input power to the output. Similar to a tuned circuit's frequency response, $Q = f_C / BW$, where f_C is the geometric center frequency = $\sqrt{(f_H \times f_L)}$, where f_H and f_L are the upper and lower limits of the frequency range being matched. The wider the frequency range, the lower Q will be.

However the value is determined, Q must also be high enough that $(Q^2 + 1) > (R_{IN} / R_{OUT})$. If these two quantities are equal, the value of X_{C2} becomes infinite (C2 = 0) and the Pi network becomes an L network. Once you've selected a value for Q, follow these steps to calculate the component values (it is assumed that $R_{IN} > R_{OUT}$):

Calculate the value of the parallel reactance, $X_{C1} = R_{IN} / Q$

Calculate the value of the parallel reactance X_{C2}

$$X_{C2} = R_{OUT} \sqrt{\frac{R_{IN} / R_{OUT}}{Q^2 + 1 - R_{IN} / R_{OUT}}}$$

Calculate the value of the series reactance X_L

$$X_L = \frac{Q R_{IN} + R_{IN} R_{OUT} / X_{C2}}{Q^2 + 1}$$

Convert the reactances to component values:

$$C = \frac{1}{2\pi f X_C} \quad \text{and} \quad L = \frac{X_L}{2\pi f}$$

If the value of Q you select results in components that are impractical, you can change Q and try again. (Large values of L or small values of C are problematic in some applications and frequencies.)

Building a Pi Network

Let's build a real Pi network that will match a 300 Ω input (such as might be present at the feed point of a folded dipole) to 50 Ω for coaxial cable on the 40 meter band. For good efficiency, we'll set f_H and f_L well outside our operating range at 7.7 and 6.6 MHz, respectively. Thus, we have

$R_{IN} = 300\ \Omega$, $R_{OUT} = 50\ \Omega$,

$f_C = \sqrt{(7.7 \times 6.6)} = 7.13$ MHz, and

BW = 7.7 − 6.6 = 1.1 MHz.

Use f_C for calculating component values.

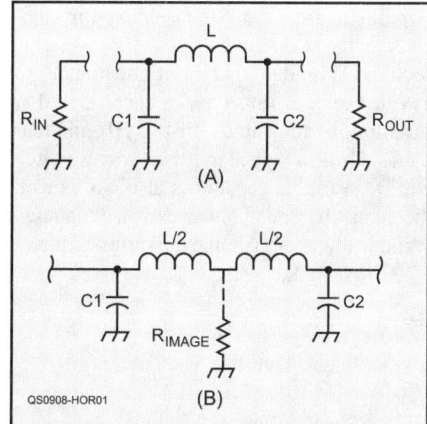

Figure 1 — The Pi network (A) is electrically identical to a pair of back-to-back L networks (B). Using a two-step transformation allows the designer to control network Q and component value. Its low pass characteristics make it popular as an amplifier output circuit.

[1]Previous Hands-On Radio columns and a complete parts list for all experiments are available to ARRL members at www.arrl.org/hands-on-radio.

[2]The ARRL Handbook for Radio Communications, 2009 Edition. Available from your ARRL dealer or the ARRL Bookstore, ARRL order no. 1018. Telephone 860-594-0355, or toll-free in the US 888-277-5289; www.arrl.org/shop; pubsales@arrl.org.

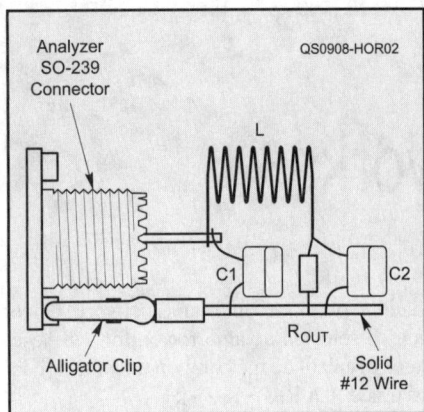

Figure 2 — The example Pi network can be built directly at the output of an SWR analyzer to minimize excess lead length.

values to see how they affect the match. (If you have some variable capacitors, you can attach them in parallel with C1 or C2 and adjust them to make a tunable matching network.) Try raising the value of Q by reducing the difference between f_H and f_L, say to 7.3 and 7.0 MHz. Rebuild the circuit and compare the variation in SWR with that of the lower Q circuit.

T Networks

The T network shown in Figure 3 creates a two-step impedance transformation, just as does the Pi network. Also, as for the Pi network, Q must be high enough that $(Q^2+1) > R_{IN}/R_{OUT}$. Unlike the Pi network, the intermediate resistance value, R_{IMAGE} is higher than either R_{IN} or R_{OUT}. This is because the L networks are "turned around" so that the parallel elements are connected across R_{IMAGE}.

With $R_{IN} > R_{OUT}$, Q, and f_C specified as before, calculating the component values for the T network requires the calculation of a pair of intermediate values, A and B, to make the equations more manageable.

Calculate the intermediate variables A and B

$$A = R_{IN}(Q^2 + 1) \text{ and } B = \sqrt{\left[\frac{A}{R_{OUT}} - 1\right]}$$

Calculate the value of the input series reactance $X_{L1} = R_{IN} \times Q$.

Calculate the value of the output series reactance $X_{L2} = R_{OUT} \times B$.

Calculate the value of the parallel reactance $X_C = A / (Q + B)$.

The version of the T network with two series capacitors and a parallel inductor is popular in impedance-matching units because of the reasonable values of components required to match 50 Ω input impedances to a wide range of impedances encountered at the input to feed lines on the HF amateur bands. Using a tapped inductor with a switch and two variable capacitors also works well mechanically and at lower expense than two variable inductors. Unlike the series-inductor

Start by calculating Q = 7.13 / 1.1 = 6.48. Check to be sure $Q^2 + 1 = 43 > R_{IN}/R_{OUT} = 300/50 = 6$, and proceed. $X_{C1} = 46.3$ Ω, so C1 = 482 pF. $X_{C2} = 20.1$ Ω, so C2 = 1109 pF. $X_L = 62.5$ Ω, so L = 1.40 μH. (A simple spreadsheet is provided on the Hands-On Radio Web site to make the calculations and allow you to experiment with the circuit design values.)

Wind your own inductor using the formulas in the Electrical Fundamentals chapter or tables in the Component Data and References chapter of *The ARRL Handbook*.[2] Eight turns of 20 gauge wire on a ½ inch diameter form over a length of ½ inch will be close to 1.4 μH. To increase inductance, squeeze the turns together. The standard values of 470 pF and 1100 pF will be fine for C1 and C2. Use a noninductive carbon composition or metal-oxide resistor for the 300 Ω load.

Figure 2 shows a suggested method of construction that allows you to build the matching network right at the output of an SWR analyzer. Many other methods of construction will work, since the frequency is relatively low and values of the components relatively large in comparison to stray inductance and capacitance.

Sweep the analyzer frequency back and forth across the band and record the variation in SWR with frequency. Vary the component

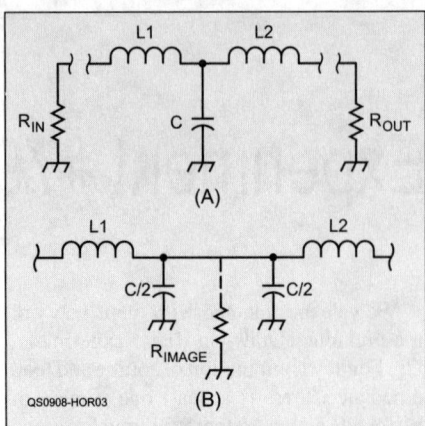

Figure 3 — The T network (A) can also be analyzed as a pair of L networks (B). The version with series capacitors and a single parallel inductor is a popular impedance-matching unit or "antenna tuner" design, although as a high-pass network, it does not offer any harmonic reduction.

Pi and T networks discussed earlier, this circuit is a high-pass network and so doesn't provide any additional harmonic reduction.

Parts List

- 470 pF and 1100 pF silvered mica or ceramic capacitors.
- 300 Ω, ¼ W carbon composition or metal-oxide resistor.
- 10 inches of 20 gauge solid hookup or magnet wire.

Recommended Reading

The three-part series "Impedance Matching" by W1DF in the March, April and May 1957 issues of *QST* remains a classic tutorial. With the online *QST* archives open to ARRL members, there's no reason not to log on to the ARRL Web site and read them!

Next Month

Battery powered accessories and gadgets abound, but which of the many battery types perform well and how can you compare one to another? Let's look at battery capacity next time.

Practical Construction

Experiment #120

Power Polarity Protection

Part of being a successful radio gadgeteer is knowing how to make sure your gadgets survive the various insults and screwups that are part of life in the electronics world. This month's and next month's columns will show you how to deal with some common gremlins that seem to live just outside your gadget's power connection.

Polarity Protection

We've all done it — accidentally connected power with plus where minus should be and vice versa, "just for a second," letting the magic smoke out of our precious components. That's why I was encouraged to see Terry Fletcher, WAØITP, publish a collection of circuits in *QRP Quarterly* that protect equipment against reversed power polarity.[1] Terry collected some circuits from various sources and published them all in one place with a comparison of each circuit's strong points and weaknesses. (I thought these were so useful, I added them to the 2013 *ARRL Handbook*, too![2]) Most of these circuits can be retrofitted into existing equipment — as long as you take note of and remember the following characteristics of each circuit.

Diode Protection

Figure 1A shows the simplest protective circuit of all — a rectifier in series with the positive lead between the power source (In) and electronic device (Out). A Schottky barrier diode is also shown next to the conventional silicon junction diode. The typical 1N4000 series junction diode rectifier has a nominal forward voltage drop of 0.7 V but at 1 A of current, the forward

---

[1]T. Fletcher, WAØITP, "Reverse Voltage Protection Overview," *QRP Quarterly*, Spring 2012, pp 12-13. Also online at **www.wa0itp.com/revpro.html**.
[2]*The ARRL Handbook for Radio Communications,* 2013 Edition. Available from your ARRL dealer or the ARRL Bookstore, ARRL order no. 4050 (Hardcover 4197). Telephone 860-594-0355, or toll-free in the US 888-277-5289; **www.arrl.org/shop; pubsales@arrl.org**.

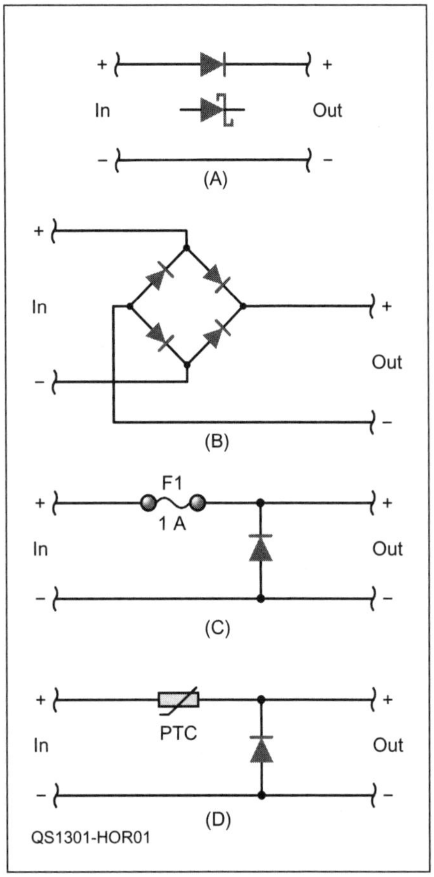

Figure 1 — Diodes connected in series and shunt can either block reverse polarity voltage or cause a protective fuse to open if voltage is reversed.

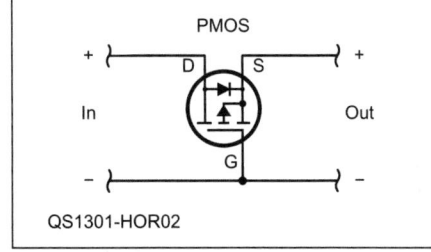

Figure 2 — A PMOS enhancement mode MOSFET conducts with a very low ON resistance with normal voltage polarity but blocks reverse polarity voltage.

voltage drop can be a bit greater than 0.9 V with nearly 1 W of power dissipated as heat. Schottky barrier rectifiers have a lower forward voltage drop: 0.75 V at 1 A for the 1N5418. Be sure your equipment will work properly with its input voltage a little bit lower than without the diode. Don't exceed the diode's average forward current rating or it will overheat and fail. The protective rectifier should have a reverse voltage rating of at least twice the supply voltage so a 1N4002 (50 V, 1 A) or 1N5418 (30 V, 1A) are suitable in 12 V systems.

The diode bridge circuit in Figure 1B not only protects the circuit but allows it to operate normally with voltage of either polarity applied. The trade off is *two* voltage drops in series with the power source so your operating voltage takes a hit of 1.5 to 2.0 V, depending on what type of diodes you use. Diode bridges can be purchased all in one package, so this is a convenient option.

Figures 1C and 1D rely on a fuse to protect the circuit. A shunt connected diode across the power source will conduct high current if voltage polarity is reversed. This will blow a metal fuse very quickly and the diode will limit any applied voltage to one negative forward drop. The fuse has no significant voltage drop under normal conditions. The fuse needs to be large enough to handle normal device current but not so large that the surge current through the reverse connected diode causes the diode to fail first. If you use this method, make sure the fuse is accessible and replaceable.

The PTC (positive temperature coefficient) fuse in Figure 1D is a *resettable* fuse that acts like a solid state circuit breaker. When its rated current is exceeded, the fuse material heats up and changes to a high resistance state. When the material cools, it changes back to the low resistance state. PTC fuses are characterized by their trip current at which the material changes resistance and their *holding current* that keeps them in the high resistance state. A typical PTC device for low current 12 V protection is the

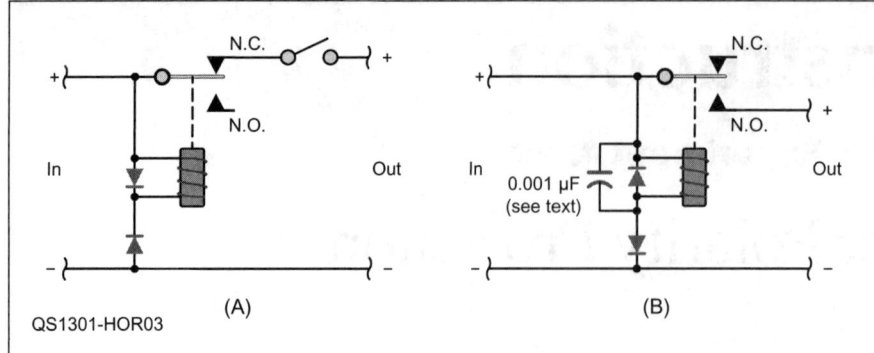

Figure 3 — Relays can be used to open (A) or not close (B) when reverse polarity voltage is applied.

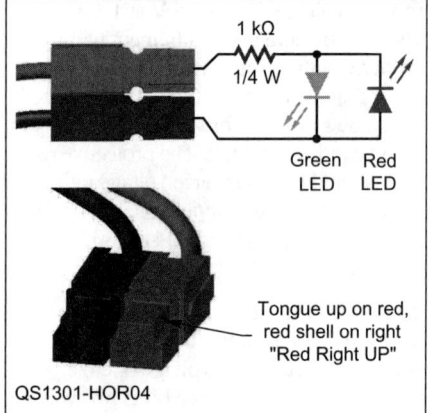

Figure 4 — A handy voltage polarity checker can be made inexpensively and added to your go-kit.

Littlefuse 60R050XPR, which has a trip current of 1 A and a holding current of 500 mA. PTC fuses do heat up a bit when in the high resistance state and they insert a small resistance in series with the power source (around 1 Ω) that can cause a voltage drop. Again make sure that the diode can handle a big surge current when needed!

Sometimes the shunt diodes in Figures 1C and 1D fail due to the high surge current. If the diode fails shorted or in a low resistance state, it will continue to blow the fuse until replaced. If the diode fails open or with a high resistance, it will no longer protect the circuit. If shunt diode protection is used and the fuse opens, check the diode to be sure it has not failed as well.

Active Device Protection

Many voltage regulators that use PNP or PMOS transistors (P channel MOSFETs), such as the LM2940, offer reverse polarity protection. A PMOS transistor will also work as shown in Figure 2. (The diode symbol represents the *intrinsic body diode* that is created due to the way the transistor is made.) The PMOS enhancement mode transistor conducts unless the drain-to-gate bias becomes negative as when the applied voltage is reversed.

PMOS devices have low ON resistance ($R_{DS(ON)}$) and high maximum current ratings. For example, the SPP08P06P (at a cost of less than a dollar) has an ON resistance of 0.3 Ω at 8.8 A of continuous drain current. Lower resistances and higher current ratings are available, as well. For more information about using PMOS and NMOS devices for polarity protection see Maxim Electronics Application Note 636, "Reverse Current Circuitry Protection."[3]

Relay Protection

Figure 3 shows a pair of circuits that can be used for polarity protection. Relay based circuits have the advantage of little to no voltage drop, even at high currents, as long as the contact ratings are sufficient. The circuits are more complex than the diode based circuits described in the preceding section but can generally handle more current and are not damaged by reverse polarity voltages. The circuits reset themselves automatically.

The circuit in Figure 3A uses the normally closed (NC) contacts of a single pole, double throw (SPDT) relay. There is no current drain through the relay coil until reverse polarity voltage is applied. There will be a few milliseconds, however, during which reverse polarity voltage gets through to your circuit if no power switch is used or the power switch is closed. This is enough time for damage to occur so this circuit is only recommended if a power switch is used to turn the equipment on and off after voltage is applied.

[3]www.maxim-ic.com/app-notes/index.mvp/id/636

The normally open (NO) contacts are used in Figure 3B. The relay contacts close and supply power to the equipment only when applied voltage has the proper polarity. The relay coil draws current continuously during normal operation. Small relays draw anywhere from 10 to 50 mA and this level of current drain may be unacceptable for low power and battery powered equipment.

Other Gotchas

WA0ITP also notes that it is very, very easy to accidentally connect the wrong terminals of a 9 V battery to its snap on clip. Make sure you protect circuits that use those batteries!

Assuming voltage polarity is okay if the power source has a red and black connection can be an expensive assumption. It's always a good idea when connecting your gadget (or radio) to an unknown power source to first check for proper voltage polarity. Figure 4 shows a simple circuit you can make and use with any connector style, although PowerPoles are shown here. Hot glue or epoxy could be used to secure the small components right on the connector making an enclosure unnecessary.

Another problem doesn't pertain to circuit protection but can occur if diodes are connected to wires near antennas — RF harmonic generation and mixing. All of these diodes are likely going to be connected to long power leads and to be in close proximity to antennas radiating a pretty strong signal. The signal is picked up by the power leads and can cause diodes that are not conducting (or heavily reverse biased) to turn ON and OFF at the frequency of the RF. This creates harmonics and if more than one signal is present, you've created a dandy mixer, as well.

The leads that pick up the signal and conduct it to the diodes are happy to take the harmonics and mixing products from the diode and radiate them. These spurious signals are right in your shack so they are very easy to receive. The easiest fix is to simply connect a small ceramic capacitor across each diode — Figure 3B shows an example. Any value from .001 to .1 μF will do. (This is a good idea for rotator controllers and other switching gadgetry in your shack, too.)

Next Month

In the following column we'll take a close look at transient protection, particularly in mobile applications. You may be shocked by what you learn!

Note that some of the solutions provided in next month's column will also protect reverse polarity at no additional cost.

Experiment #121

Transient Protection

Last month's column dealt with ways to prevent damage to equipment if power were applied with voltage polarity reversed.[1] That is a pretty common problem, so it's good to protect against it — called *mitigation* — whether from a human caused accident or a wiring fault. Another class of problems also exists but isn't talked about much within the ham community — *transients*.

A transient, by definition, "…refers to momentary overvoltages or voltage reductions in an electric power system…" (*CRC Electrical Engineering Dictionary*) Just because a transient is momentary doesn't mean it can't cause harm. As you'll see in the following description of transients found in vehicles, they can be plenty damaging.

Automotive Transients

You might think with power supplied by a battery, a vehicle's dc power system is reasonably well behaved. Unfortunately from an electronics view, that's not true, as you can see in Figure 1. The voltage in vehicles that use a *starting battery* as the energy source varies between 10.5 V (a discharged battery) to more than 15 V during heavy charging. The typical 13.8 V ±15% input voltage specification corresponds to a fresh battery during normal charging. The subject of this column, though, is not the dc supply voltage.

There are numerous ac and transient signals superimposed on that dc supply voltage — for example, the rapid current switching in the vehicle's alternator creates sharp (short duration) transients on the *power bus* (the heavy wiring that distributes power within the vehicle). If not adequately filtered by the radio, the result is a high pitched *alternator whine,* which follows the engine speed and is added to both receive and transmit signals.

SAE (Society of Automotive Engineers) Standard J1113, "Immunity to Conducted Transients on Power Leads," describes transients encountered on a vehicle's power bus such as these common occurrences:

- *Load Dump* — occurs when a loose battery connection opens up during charging and the alternator's energy is "dumped" on the power bus with no battery to hold down the voltage.

- *Alternator Field Decay* — occurs every time the vehicle is turned off and the alternator's stored energy has to be dissipated via the power bus.

- *Inductive Load Switching* — the kick back voltage from an inductive load (such as an electric window motor) being turned off.

- *Mutual Coupling* — transient energy that is coupled between conductors in a wiring harness.

Table 1 summarizes the electrical characteristics of these transients. There are quite a few more transients that are described in SAE Standard J1113 but these are common and cover the range of voltage and energy amplitudes that vehicle electronics have to deal with.

Obviously, some of those transients are pretty severe — such as the big load dump transient in Figure 2. It's a wonder anything survives being plugged into a vehicle. Your electronics are not going it alone, however, as the vehicle's manufacturer has built in some transient protection for you. As explained in the Littelfuse application note *Suppression of Transients in an Automotive Environment*, vehicle electronics are already protected by a *central suppressor* in the vehicle, usually located as close to the master control computer module as possible.[2] There are usually suppressors in other modules around the vehicle, too. This helps limit the voltage and energy levels to which your gadgets and radios are exposed. Nevertheless, it's a good idea to provide some limited protection.

Electrostatic Discharge and Lightning

Another source of transients, well known to inhabitants of areas with dry weather, is *electrostatic discharge* or *ESD*. A sudden discharge of static electricity by a spark,

Table 1
Typical Vehicle Transients

Type	Voltage (V)	Energy (Joules)	Duration	Occurs
Load Dump	<125	>10	200-400 msec	Infrequently
Field Decay	−100 to +40	<1	200 msec	At turn off
Inductive	−300 to +80	<1	<320 μsec	Often
Mutual Coupling	<200 V	<1	1 msec	Often

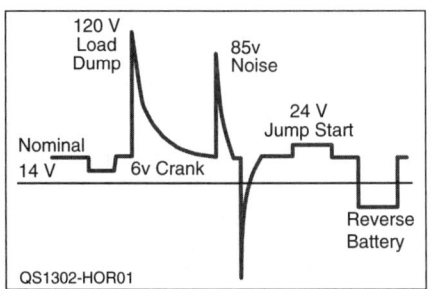

Figure 1 — Typical vehicle power system voltage levels. (Data courtesy of Littelfuse Corp.)

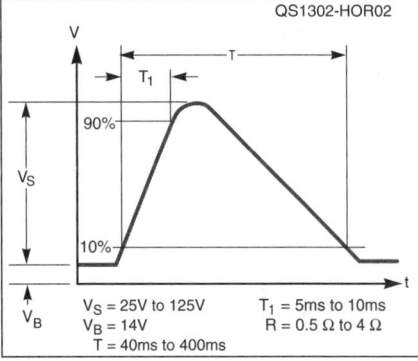

Figure 2 — The load dump transient that occurs when the connection between a battery and the alternator opens during charging. (Data courtesy of Littelfuse Corp.)

such as from walking across a carpet then touching a grounded surface, is a typical example of ESD. In fact, the standard ESD test generator uses a finger shaped probe.

There can be enough energy in an ESD to destroy semiconductors or scramble the operation of a circuit. A typical ESD transient lasts for less than 50 μsec but can generate voltages up to 15 kV. ESD transients can appear on power and signal wiring, connectors, controls and switches, displays — anything a finger can touch, even metal enclosures.

Lightning can obviously generate pretty significant transients though we aren't addressing protection from a direct strike in this column. An indirect or nearby strike can generate fast rising transients with energy levels equal to or larger than the automotive inductive transient and with voltages nearly as high as an ESD pulse.

Protective Components

There are several ways to protect electronics against transients — block them, route them away from the circuitry or dissipate their energy as heat. The goal is to limit the resulting voltage to levels the electronics can handle. Several different types of protective components have been developed to handle transients:

- *MOV (metal oxide varistor)* — partially conductive powder pressed into a disc or cylinder so that it is nonconductive up to its *clamping voltage*.[3] At voltages (of either polarity) higher than the clamping voltage, its resistance drops, limiting the voltage by dissipating energy as heat. MOVs are generally connected between the circuit being protected and ground so that the lower resistance causes the MOV to absorb the transient's energy and keep voltage at a safe level. After repeated transients, MOVs generally fail in a low resistance state.

- *Transient Voltage Suppressor (TVS) Diode* — a special type of Zener diode that offers more precise clamping action.[4] Two TVS diodes back to back in a single package make a bipolar clamp that has similar characteristics to an MOV. Unless overloaded, TVS diodes can handle repeated transients without changing their characteristics.

- *Diode Clamps* — Most effective for ESD and fast transients, clamping diodes route energy away from the protected circuit into the power supply where it is absorbed by the filter components.[5,6] This limits the circuit voltage to one forward voltage drop beyond the power supply voltage.

- *RC Pulse Filtering* — Useful for both RFI and ESD, a series resistor followed by a capacitor to ground act as a low-pass filter, smoothing out transient voltages and dissipating some of the energy as heat.

Testing and Using Protective Components

You can observe the protective action of these components and methods on the workbench by constructing the test circuit in Figure 4. The input test signal is either variable dc (0 to 12 V) or a pulse train (square wave generator with a series dc blocking capacitor). You will also need a +6 V dc supply which can be made from batteries if you don't have a second power supply. The 1 kΩ resistor limits the current through the protective components.

Here are some typical components to use — feel free to substitute:

- MOV: V8ZA05P — clamps at 6 V.
- TVS: 1.5KE8.2CA — clamps at ±8.2 V (bipolar).
- A 1N4001 general purpose silicon diode can be used for the clamping diode in this test.

Begin by connecting the MOV to the 1 kΩ resistor and monitor V_{OUT} with a voltmeter. Slowly increase the input dc voltage until the MOV begins to conduct, at which point the output voltage will increase more slowly than the input voltage. Take several measurements and make a graph of V_{OUT} versus V_{IN}. Replace the MOV with the TVS diode and repeat the experiment, plotting the input and output voltages on the same graph. Disconnect the TVS diode and connect the clamping diode from V_{OUT} to the 6 V supply and repeat the measurement and graph of input and output voltage. Observe and compare the different shapes of the clamping action by each device.

Disconnect the input dc voltage and the clamping diode. Create a pulse generator by connecting a series 0.001 μF capacitor in series with the output of a square wave generator. Set the square wave generator to approximately 1 kHz and a peak to peak output of at least 1 V. Connect the pulses to the circuit's input and observe them at V_{OUT}. Now connect a pulse filtering capacitor (use 0.001, 0.01, 0.1 and 1 μF capacitors) to the resistor and observe the effects of each capacitor in limiting the pulse peak voltage and lengthening it.

Summary

This short discussion just scratches the surface of transient protection. The referenced application notes can provide a lot more information — Littelfuse offers many application notes on its website and references such as Standler's book on transients is excellent and detailed.[7] In the meantime, considering adding protection to your circuits and equipment, especially mobile stations.

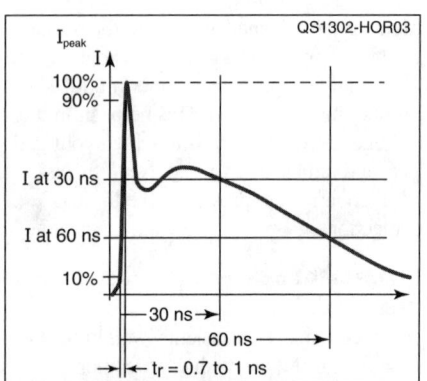

Figure 3 — ESD Transient waveform. (Data from standard IEC 61000-4-2.)

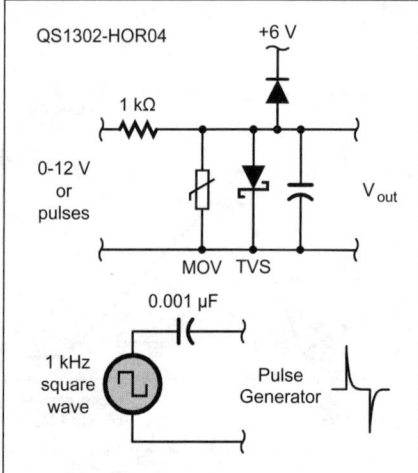

Figure 4 — Test circuit for observing the voltage clamping effects of different protection components.

References

[1] All previous Hands-On Radio experiments are available to ARRL members at **www.arrl.org/hands-on-radio**.
[2] Littelfuse, *Suppression of Transients in an Automotive Environment*, AN9312.5, 1999, **www.littelfuse.com/data/en/Application_Notes/an9312.pdf**.
[3] **www.resistorguide.com/varistor**
[4] **www.semtech.com/images/promo/What_are_TVS_Diodes.pdf**
[5] Littelfuse, *ESD Protection Design Guide*, **www.littelfuse.com/data/en/Applications/Littelfuse_ESD_System_Level_Guide.pdf**
[6] Littelfuse, *Tips for Enhancing ESD Protection*, **www.littelfuse.com/data/en/Application_Notes/Littelfuse-Tips_for_Enhancing_ESD_Protection.pdf**
[7] R. Standler, *Transient Protection of Electronic Circuits*, Dover.

Experiment 90 — Construction Techniques

Back to the Bench

This month we're reacquainting ourselves with our toolboxes and soldering irons, having neglected them in favor of computer screens lately. Building the circuit tests your model and allows you to complete the four step cycle — *design, simulate, build, compare*. Clean off your workspace and let's dig in!

Dead Bug

It's hard to beat *dead bug* or *ugly style* construction for speed and convenience if you're building an initial prototype. No special tools or parts are required — just grab a handy piece of unetched printed circuit board (PCB) material and start soldering. The surface of the PCB acts as a common connection for all parts — often ground potential. If a component lead is to be connected to circuit common, it's soldered directly to the PCB surface. This mounts the component to the PCB, which then also acts as the mechanical base. Components are held in place by soldering their leads together directly or by using wire jumpers if they don't reach.

Figure 1 shows a typical candidate for dead-bug construction — a three-terminal voltage regulator circuit based on a 7800 series IC. You can find out more about using three terminal regulators in Hands-On Radio experiments #8 and #70 and in Section 3.8 of the *2010 ARRL Handbook*.[1,2] The copper surface of the PCB does double duty in this circuit, providing both a ground plane and a small amount of heat sinking. (If your voltage regulator dissipates more than a few hundred milliwatts, you should use a real heat sink.)

Start by carefully bending the pins of the regulator IC. Pins 1 and 3 can be bent *up* (away from the PCB surface) to act as tie points for the input, output and bypass capacitors. Pin 2 takes a bit more technique, since it will need to lie flush along the surface of the PCB. First, bend the thin part of the lead *down* (toward the PCB surface) at about a 45° angle. At the point at which the lead would intersect the PCB surface, bend it back up so that the outer part of the lead is parallel to the surface. This may take a couple of tries if you haven't done it before. DIP ICs mount upside-down against the PCB, held in place by grounded pins soldered to the PCB.

It's important that the pin lay flat against the board so that the IC's metal tab can also make good contact with the copper. Thermal contact between the IC's tab and the PCB surface is insured by holding the IC to the board with a machine screw with a small amount of thermal compound or a thermal pad between the IC and PCB surfaces.

Once the IC is mounted, solder the bypass capacitors between pins 1 and 3 to the PCB surface as ground. In Figure 1, you can see that I've drilled a hole through the board and used a knot to hold the input and output wiring in place — cheap but effective.

Done carefully, dead-bug construction is RF-friendly up into the low UHF range. The PCB's surface also acts as a *ground plane*, proving low inductance ground connections and helping reduce coupling between different parts of the circuit. To get good RF performance, leads must be kept short and direct. Remember to keep inputs and outputs

[1]All previous Hands-On Radio experiments are available to ARRL members as downloadable PDF files at **www.arrl.org/hands-on-radio**.
[2]*The ARRL Handbook for Radio Communications,* 2010 Edition. Available from your ARRL dealer or the ARRL Bookstore, ARRL order no. 1448 (Hardcover 1462). Telephone 860-594-0355, or toll-free in the US 888-277-5289; **www.arrl.org/shop**; **pubsales@arrl.org**.

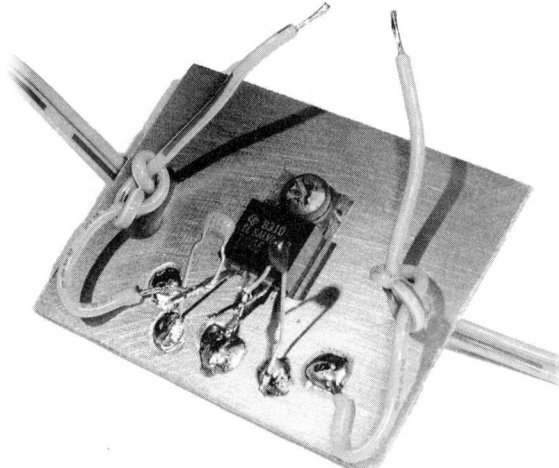

Figure 1 — Dead bug or ugly-style construction uses the surface of unetched PCB material as the base for circuit assembly. Component leads are soldered directly to the PCB surface or to other leads. (Photos and graphics reprinted with permission from *Circuitbuilding Do-It-Yourself for Dummies*, Wiley Publishing.)

Figure 2 — Small pads made from PCB material are used as isolated connecting points in Manhattan style construction. Pads are either soldered or glued to the PCB surface.

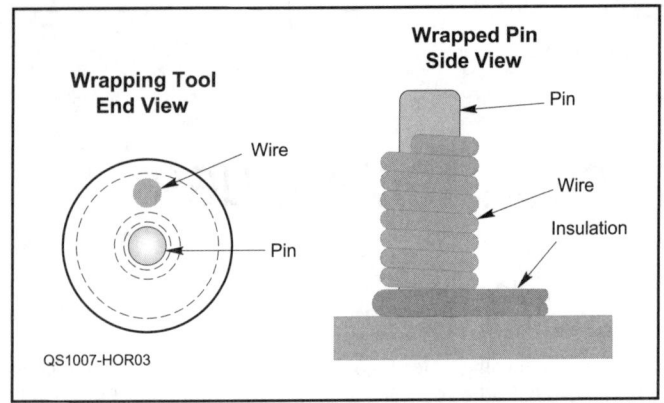
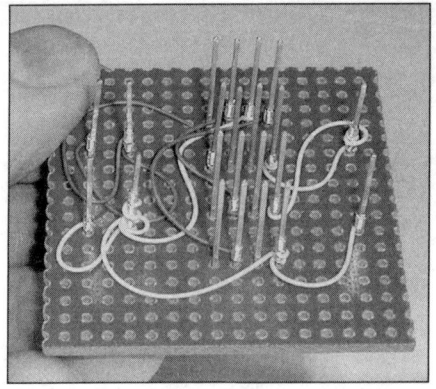

Figure 3 — A wire wrap connection is made by using a wire wrap tool to coil the solid wire around the sharp corners of the wire wrap pin. A wrap of insulation at the bottom of the pin provides stress relief for the wire and prevents breakage.

Figure 4 — The wire wrapped circuit's components are soldered to pins or inserted into sockets that extend through the insulating board. Individual wires make the connections.

away from each other, to the extent feasible.

The dead bug procedure begins with cleaning the PCB surface with steel wool or a scrubbing pad. Then solder components with common connections to the PCB. When those components are all mounted, solder the other components between the remaining leads. Because it can be difficult to make large changes in the circuit layout, dead-bug construction is best for simple circuits or designs in which you are confident. If using a multi pin IC, be careful to note that you will be working with the pins in a reversed layout from the normal view. It helps to use a pencil or label to indicate pin numbers. Don't ask me how I learned this.

Manhattan Style

A technique that is closely related to dead bug style is *Manhattan style*. This technique uses pads of PCB material to provide isolated mounting points for circuit connections. The individual pads are hot glued or soldered to the PCB surface. Components are then soldered to the pads. Because the pads are often square, the resulting resemblance of the completed circuit to a city street grid gives the technique its name. Grounded component leads are still soldered directly to the PCB surface as in dead bug construction.

To make pads of PCB material, remove a strip of PCB about ¼ inch wide. Use a saw or score the material with a sharp utility knife and straightedge, then bend the piece over the edge of a work surface to break it loose. Use heavy wire cutters (and eye protection) to cut the strip into small squares.

Figure 2 shows a circuit assembled using Manhattan style construction. Connection pads were attached to the PCB surface with hot glue. Gluing works for pads made from both single and double sided PCB material. If you use glue, you'll have to work fast to keep your soldering from melting the glue, allowing the pad to shift position. Some glues, including acrylonitrile based products, may not hold well while heated. If you use double sided PCB pads, you can solder the pads to the board by tinning the pad and PCB, then placing the pad on the PCB and reheating the PCB until the solder melts on both surfaces. You'll have to hold the pad against the PCB surface with a small screwdriver or sturdy toothpick during this procedure.

Manhattan style construction creates a larger (and easier) connection point for ungrounded connections. The combination of using the PCB surface and the pads allows for easier circuit layout, but the resulting circuit is usually somewhat larger than if constructed purely using the dead bug technique. Nevertheless, it's also easier to make changes to the circuit if isolated pads are used. Because leads are generally a bit longer than in dead-bug technique, Manhattan style construction is generally used at HF and below.

Wire Wrap

Our final technique is an old standby, particularly in digital electronics — *wire wrapping*. In this technique, individual pins or sockets with pins are press-fit into holes in an insulating material called *perforated board*. Connections are made by tightly wrapping solid, tinned wire around square pins so that the pin's corners bite into the wire for a secure connection.

Figure 3 shows the end view of a wire wrapping tool and a side view of a properly wrapped connection. You can either use a manual tool, about the size of a small screwdriver, or a motorized tool that resembles a soldering gun. For small projects, a manual tool is fine. One end is used for wrapping, the other end is an unwrapping tool. A wire stripper is usually in the middle.

A special type of wire is used for wire wrapping. The most common is #30 AWG with Kynar insulation that is easy to strip. Be sure to get the right gauge wire for your stripping tool or the stripping process may leave nicks, causing the wire to break easily. Using three or four colors of insulation will also help you keep the connections straight in the resulting nest of wires. Wire wrapping tools, wire and pins are sold by electronics parts and tool vendors. You may also be able to buy prestripped wires, although this is relatively expensive for the casual builder.

Figure 4 shows a finished circuit that was assembled using the wire wrap technique. Components and input/output wiring are soldered to pins pressed into the board. (I used a miniature screwdriver.)

The bottom side of the circuit shows how wires are wrapped onto the pins that extend through the perforated board. Two colors of wire insulation are used here — the darker is used for power connections and the lighter for signal connections. Where more than one connection is made to a pin, the wires are wrapped onto the pin one above the other, not wrapped directly over the previous wrap.

Wire wrapping was originally invented for digital circuitry with clock frequencies of few MHz or lower. It can be used for audio and dc circuits. Because the wires loop and cross frequently, wire wrapping is not good for circuits that are sensitive to coupling between signal paths. The thin wires and wrapped connections are not well suited to carry currents of more than 100 mA. It's very easy to create and modify circuits, however, so wire wrapping remains a common technique.

[3]Ward Silver, *Circuitbuilding Do-It-Yourself for Dummies*, Wiley Publishing, 2007. Available from your ARRL dealer or the ARRL Bookstore, ARRL order no. 0015. Telephone 860-594-0355, or toll-free in the US 888-277-5289; **www.arrl.org/shop**; **pubsales@arrl.org**.

Experiment #89 — Overvoltage Protection

Every summer hams head to the hills, operating from all manner of power sources. Some are well-regulated, but away from the relative stability of the utility power grid, all sorts of things can go wrong. One of the most common problems is excessive supply voltage, or *overvoltage*, caused by sudden drops in load, failure of a regulator, or short- or open-circuits in the power distribution wiring. Regardless of why excessive voltage occurs, the effect on connected equipment can range from a blown fuse to "fire in the hole" — even if the equipment is switched OFF at the time!

Just in time for ARRL Field Day and other summer adventures, we're going to review overvoltage protection circuits at the heart of two projects in the "Power Supplies" chapter of *The 2010 ARRL Handbook.*[1] The first is a *crowbar* circuit for dc power sources, such as a linear or switchmode supply. The second is intended for ac power, such as from a generator or inverter.

Commercial equipment is specified to operate from a range of input voltages, such as 9 to 15 V, or at a nominal input voltage with some tolerance, such as 12 V ±5%. Any voltage above those specified ranges (>15 V or >12.6 V) can be considered overvoltage.

Dropping the Crowbar

DC power supplies can create output overvoltage for many reasons. In a linear power supply, it is not uncommon for a *pass transistor* that regulates the flow of output current to fail as a short circuit or low resistance, placing the full rectifier output voltage on the power supply output terminals. For the common 12 V supply this can be 18 V or more, placing severe stress on attached equipment. Regulator circuits can fail in either linear or switchmode supplies, as well.

Since it is assumed that the power supply is unable to control its output voltage, an external device must limit the voltage and remove power from the supply. The crowbar circuit does this by placing a short circuit directly across the output terminals whenever excessive voltage is detected. This is intended to blow fuses or trip circuit breakers in the supply and to discharge any filter capacitors in the supply's rectifier or output circuits.

The usual choice for the device that does the discharging is a silicon controlled rectifier (SCR). As described in Experiment #10, an SCR acts as an open circuit between its anode and cathode until it is triggered by current flowing from the gate to the cathode.[2] It then conducts current from the anode to the cathode until the current level drops below the *holding current*, I_H, even if gate current is interrupted. Once anode-to-cathode current is halted, it is necessary to retrigger the SCR.

While conducting, SCRs have a low forward voltage drop and can safely handle high currents of dozens of amps for short periods, even without heat sinks. This is enough to cause most amateur power supplies to blow their input power or rectifier fuses.

An SCR Overvoltage Protector

Figure 1 shows the circuit for the SCR based crowbar found in section 17.16.8 of *The 2010 ARRL Handbook*. Simpler circuits use current through a Zener diode to trigger the SCR — when the input voltage increases to a sufficient level, the Zener begins to conduct and the current flows through the SCR gate, triggering the SCR. The trigger level of these circuits drifts with temperature and Zener voltage. Using the MC3423 IC allows much better control of the input voltage at which the SCR fires. Note: If you can't locate an MC3423, an NTE7172 is a direct replacement.

Begin by browsing to **www.onsemi.com**. Enter MC3423 into the SEARCH window then download the MC3423 PDF datasheet. Find Figure 2, the representative block diagram. The key item in Figure 2 is found at the lower left — the 2.6 V internal voltage reference. (Some older versions of this chip have a 2.5 V reference.) This is the MC3423's activation threshold. Whenever the voltage at the SENSE 1 input is *less* than 2.6 V, the output of the left hand comparator is HIGH, turning on the transistor whose collector is connected to pin 4. Note that in the *Handbook* circuit, pin 4 is connected directly to pin 3, the SENSE 2 input to the right hand comparator. If the transistor is ON, the right hand comparator's output is LOW and the output circuitry connected to pin 8 is OFF. (Comparators are discussed in Experiment #11.)

If the voltage at SENSE 1 is *greater* than 2.6 V, however, the left hand comparator output goes LOW, turning the transistor OFF, and the right-hand comparator output goes HIGH, turning the output circuitry ON. In our circuit, this triggers the SCR.

The circuit is based on Figure 5 of the

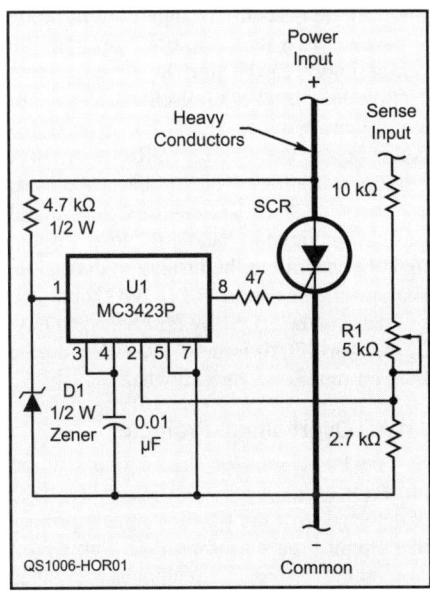

Figure 1 — Schematic diagram and parts list for the overvoltage crowbar circuit. Unless otherwise specified, resistors are ¼ W, 5% tolerance carbon composition or film units. A PC board is available from FAR Circuits (www.farcircuits.net).
D1 — Zener diode, ½ W, 6 to 9 V for a 12 V supply.
R1 — 5 kΩ, PC mount trimmer potentiometer.
SCR — C38M stud mount (TO-65 package).
U1 — MC3423P integrated circuit, 8 pin DIP.

---

[1] *The ARRL Handbook for Radio Communications,* 2010 Edition. Available from your ARRL dealer or the ARRL Bookstore, ARRL order no. 1148). Telephone 860-594-0355, or toll-free in the US 888-277-5289; **www.arrl.org/shop; pubsales@arrl.org**.

[2] All previous Hands-On Radio experiments are available to ARRL members as downloadable PDF files at **www.arrl.org/hands-on-radio**.

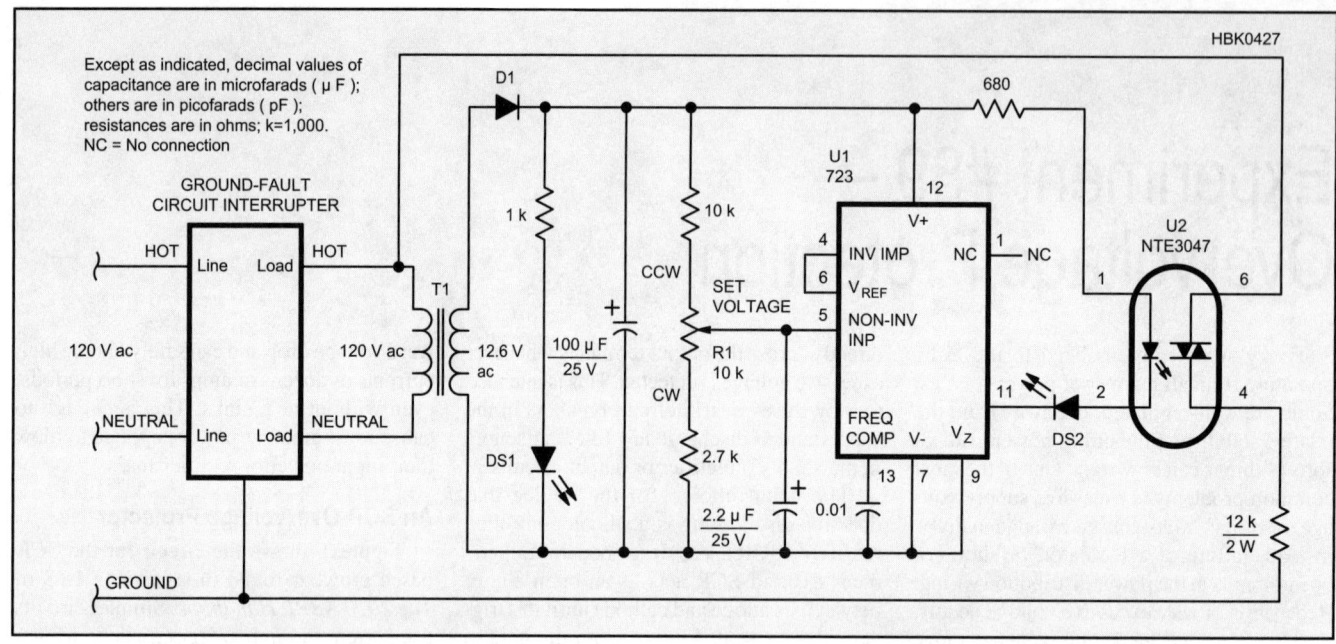

Figure 2 — Schematic diagram and parts list for the Field Day equipment overvoltage protection circuit. The circuit must be used with a separate ground fault interrupter at each station. DS2 illuminates briefly on overvoltage. Unless otherwise specified, resistors are ¼ W, 5% tolerance carbon composition or film units. A PC board is available from FAR Circuits (www.farcircuits.net).

D1 — 200 PIV, 1 A diode, 1N4003 or equivalent.
DS1, DS2 — Small LEDs.
R1 — 10 kΩ, PC mount multi-turn trimmer potentiometer.
T1 — 12.6 V transformer.
U1 — 723 adjustable voltage regulator integrated circuit.
U2 — Optoisolator with TRIAC output, NTE3047 or equivalent.

MC3423 datasheet. The combination of the 4.7 kΩ resistor and D1, the Zener diode, is *not* a sensing input. It is a power supply for the IC. Why? Because if the power supply and sensing input are the same, when the crowbar fires, the chip loses its power supply! This is okay in most cases because the SCR will stay on until it has discharged the power supply and current falls below I_H. By separating the chip's power supply from the sensing input, the crowbar circuit can be applied in many configurations.

The SENSE 1 input is pin 2 and is connected to the voltage divider adjusted by R1. The values in the string of resistors result in a voltage of 2.6 V at SENSE 1 when the sense input voltage is 15 V with the 5 kΩ potentiometer at mid travel. Figure 4 of the datasheet shows how to calculate the resistor values for any desired threshold voltage.

As shown in Figure 6 of the datasheet, the 0.01 μF capacitor at the junction of pins 3 and 4 serves to delay triggering, both to reduce susceptibility to RFI and to keep the circuit from triggering on very short transients. This value requires the overvoltage condition to exist for $12 \times 10^3 \times 0.01 \times 10^{-6} = 120$ μs before the circuit triggers.

Why all the complexity? The IC's cascaded comparators allow the circuit designer to monitor more than one input signal or to use other signals to inhibit or enable the circuit. Note also that pin 6 (Indicator Output) can sink current through an LED or lamp to provide a visual indication that the circuit has activated.

How does one choose an SCR for a crowbar circuit? The *Handbook* circuit uses a C38M SCR rated at 35 A forward current, but the peak non-repetitive surge current rating (the value used for a one-time shorting of a power supply) can be much higher, depending on how long it takes to discharge the power supply and blow its fuses. For most transceiver power supplies, an SCR rated at 30 A or more will suffice. Note that the power leads are shown as thick lines — this wire should be heavy enough to handle the load without melting or igniting the insulation during the short discharge period — #18 AWG or larger.

See also the article by Phil Salas, AD5X, in the April 2010 issue of *QST* that uses a high current Zener as a crowbar.[3]

An AC Overvoltage Protector

The basic idea of the ac overvoltage protector is the same — when overvoltage is detected, trip the power source's protective circuit breaker and remove power. You wouldn't want to short circuit most ac sources because of the high surge currents that would result. In the case of the *Handbook* project in section 7.16.7, a ground fault circuit interrupter (GFCI) circuit breaker handles the high current instead of a semiconductor device.

Figure 2 shows the overvoltage protector circuit. The GFCI breaker is wired to supply both the overvoltage protector circuit and whatever ac powered equipment it protects. The common 723 voltage regulator IC is used

[3]P. Salas, AD5X, "Compact Voltage Protector and Fuse Assembly for 100 W Transceivers," *QST*, Apr 2010, pp 30-32.

as the voltage sensor. (Its datasheet is available for download at **www.national.com/ds/LM/LM723.pdf**.) A 12.6 V transformer steps the input voltage down and D1 and the 100 μF capacitor form a half-wave rectifier to power the circuit. The voltage threshold is set by the adjustable voltage divider labeled SET VOLTAGE and connected to pin 5 of the IC. The 723 uses a 7.15 V internal reference to control a single *error amplifier* that takes the place of the comparators in the MC3423. The circuit configures the 723 so that if the input voltage exceeds the threshold, the V_C output (pin 11) conducts current to ground.

Instead of an SCR placing a short circuit across power supply terminals, the ac overvoltage protector uses an optically isolated TRIAC to trip the GFCI circuit breaker. (Opto-isolators are described in Experiment #14.) Note the connection between the HOT terminal of the breaker, through the TRIAC, through a 12 kΩ resistor that limits current to about 10 mA to safety ground. When the TRIAC conducts (when the V_C output of the 723 sinks current to ground) the GFCI senses the resulting imbalance between HOT and NEUTRAL currents and trips, removing power from the overvoltage protection circuit and anything else connected to its LOAD terminals. (GFCI breakers are designed to trip at fault currents of 5 mA.)

As the *Handbook* notes, you'll need one overvoltage protector circuit and GFCI for each ac-powered circuit. Be extra careful when wiring the TRIAC and be sure to follow the regular color code of black for HOT and white for NEUTRAL — safety first!

Experiment #80 — Battery Capacity

What makes a *good* battery? The most common answer is battery capacity, or as it is commonly evaluated, battery *life*. What battery life really means is *the time over which a battery can operate a specific piece of equipment in a specific manner*. The life of a battery can vary greatly, depending on how the equipment is used, so battery lifetime in hours or days isn't really very useful. What determines lifetime is battery capacity.

The units of battery capacity are in amp-hours, abbreviated Ah and represented by the letter C in equations. This is a measure of how much energy the battery stores, similar to watt-hours or kilowatt-hours that measure energy in your home. Large batteries are rated in amp-hours while smaller batteries, like AA or coin cells are rated in milliamp-hours or mAh.

Energy and Charge

There is something a little strange about Ah. Amperes (coulombs / second) times hours (3600 seconds / hour) yields coulombs, not joules. How do you get energy from charge? Amp-hours is really a measure of charge. It tells you how many electrons are available to deliver energy from the battery as they flow through your circuit.

The energy carried by any one of those electrons is determined by the electro-motive force (EMF) that pushes it through the circuit. You can measure the EMF as the open circuit battery terminal voltage, V_{OC}, with no load attached. For example, a high-impedance voltmeter attached to the terminals of an alkaline cell might read 1.55V. The current drawn by the meter is very small and its effects can be neglected. That means there is 1.55 V of EMF available to push the electron from wherever it resides in the battery, out through the negative terminal, through the external circuit, and back into the battery. (Actually, that specific electron may never make it all the way around the circuit. Even at 1 A of current flow through a 20 gauge wire, individual electrons are only moving at about ⅕ mm/s.)

Since Ah or mAh gives you charge, the number of available electrons, to find the total amount of energy available, multiply V_{OC} by the Ah rating by 3600 sec / hr. A typical AA battery is thus rated to store a theoretical maximum of 1.55 V × 2 Ah × 3600 s / hr = 11,160. That's a lot of energy and a good reason to respect batteries!

Chemistry Types

Where does the 1.55 V come from? Why not 1.2 or 1.8 V? The battery's open circuit voltage is determined by its internal chemistry, meaning the types of chemicals stored in the battery, and the chemical reaction that occurs as electrons are allowed to flow between the battery terminals. Due to the differing affinity for electrons by the atoms making up the chemicals, electrons will flow from one atom to another if there is a path between the atoms. The media that makes up the path is called the *electrolyte* and it is usually a liquid, paste or gel.

The voltage that drives the electrons from one atom to another is called the *electropotential* and it depends on the exact chemicals used. Table 1 lists the chemistry of several common battery types and the nominal voltage you can expect from a fully charged battery supplying a light load. Table 2 compares the capacities of different types of battery chemistry and cell sizes.

Discharge Curves

Battery capacity is a useful number, but it does not represent the more complex behavior of a battery in actual use. The calculation of stored energy assumes that battery voltage is constant until the battery is discharged. Battery voltage is not constant, however; rather, it slowly drops as more and more energy is delivered because the internal resistance of the battery increases as the chemicals supplying the electrons are used up. Battery voltage

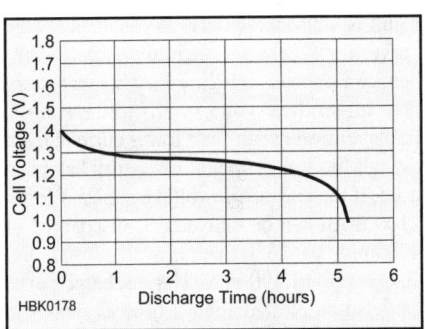

Figure 2 — Typical NiMH battery discharge characteristic. The load on the battery must also be specified for actual discharge curves.

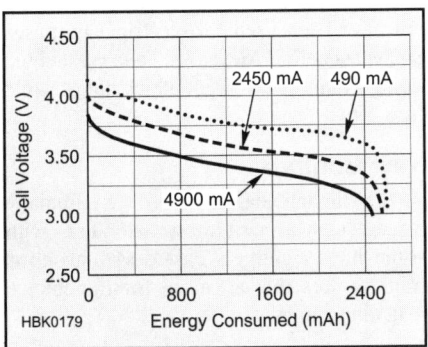

Figure 1 — Typical Li-ion battery discharge curves at three discharge rates.

Table 1
Battery Types

Chemistry	Nominal Voltage (V)
NiCd	1.2
NiMH	1.2
Carbon zinc	1.5
Alkaline	1.5
Lead acid	2.1
Li-ion	3.5-3.7

Table 2
Typical Battery Capacity

Cell Size	Chemistry	Nominal Capacity (mAh)
AAA	Alkaline	1000-1300
AAA	NiCd	250-350
AAA	NiMH	800-1200
AA	Alkaline	2000-3000
AA	NiCd	600-1000
AA	NiMH	1700-2700
C	Alkaline	6000-8000
C	NiCd	2000-3500
C	NiMH	3000-5500
D	Alkaline	12000-20000
D	NiCd	5000
D	NiMH	3000-12000
9 V	Alkaline	500-650
9 V	NiCd	120
9 V	NiMH	200-350

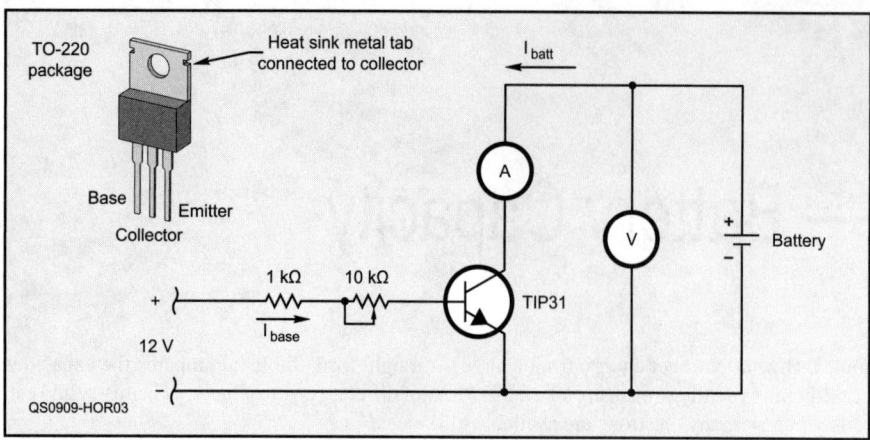

Figure 3 — This circuit draws a constant value of battery current, controlled by the 10 kΩ potentiometer.

eventually drops to a level at which the equipment it is supplying no longer works and the battery is considered dead. Even though there may be additional energy left in the battery, it can't be delivered at a high enough voltage to operate the equipment. This is the battery's end-of-life voltage.

If the battery's terminal voltage under load is plotted versus the amount of energy used, the result is a discharge curve. Typical discharge curves for a rechargeable lithium ion (Li-ion) battery are shown in Figure 1. At higher battery current, terminal voltage falls to the end-of-life level faster than for a lower current, even though the same amount of energy has been used. If the end-of-life voltage is 3.25 V, only 2100 mAh can be delivered if discharged at 4900 mA, but 2550 mAh is available at a discharge rate of 490 mA. The discharge curve can also be plotted against time as shown in Figure 2. Different types of batteries can be compared by drawing identical currents from them and comparing the times at which they reach their end of life voltage.

Battery Up!

We're going to do two battery test experiments. The first will illustrate how a specific type of battery performs at different current levels. The second will compare different types of batteries. Start by acquiring a fresh trio of alkaline AA cells to sacrifice in the name of science. To control current through the batteries, connect the circuit in Figure 3 using a battery holder or clip leads and a rubber band to make the battery connections. Use a pair of locking pliers as a heat sink for the transistor as in Experiment #381.[1] Set the meter reading current to a scale appropriate for measuring about 100 mA.

Measure and record battery voltage. Note the time, set the 10 kΩ pot to maximum resistance, and apply power. Reduce the pot's resistance until current is about 200 mA — you can fine tune with the power supply voltage if the adjustment is too sensitive. Adjusting the TIP31 base current to keep battery current approximately constant, record battery voltage and current once per hour until battery voltage drops to 1.0 V — our end of life value. For a 2000 mAh battery, it should take no more than

[1]Previous Hands-On Radio columns and a complete parts list for all experiments are available to ARRL members at www.arrl.org/hands-on-radio.

2000 mAh / 200 mA = 10 hr. (You can suspend the test by turning off the power supply.) Graph the results as in Figure 2.

Reset the pot to maximum resistance and swap in a fresh battery. As before, note the battery voltage, then adjust base current until the battery current is 500 mA and record battery voltage every 30 minutes or so until battery voltage reaches 1.0 V. Repeat with a battery current of 1 A. You should now have three discharge curves that look somewhat like those in Figure 2.

Why not just hook up a fixed value resistor and measure voltage versus time? Because battery voltage changes as the battery is discharged, the amount of battery current would also change according to Ohm's law. A constant current is a better and more repeatable test.

Now find some different types of AA batteries, such as carbon zinc, NiCd, and NiMH and repeat the 500 mA current test. Compare the discharge curves. Which battery type is better?

Parts List

- Transistor, TIP31 or equivalent.
- Potentiometer, 10 kΩ, preferably linear taper.
- Resistor, 1 kΩ, ¼ W.
- Batteries, three each, size AA — alkaline, carbon zinc, NiCd and NiMH.

Recommended Reading

Here's a treat for readers interested in electro history. Check out the Wikipedia (www.en.wikipedia.org) entry on the *Baghdad Battery*, an ancient artifact that may or may not have been used for electrical purposes 800 years before Alexander Volta created the first galvanic pile and got a kick out of some frog legs.

Next Month

It's still antenna season out there in many places, so let's revisit transmission lines with some handy gadgets called synchronous transformers that are used for impedance matching chores.

Experiment #67 — The Return of the Kit

While happily ambling the aisles of Hara Arena and plodding the pathways of the outside flea market of the Dayton Hamvention, I realized that a renaissance had occurred — the return of the kit. Vendors of kits or sellers of radio stuff built from kits were everywhere! This certainly runs counter to the perception that "hams aren't building." To my eye, there have never been more kits available in such variety. Hands-On Radio readers can feast on that cornucopia!

Before we start, I must report that the *Dual Function Generator* and *Spectrum View* software used in the past few experiments are *not* free with *The 2008 ARRL Handbook* as I indicated. They were included as part of a special introductory offer. Nevertheless, those programs remain available as part of the *Software Library for Hams 2.0* CD from the ARRL — still a bargain at $20, but not quite free as reported.[1]

Kit History

Hams who were licensed before the mid-1980s will recall the kit vendor that dominated amateur building for 30 years: Heathkit (**www.heathkit.com**) is still in business, providing education and training for a variety of technologies. It is the familiar green and gray cabinets of amateur rigs, test equipment and home electronics that had the biggest impact on hams, however. A quick browse through Heathkit Museum Web site (**www.heathkitmuseum.com**) will provide glimpses of the colorfully named gear that was the core of many ham stations. How could you not be proud and confident as the owner of a *Marauder* transmitter? Many Heathkit rigs, amplifiers and accessories are still making contacts today. When they need service, a Heathkit meter or generator may even be used to set them right again. But enough nostalgia — what about today's kits?

Kit Building Today

Heathkits were known for their detailed,

[1] *Dual Function Generator* and *SpectrumView* are part of the *ARRL Software Library for Hams, Vol 2.0*. Available from your ARRL dealer or the ARRL Bookstore, ARRL order no. 9825. Telephone 860-594-0355, or toll-free in the US 888-277-5289; **www.arrl.org/shop/**; **pubsales@arrl.org**.

Figure 1 — FAR Circuits, shown here in their Dayton Hamvention tent, stocks printed-circuit boards for many of the projects described in *QST* and other amateur magazines.

lavishly illustrated, step-by-step instruction manuals. In fact, the Heath motto was, "We won't let you fail!" Their customer service was exceptional. However, that level of support became commercially unsustainable and today's kit builder shouldn't expect it. Nevertheless, the manuals that come with current kits are usually sufficient for the homebrewer to be successful.

Kits available today for ham radio range from electronic gadgets that cost a few dollars to sophisticated transceivers. The Elecraft K2 HF transceiver (**www.elecraft.com**) is the most sophisticated amateur kit ever. (The Elecraft K3 is a modular, no-solder kit and actually requires less assembly than the K2.) In between lies a profusion of kits for test equipment, accessories and radios. Let's start with the simplest form and work our way up.

Kits from Magazine Articles

Construction articles in ARRL publications such as *QST* and in other magazines are often based on a printed-circuit board (PCB). Most kit builders are not equipped to make their own PCBs and would have to resort to other less-suitable building techniques if the PCB were not available for purchase. Enter FAR Circuits (**www.farcircuits.net**). About half of their traveling inventory can be seen in Figure 1, taken in the Hamvention flea market area. These aren't complete kits, which would have all of the electronic parts and an enclosure. What you get from FAR is a copy of the original article and a single or double-sided PCB. You add your own parts to complete the kit. This is a very cost-effective way to build something you've seen in a magazine if you have a stock of parts or are willing to order them.

If FAR doesn't carry a circuit-board for the article, it's possible that the author has a small stock. Check the notes at the end of the article to see if kits of parts or a PCB is available. Unless the kit is very popular, however, the author is unlikely to maintain a stock of PCBs for long and once sold out they are rarely available.

Test Equipment Kits

Building one's own test equipment is a great way both to learn about electronics and to save a lot of money compared to purchasing lab-quality gear. For example, a used commercial function generator will cost a minimum of $100 from a dealer (plus shipping), but the simple sine-square wave generator kit from Ocean State Electronics (www.oselectronics.com) will only set you back $40 and you get to practice surface-mount soldering, too.

You will find that kits are available for many mid level test and measurement functions. Don't expect to find voltmeter kits, since fully assembled meters are available for under $20. Nor will you find expensive high performance oscilloscopes or signal generators in kit form. But there are plenty of kits for power supplies, voltage regulators, component testers, frequency counters and others at reasonable prices.

Several vendors also sell training kits for soldering and surface-mount techniques. This is a great way for a beginner or student to get started. For example, the Elenco AK-100 "Learn to Solder" kit, available from many vendors, costs less than $20, and comes with solder, soldering iron and wire cutters that can be used to build other kits.

Transceiver Kits

The thrill of building your own radio and making contacts with something built yourself is unmatched in Amateur Radio. Your editor's first radio was a Heathkit HW-16, painstakingly assembled at his bedroom workbench and helpfully troubleshot by WBØDYU. Today, there are dozens of receiver, transmitter and transceiver kits.

For your first attempt at building your own radio, it's best to start with a simple CW QRP transceiver kit. The Ten-Tec 1340 (radio.tentec.com/kits/transceivers/1300) comes with an extensive instruction book and a customer service department for help. As you become more skilled at building, you may want to try your hand at building a radio from separate receiver and transmitter modules. A PCB for the *QST* classic Tuna Tin 2 transmitter by the late Doug DeMaw, W1FB, is available from FAR Circuits and a complete kit (along with several amusing variations) from the Maine QRP Club (www.qrpme.com).[2]

Advanced builders will be comfortable tackling one of the Hendricks QRP Kits transceivers as shown in Figure 2 (www.qrpkits.com), a Northern California QRP Club (www.norcalqrp.org) design, or one of the Elecraft kits mentioned previously. You will find the QRP clubs to be hotbeds of building activity — probably the most active builders in Amateur Radio!

Kit Building

Your kit building adventures will be suc-

[2]D. DeMaw, W1CER, "Build A Tuna Tin 2," *QST*, May 1976, pp 14-16.

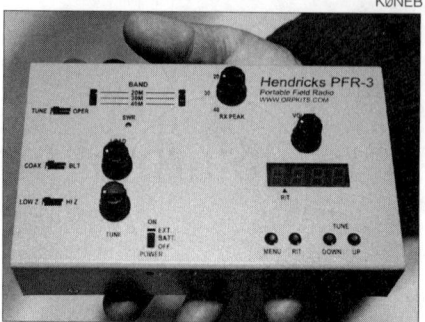

Figure 2 — Advanced kit builders can take on projects such as this 40-30-20 meter, 5 W, QRP CW rig from Hendricks QRP Kits.

Figure 3 — The Elenco SM200K Surface-Mount Soldering Kit is a good introduction to SMT building techniques. A magnifying glass and a pair of tweezers are the only special tools required.

cessful if you are careful and follow instructions. You also need to have the basic skills needed for the kit. For your first few kits, try one that uses *through-hole components* with leads that are inserted into PCB holes rather than surface mount devices (SMDs). These are easy to build, with large components and soldering surfaces. A regular soldering iron of 25 to 50 W and 60-40 0.032 inch diameter rosin-core solder will do nicely. For tools you'll only need a small pair of needle-nosed pliers and wire cutters.

If you haven't had any experience with *surface-mount technology* (SMT) in which the components are soldered directly to the PCB pads without leads, you should try a training kit. The Elenco SM200K "Learn Surface-Mount Soldering" kit shown in Figure 3 is widely available. You will need some kind of magnifier, especially if your vision isn't sharp. Lightweight, head-mounted magnifiers are available at sewing and craft stores for much less than at electronics and tool outlets. Make sure your work surface is brightly illuminated, too.

Once you're ready to build your kit, start by doing a complete check of the parts against the kit's part list. This is the time to find out that you have extra or missing components! Building a kit is a perfect time to start learning the resistor color code and parts marking conventions. (*The ARRL Handbook* has a section on component marking.[3]) Sort the parts out into groups so that you can find them easily. Egg cartons and muffin tins are simple and inexpensive for use as parts holders.

Most kits will have step-by-step instructions and it can't be emphasized too much that it is important to follow them. Read through the whole manual first to get a feel for how the project will go. Check off each completed step so that if you're interrupted you won't forget where you stopped. (It's not always obvious.) If the instruction doesn't seem to correspond with the kit's layout or available components, stop! Don't proceed until you have figured it out. Troubleshooting is a lot harder than doing it right the first time.

As you build the kit, make notes in your electronics notebook. You do keep a notebook, don't you? A simple, spiral-bound notebook should be your constant companion at the workbench. Record assembly notes, test results, any discrepancies between the instructions and the kit, and any good ideas that occur to you as you build. If you have to repair or decide to modify the kit later, your notes may prove invaluable.

Finding Kits

An excellent way to find kit vendors is to use the ARRL Technical Information Service (www.arrl.org/tis). Click TISFIND and then enter "kits" into the SEARCH ON-LINE FOR: window. You'll find dozens of kit vendors. Paper catalogs are available from vendors such as Ramsey Electronics (www.ramseyelectronics.com), Jameco (www.jameco.com) and Vectronics (www.vectronics.com). If you are looking for a specific type of kit, an Internet search engine is probably a better method.

Recommended Reading

Your job is to use the ARRL TIS and browse through some of the kit vendor Web sites. Take a look at the gear in your shack and on your workbench, then review the Web sites to see if any kits might fill one of those smaller holes.

Next Month

We'll return to poring over the workbench next month as we start a multi-part experiment featuring phase-locked loops (PLLs). Ready, aim, lock!

[3]*The ARRL Handbook for Radio Communications*, 2008 Edition. Available from your ARRL dealer or the ARRL Bookstore, ARRL order no. 1018. Telephone 860-594-0355, or toll-free in the US 888-277-5289; www.arrl.org/shop/; pubsales@arrl.org.

Experiment #71 — Circuit Layout

Success with radio electronics has a lot to do with knowing how to construct a circuit. As frequency increases, this sort of "radio know-how" becomes increasingly important. This month's experiment will present an example of the effects of circuit layout on its performance at different frequencies. (This experiment is based on a column by Dave Kelley, ND3K, professor of Electrical Engineering at Bucknell University and a frequent reader of Hands-On Radio.)[1]

The Basic Filter

The circuit we're going to build — in three different ways — is a low-pass filter with a cutoff frequency of around 28 MHz. The schematic is shown in Figure 1. The filter is designed to be used with a 50 Ω source and load.

Figure 1 also shows (in red) the *parasitic reactances* associated with each of the *ideal* components. For example, the *interturn capacitance* of L1 creates C_{w1}, typically a few pF for an airwound coil of the size you'll wind. The *lead inductance* of C1 appears as L_{s1} and is about 12 nH/inch for 20 gauge wire. More parasitic capacitance appears as C_{io}, the capacitance between the input and output connections. The size and placement of the components affects each of these parasitic values.

Sans Solder

The first version of the circuit is built on a solderless breadboard, as seen in Figure 2. To wind the inductors, start with 16 inches of solid 20 gauge wire. (Scrape the enamel coating off the ends of the wire with a knife or file.) On a ½ inch form (such as a drill bit or dowel), wind nine turns over a length of about ⅞ inch. Leave about ¾ inch of wire on each end to make leads.

Plug the inductors into the breadboard at an angle to each other. If the inductors are placed end-to-end in a straight line, their magnetic fields will *couple* and upset circuit performance quite a bit. Connect the capacitors from the inductors to the breadboard's ground rail. It's not necessary to minimize lead length or arrangement — yet.

Now comes the fun part — measuring how the circuit responds at different frequencies. As a signal source, you'll need a signal generator or an SWR analyzer, such as an MFJ-259. Any stable source that can output more than a few hundred millivolts at frequencies up to 50 MHz or higher will do. You'll also need an oscilloscope with a bandwidth of at least 20 MHz. To connect the signal source to the circuit, find a 2 to 3 foot piece of coaxial cable with a connector on one end. Attach a pair of alligator clips to the end without the connector, or just solder the braid and center conductor to short pieces of wire that can be plugged into the breadboard.

[1]D. Kelley, "The Good, the Bad, and the Ugly: Demonstrating Basic Circuit Layout and Measurement Concepts," *IEEE Antennas and Propagation Magazine*; Vol 49, No 6, Dec 2007, p 153.

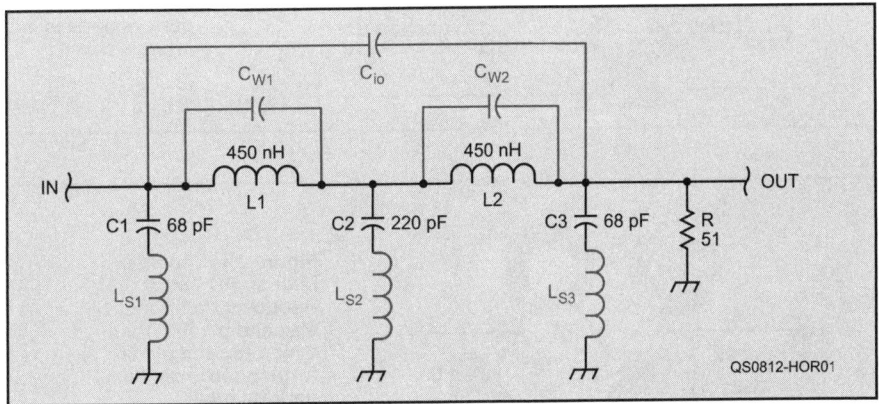

Figure 1 — Performance of the two-section low-pass filter is affected by parasitic reactances (shown in red).

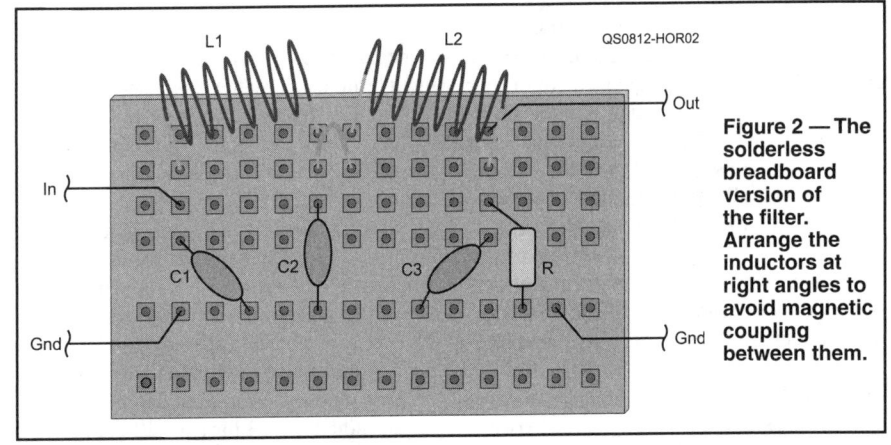

Figure 2 — The solderless breadboard version of the filter. Arrange the inductors at right angles to avoid magnetic coupling between them.

Set the source to about 1 MHz and connect it to the filter. Use the 'scope to measure the filter input and output signals in V_{P-P}. Increase the source frequency until the output voltage, $V_{OUT} = 0.707\ V_{IN}$. This is the filter's *cutoff frequency*, f_0, and it should be somewhere above 20 MHz and less than 35 MHz. If you have a 'scope with a bandwidth of 20 MHz or less, you can still use it if it can provide an indication, because we are taking *relative* measurements between input and output.

Now sweep the source from 1 MHz to 100 MHz or as high as the source will go. Watch the output signal amplitude. Is it steady through the *passband* below f_0 or does it vary? Does the filter's *amplitude response* (the ratio of output to input) de-

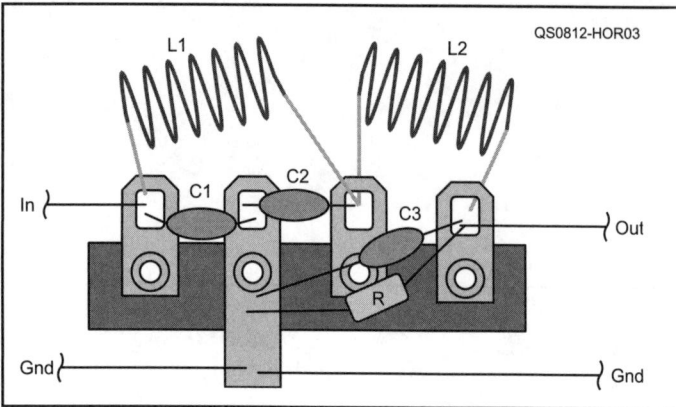

Figure 3 — The filter built on a terminal strip. By reducing lead length and capacitance between terminals, high-frequency performance is improved.

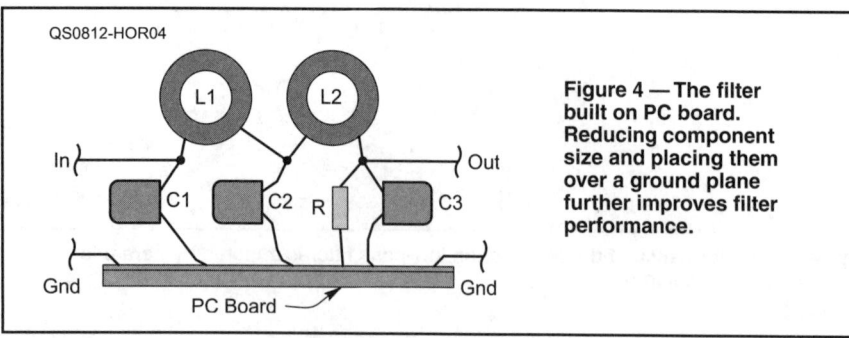

Figure 4 — The filter built on PC board. Reducing component size and placing them over a ground plane further improves filter performance.

crease smoothly (*roll-off*) up to 3 × f_0 or are there peaks and notches and abrupt or erratic changes? Increase the frequency still further and see if the filter's attenuation levels off or even begins to decrease. Graph the filter's frequency response in dB by downloading the spreadsheet available on the Hands-On Radio Web site under Experiment #18.[2]

You'll probably see a pretty uneven response at and above f_0 due to the effects of the parasitic reactances. These effects are quite undesirable and hard to predict. Rearrange the components to see changes in the frequency response.

The parasitic capacitance present between the connector strips of the breadboard affect filter performance above a few MHz, as well. All in all, it's easy to see why this construction technique is not recommended for radio frequency projects.

Terminal Strip

The parasitics inherent to the breadboard are avoided in a style of construction popular before printed-circuit boards became the norm. The *terminal strip* is a good way to connect simple circuits, particularly for parts with wire leads.

Rebuild the filter on a terminal strip as shown in Figure 3. The terminal strip can be screwed to a scrap of wood, or soldered to a piece of printed circuit board scrap. Any solid mounting method will work.

Reattach the source by soldering the coax center conductor and shield directly to the input terminals. Use the spreadsheet to graph the circuit's frequency response. Compared to the frequency response of the breadboard filter, the terminal strip version will show a much smoother frequency response with higher attenuation at high frequencies. This is because the component leads are shorter (less inductance) and there is less capacitance between adjacent contacts. Additionally, all the grounds are connected at a single point, minimizing inductance in ground connections.

Now replace the air-wound inductors with toroid inductors: 13 turns of 20 gauge wire spaced evenly around a T-30-10 powdered-iron core. (*The ARRL Handbook* shows how to wind a toroid inductor.)[3] About 1 foot of enameled wire will do the job. (Don't forget to remove the enamel from each end of the winding!)

Repeat the frequency response measurements of the filter and note any changes in the response. With the toroids, you'll probably see even smoother pass-band and roll-off characteristics. The *ultimate attenuation* at higher frequencies will likely be higher since there is less input-to-output parasitic capacitance.

PC Mount

Finally, obtain a scrap of printed-circuit board a couple of inches square. Clean the surface with steel wool or a scrubbing pad and build the filter as shown in Figure 4, using the toroid inductors. Reattach the source by soldering the coax braid to the circuit board and the center conductor directly to the filter input. Repeat your frequency response measurements.

Compare the sets of measurements you've made with all three types of construction and the two types of inductors. As the effects of circuit construction and component parasitics are reduced, the response of the filter becomes closer to the ideal of a flat passband, smooth roll-off, and even attenuation at high frequencies. When you build your next RF project, remember the effects of construction!

Parts List

- Terminal strip (3 terminals plus ground terminal).
- Capacitors. Two each 68 pF and one 220 pF silver mica.
- Powdered iron toroid cores; two type T-30-10.
- Resistor, 51 Ω, carbon composition or film.
- Enameled wire, 6 feet of 20 gauge.
- PC board scrap.

If you have been wondering where to get parts for the experiments in the Hands-On Radio columns, the ARRL (**www.arrl.org/shop**) and Kanga US (**www.kangaus.com**) have teamed up to provide a kit with all of the electronic parts for experiments #1 through #61. For columns #62 and later, check the parts list spreadsheet on the Hands-On Radio Web site.

Recommended Reading

For more information on how components behave at high frequencies, *The ARRL Handbook* is a good place to start. Chapters 5 and 6 provide good information on the types of components and their RF characteristics. *Experimental Methods in RF Design* provides lots of examples of how to build RF circuits properly.[4]

Next Month

While most hams are familiar with the idea of standing wave ratio (SWR), not many delve into how it is calculated and fewer are familiar with the measurement preferred by professional RF designers — return loss. Next month, we'll reflect on those and other concepts.

---

[2]All Hands-On Radio experiments and an extensive FAQ are available at **www.arrl.org/tis/info/HTML/Hands-On-Radio**.

[3]*The ARRL Handbook for Radio Communications*, 2009 Edition. Available from your ARRL dealer or the ARRL Bookstore, ARRL order no. 0261 (Hardcover 0292). Telephone 860-594-0355, or toll-free in the US 888-277-5289; **www.arrl.org/shop/**; **pubsales@arrl.org**

[4]W. Hayward, W7ZOI, R. Campbell, KK7B, and B. Larkin, W7PUA, *Experimental Methods in RF Design*. Available from your ARRL dealer or the ARRL Bookstore, ARRL order no. 8799. Telephone 860-594-0355, or toll-free in the US 888-277-5289; **www.arrl.org/shop/**; **pubsales@arrl.org**

Parts List

The following components are used for the experiments shown in this book, updated through Experiment 121

	Max Qty	Experiment	Notes
1/4 WATT RESISTORS (All values are ohms)			
10	2	2, 13, 15, 16, 46	
27, non-inductive	1	21	
39	2	13	
47	1	10	
51	1	13, 26, 71	
68	2	45	
75	2	13	
100	several	2, 13, 15, 16, 24, 29, 37, 46, 48, 74, 77, 99, 100, 103	Special batch requirements for Experiment 100.
150, non-inductive	1	21	
220	1	48, 77	
220, non-inductive, such as carbon composition	2	81	
270	1	1, 28, 45, 71, 76	
300	1	79	
330	1	6, 28, 29, 38	
390	1	48	
470	7	8, 28, 29, 37, 69, 77	
510	1	9, 55	
910	1	71	
1k	10	2, 3, 8, 12, 28, 29, 31, 33, 36, 37, 39, 40, 55, 73, 77, 80, 101, 109, 111, 121	
1.2k	1	2	
1.5k	1	1, 28, 77, 93	
2.2k	2	4, 8, 11, 14, 77, 78	
2.7k	1	27	
3.3k	1	93	
3.9k	1	3, 6	
4.7k	2	5, 8, 9, 45, 54, 68, 77, 93, 97, 109	
5.1k	1	6	
6.8k	1	1, 28, 55, 77	
10k	4	2, 3, 4, 7, 17, 18, 30, 31, 40, 41, 46, 54, 69, 73, 99, 109	
11k	1	109	
15k	1	2	
16k	4	42	
22k	1	2, 4, 11, 41, 45	
27k	1	11, 99	
39k	1	1, 5, 28, 41, 77, 93	
47k	1	4, 31, 103	
56k	1	41	
62k	1	5	
75k	1	41	
91k	1	5	
100k	1	99	
120k	1	39	
220k	1	11	
270k	1	38	
330k	1	97	
470k	1	31	
680k	1	38	
1M	several	43, 55, 99	
4.7M	1	97	
POWER RESISTORS (All values are ohms)			
5, 5W	1	24	Resistors with larger power dissipations can be used. Also okay to combine smaller units in series or parallel.
50, 10W	2	9(2)	
100, 1W	1	10, 32	

POTENTIOMETERS (All values are ohms)

100	1	28, 42	
1k	1	3, 4, 14	
5k, panel-mount	1	78	
5k, PCB mount	1	109	Bourns series 3386F.
10k	2	3, 8, 11, 12, 14, 54, 55, 68, 69, 80	
20k	1	9	
50k	1	10, 103	
100k	1	1, 2, 38, 41, 92	
1M	1	17, 39, 40, 109	

CAPACITORS

50 to 200 pF fixed-value or air-variable	1	21, 43	
500 pF variable	1	102	Maximum value not critical.
2.7 pF, ceramic	1	43	
68 pF, silver mica	2	71	
100 pF, polystyrene or ceramic	2	45, 46, 97	
200 pF, polystyrene or silver mica	1	43, 46	
220 pF, silver mica	1	71	
270 pF, film or ceramic	1	97	
390 pF, polystyrene or silver mica	2	46	
470 pF, silver mica	1	79	
470 pF, film or ceramic	1	97	
1 nF, 50V ceramic	3	17, 37, 74, 93, 103, 121	
1.2 nF (1200 pF), silver mica	1	79	
1.5 nF, film or ceramic	1	18	
2.7 nF, film or ceramic	1	4	
5.6 nF, film or ceramic	1	4	
10 nF, film or ceramic	4	5, 25, 42, 45, 46, 71, 93, 109, 121	
16 nF, film or ceramic	2	41	
22 nF, film or ceramic	1	68	
33 nF, film or ceramic	2	4	
56 nF, film or ceramic	1	4	
0.1 µF, 50 V ceramic	6	8, 12, 17, 24, 26, 27, 38, 39, 40, 43, 68, 69, 71, 99, 121	
1 µF, 25 V electrolytic	4	7, 40, 69, 93, 121	
1 µF, 35 V tantalum	1	6, 8	
10 µF, 25 V electrolytic	3	1, 2, 3, 4, 5, 7, 17, 28, 31, 68, 77	
33 µF, 25 V electrolytic	1	109	
47 µF, 15 V tantalum	1	15, 16	Changed from 100 µF in original article.
100 µF, 35 V electrolytic	2	10, 15, 16, 30, 31	Exp 15/16 input capacitor changed to electrolytic.
4700 µF, 15V electrolytic	2	10, 38	

INDUCTORS AND CORES

Component	Qty	Page(s)	Notes
2 µH	1	43	Hand wound, 8 turns, 1½" dia, 1" long.
100 µH, 1 amp	1	15, 16, 118	
22 µH	1	46	Should be rated for 50 mA.
1 mH	1	26, 74	
FT37-43 (ferrite) or equivalent	1	45	
FT240-61 (ferrite) or equivalent	1	48	FT240 balun kit; Amidon Associates.
T50-6 (powdered iron) or equivalent	1	46	Original shopping list had FT50-6.
T30-10 (powdered iron) or equivalent	2	71	
Ferrite snap-on split core	1	103	RadioShack 273-105 or type #31 or #43 mix.
T-157-2 (powdered iron) or equivalent	1	103	

TRANSFORMERS

Component	Qty	Page(s)
115 V Pri, 12.6 V Sec, RadioShack 273-1365 or similar	1	10, 33
1000 ohm - 8 ohm audio transformer, RadioShack 273-1380 or similar	1	33

TRANSISTORS

Component	Qty	Page(s)	Notes
2N3904	2	1, 2, 19, 20, 25, 28, 31, 40, 45, 46, 77	
2N3906	1	19, 25, 40, 109	
2N4401	1	8	
2N4416	1	43	
2N7000	1	39, 40	
J310 or MPF102 JFET	2	99	
TIP31	1	9, 80	
TIP42	1	38	
IRF510 MOSFET	1	9, 12, 15, 16	
SCR, 100 V, 8 A Digi-Key MCR218-004OS-ND	1	10	Original RadioShack part number no longer available.

DIODES AND RECTIFIERS

Component	Qty	Page(s)
1N34	1	76, 102, 103
1N4001	1	71, 76, 77, 109, 121
1N4148	12	6, 7, 14, 25, 26, 30, 39, 40, 42, 54, 55, 76, 97
1N4732A	1	76
1N4733A	1	6, 38
1N5234B	1	109
1N5817	1	76
1N5819	1	15, 16

INTEGRATED CIRCUITS

Component	Qty	Page(s)
78L-05, -08, -12, or -15 voltage regulator	1	25
4N35 Optocoupler	1	14
741 op amp	2	3, 4, 8, 17, 19, 42, 54, 55, 73
555 timer	2	5, 69, 93
565 Phase-locked Loop	1	68
LM317 op amp	1	8, 71
LM311 Comparator	1	11, 38
LM324 quad op amp	1	17
LM393 dual comparator	1	109
CD4013 dual-D -type flip-flop	1	101
CD4027 dual JK -type flip-flop	1	97
CD4028 BCD-to-		

Item	Qty	Page(s)	Notes
Decimal Decoder	1	37	
CD4029 Up/Down Counter	1	36	
CD4511 BCD-to-7-Segment Decoder	1	37	
74HC4001 Quad NOR gate	1	97	
74HC4040 12-stage ripple counter	1	97	

LIGHTS AND DISPLAYS

Item	Qty	Page(s)	Notes
Bulb, 12 V RadioShack 900-2665 or similar	1	10	
LED, Red RadioShack 276-026 or similar	5	11, 36, 39, 40, 93, 101, 109	
LED, Green RadioShack 276-022 or similar	1	39	
7-Seg Display, Common-Cathode, Jameco 17187 or equiv.	1	37	

MISCELLANEOUS

Item	Qty	Page(s)	Notes
40-meter crystal in HC-8 or HC-16 holder	1	46	7030-7045 kHz recommended
#12 AWG solid, bare wire	6 feet	96	Use scrap house wiring and remove insulation.
#14 AWG solid, enameled wire	14 feet	48	FT240 balun kit from Amidon Associates includes wire.
#20 or #22 AWG solid wire	16 feet	21, 71, 79, 102	Total wire length for #46 & #46 and #71.
#24 to #28 AWG solid wire	3 feet	45, 46, 47	Total wire length for #46 & #46; use scrap for #47.
PL-259 connector	2	116	
SO-239 connector	1	21, 48	
UHF double-male adapter	1	21	
UHF tee connector	1	103, 116	
Plastic film can or pill bottle	1	22	
SPST or SPDT switch	1	22	
SPST momentary NO pushbutton	1	109	Omron 6mm for the PCB layout.
RG-58 coaxial cable	25 feet	22, 111, 116	
RG-59 coaxial cable	20 feet	81	
Wooden spring-loaded clothespins	23		
Scraps of 2x4 wood	23		
Thick cardboard	23		
¼-inch diameter dowel	23		
Glue	23		
Wood screws	23		
#6-32 machine screws	23		
#6-32 nuts	23		
#6 flat washers	23		
DPST switch	2	30	
TO-220 heat sink	1	24	
Mounting hardware for TO-220 heatsink	1 set	24	
1.5 V AAA or 9 V battery	1	32	
Terminal strip, 3 terminals plus ground	1	71	
Crimp-on ring terminals, #10 holes, 16-14 AWG wire	2	116	
PC board scrap	1	70, 71	
5A inline fuse and holder	1	71	

Item	Qty	Ref	Notes
Mechanical quadrature encoder	1	101	Bourns PEC16 series recommended.
Microammeter	1	102, 103	50 to 250 µA max value.
SPST Relay, 12 V coil	1	109	Potter & Brumfield T90N1D12-12 for the PCB layout.
V8ZA05P 6V MOV	1	121	
1.5KE8.2CA 8.2V bipolar TVS	1	121	

TEST EQUIPMENT ITEMS

Item	Qty	Ref
⅛" stereo phone plug	1	64
⅛" Mono phone plug-to-stereo phone jack adapter	1	64
3' male-to-male stereo audio jumper cable	1	64
⅛" stereo headphone splitter	1	64

Quick References

Resistor Color Code

Color	Value	Multiplier	Tolerance
Black	0	x 1 (10^0)	
Brown	1	x 10 (10^1)	1%
Red	2	x 100 (10^2)	2%
Orange	3	x 1000 (10^3)	
Yellow	4	x 10,000 (10^4)	
Green	5	x 100,000 (10^5)	0.5%
Blue	6	x 1,000,000 (10^6)	0.25%
Violet	7	x 10,000,000 (10^7)	0.1%
Gray	8	x 100,000,000 (10^8)	0.05%
White	9	x 1,000,000,000 (10^9)	
Silver		x 0.1 (10^{-1})	5%
Gold		x 0.01 (10^{-2})	10%
No color			20%

For more information on resistor markings:
www.radio-electronics.com/info/data/resistor/resistors-colour-code-coding-chart.php

Capacitor Value Markings

###L (Three numbers and a letter)
Numbers 1 and 2 are value digits
Number 3 is a multiplier; 0 - x1, 1 - x10, 2 - x100, 3 - x1000, 4 - x10,000
Letter denotes tolerance; J - 5%, K - 10%, L - 20%

##p or ##n
Numbers 1 and 2 are value digits
p denotes pF, n denotes nF

For more information on capacitor markings: www.radio-electronics.com/info/data/capacitor/capacitor-markings.php

Drill Sizes Commonly Used in Electronics

Size Number	Diameter	Next Largest Fractional Size	Clears Screw Size	For Self-tapping Screw Size
11	0.191"	13/64"		10
19	0.166"	11/64"	8	
21	0.159"	11/64"		10-32
25	0.149"	5/32"		10-24
28	0.140"	9/64"	6	
29	0.136"	9/64"		8-32
33	0.113"	1/8"	4	
36	0.106"	7/64"		6-32
43	0.089"	3/32"		4-40
44	0.086"	3/32"	2	
50	0.070"	5/64"		2-56

Complete drill bit table: en.wikipedia.org/wiki/Drill_and_tap_size_chart

Metric System of Units

Prefix	Symbol	Multiplication Factor
tera	T	10^{18}
giga	G	10^{9}
mega	M	10^{6}
kilo	k	10^{3}
centi	c	10^{-2}
milli	m	10^{-3}
micro	μ	10^{-6}
nano	n	10^{-9}
pico	p	10^{-12}

Dimension Conversions

25.4 mm/inch   0.0393 inch/mm
2.54 cm/inch   0.393 inch/cm
30.48 cm/foot   0.0328 foot/cm
0.305 meter/foot   3.28 foot/meter
0.914 meter/yard   1.094 yard/meter

Voltage Conversions

Sine or square wave   VPEAK-TO-PEAK = 2 x VPEAK
Sine wave   VRMS = 0.707 x VPEAK, VPEAK = 1.414 x VRMS
Square wave   VRMS = VPEAK

Decibel Conversions

Power to Decibels   dB = 10 log10 (Power 1 / Power 2)
Voltage to Decibels   dB = 20 log10 (Voltage 1 / Voltage 2)
Decibels to Power   Power 1 = Power 2 x antilog10 (dB / 10)
Decibels to Voltage   Voltage 1 = Voltage 2 x antilog10 (dB / 20)

The antilog or inverse log function is often labeled log-1 on calculators.

Where to Find Component Data Sheets

Source	Web Site
Datasheet Archive	www.datasheetarchive.com
NTE (manufacturer of cross-reference parts)	www.nteinc.com
Open Directory Project	www.dmoz.org/Science/Technology/Electronics/

Reference/Application_Notes_and_Data_Sheets
Datasheet Café (directory of manufacturer datasheet sites)   www.datasheetcafe.com

Vendors for Electronic Components & Supplies

All Electronics	allelectronics.com
Digi-Key Electronics	www.digikey.com
Jameco	www.jameco.com
Mouser Electronics	www.mouser.com
MCM Electronics	www.mcmelectronics.com
Marlin P Jones & Associates	www.mpja.com
RadioShack	www.radioshack.com
Ramsey Kits	www.ramseyelectronics.com
Tower Electronics	www.pl-259.com
Velleman	www.vellemanusa.com/

Technical Reference and Tutorial Web Sites

All About Circuits	www.allaboutcircuits.com
American Radio Relay League's Technical Information Service	www.arrl.org/technical-information-service
ARRL Hands-On Radio	www.arrl.org/hands-on-radio
Arrick Robotics	www.robotics.com/robots.html
Battery University	www.batteryuniversity.com/
Discover Circuits	www.discovercircuits.com
Op-Amp Electronics	www.opamp-electronics.com
US Navy Electricity and Electronics Training Series	www.phy.davidson.edu/instrumentation/NEETS.htm
Radio-Electronics	www.radio-electronics.com/
Williamson Labs	www.williamson-labs.com

Books and Magazines

Active Filter Cookbook, by Don Lancaster
The ARRL Handbook
The Art of Electronics, by Horowitz and Hill
Circuit Cellar, **www.circuitcellar.com**
Circuitbuilding for Dummies, by Ward Silver NØAX
CMOS Cookbook, by Don Lancaster
Electronic Circuits 1.1 and Electronic Circuits 1.2, by Intellin Organization
Electronic Formulas, Circuits, and Symbols, by Forest Mims
Electronics for Dummies, by McComb and Boysen
Electronic Projects for Dummies, by Boysen and Muir
Tab Electronics Guide to Understanding Electricity and Electronics by Randy Slone
Nuts and Volts Magazine, **www.nutsandvolts.com**
Op-Amp Cookbook, by Walter Jung
Power Supply Cookbook, by Marty Brown
Timer, Op-Amp, and Optoelectronic Circuits, by Forrest Mims
Understanding Basic Electronics, by Larry Wolfgang WR1B

Semiconductor Diode Specifications[†]
Listed numerically by device

Device	Type	Material	Peak Inverse Voltage, PIV (V)	Average Rectified Current Forward (Reverse) IO(A)(IR(A))	Peak Surge Current, I_{FSM} 1 s @ 25°C (A)	Average Forward Voltage, VF (V)
1N34	Signal	Ge	60	8.5 m (15.0 μ)		1.0
1N34A	Signal	Ge	60	5.0 m (30.0 μ)		1.0
1N67A	Signal	Ge	100	4.0 m (5.0 μ)		1.0
1N191	Signal	Ge	90	15.0 m		1.0
1N270	Signal	Ge	80	0.2 (100 μ)		1.0
1N914	Fast Switch	Si	75	75.0 m (25.0 n)	0.5	1.0
1N1183	RFR	Si	50	40 (5 m)	800	1.1
1N1184	RFR	Si	100	40 (5 m)	800	1.1
1N2071	RFR	Si	600	0.75 (10.0 μ)		0.6
1N3666	Signal	Ge	80	0.2 (25.0 μ)		1.0
1N4001	RFR	Si	50	1.0 (0.03 m)		1.1
1N4002	RFR	Si	100	1.0 (0.03 m)		1.1
1N4003	RFR	Si	200	1.0 (0.03 m)		1.1
1N4004	RFR	Si	400	1.0 (0.03 m)		1.1
1N4005	RFR	Si	600	1.0 (0.03 m)		1.1
1N4006	RFR	Si	800	1.0 (0.03 m)		1.1
1N4007	RFR	Si	1000	1.0 (0.03 m)		1.1
1N4148	Signal	Si	75	10.0 m (25.0 n)		1.0
1N4149	Signal	Si	75	10.0 m (25.0 n)		1.0
1N4152	Fast Switch	Si	40	20.0 m (0.05 μ)		0.8
1N4445	Signal	Si	100	0.1 (50.0 n)		1.0
1N5400	RFR	Si	50	3.0 (500 μ)	200	
1N5401	RFR	Si	100	3.0 (500 μ)	200	
1N5402	RFR	Si	200	3.0 (500 μ)	200	
1N5403	RFR	Si	300	3.0 (500 μ)	200	
1N5404	RFR	Si	400	3.0 (500 μ)	200	
1N5405	RFR	Si	500	3.0 (500 μ)	200	
1N5406	RFR	Si	600	3.0 (500 μ)	200	
1N5408	RFR	Si	1000	3.0 (500 μ)	200	
1N5711	Schottky	Si	70	1 m (200 n)	15 m	0.41 @ 1 mA
1N5767	Signal	Si		0.1 (1.0 μ)		1.0
1N5817	Schottky	Si	20	1.0 (1 m)	25	0.75
1N5819	Schottky	Si	40	1.0 (1 m)	25	0.9
1N5821	Schottky	Si	30	3.0		
ECG5863	RFR	Si	600	6	150	0.9
1N6263	Schottky	Si	70	15 m	50 m	0.41 @ 1 mA
5082-2835	Schottky	Si	8	1 m (100 n)	10 m	0.34 @ 1 mA

Si = Silicon; Ge = Germanium; RFR = rectifier, fast recovery.
[†]For package shape, size and pin-connection information see manufacturers' data sheets. Many retail suppliers offer data sheets to buyers free of charge on request. Data books are available from many manufacturers and retailers.

Suggested Small-Signal FETs

Device	Type	Max Diss (mW)	Max V_{DS} (V)	$V_{GS(off)}$ (V)	Min gfs (μS)	Input C (pF)	Max ID (mA)[1]	f_{max} (MHz)	Noise Figure (typ)	Case	Base	Applications
2N4416	N-JFET	300	30	−6	4500	4	−15	450	4 dB @400 MHz	TO-72	1	VHF/UHF amp, mix, osc
2N5484	N-JFET	310	25	−3	2500	5	30	200	4 dB @200 MHz	TO-92	2	VHF/UHF amp, mix, osc
2N5485	N-JFET	310	25	−4	3500	5	30	400	4 dB @400 MHz	TO-92	2	VHF/UHF amp, mix, osc
2N5486	N-JFET	360	25	−2	5500	5	15	400	4 dB @400 MHz	TO-92	2	VHF/UHF amp. mix, osc
3N200 NTE222 SK3065	N-dual-gate MOSFET	330	20	−6	10,000	4-8.5	50	500	4.5 dB @400 MHz	TO-72	3	VHF/UHF amp, mix, osc
3N202 NTE454 SK3991	N-dual-gate MOSFET	360	25	−5	8000	6	50	200	4.5 dB @200 MHz	TO-72	3	VHF amp, mixer
MPF102 NTE451 SK9164	N-JFET	310	25	−8	2000	4.5	20	200	4 dB @400 MHz	TO-92	2	HF/VHF amp, mix, osc
MPF106 2N5484	N-JFET	310	25	−6	2500	5	30	400	4 dB @200 MHz	TO-92	2	HF/VHF/UHF amp, mix, osc
40673 NTE222 SK3050	N-dual-gate MOSFET	330	20	−4	12,000	6	50	400	6 dB @200 MHz	TO-72	3	HF/VHF/UHF amp, mix, osc
U304	P-JFET	350	−30	+10	27		−50	—	—	TO-18	4	analog switch chopper
U310	N-JFET	500 300	30 30	−6	10,000	2.5	60	450	3.2 dB @450 MHz	TO-52	5	common-gate VHF/UHF amp,
U350	N-JFET Quad	1W	25	−6	9000	5	60	100	7 dB @100 MHz	TO-99	6	matched JFET doubly bal mix
U431	N-JFET Dual	300	25	−6	10,000	5	30	100	—	TO-99	7	matched JFET cascode amp and bal mix
2N5670	N-JFET	350	25	8	3000	7	20	400	2.5 dB @100 MHz	TO-92	2	VHF/UHF osc, mix, front-end amp
2N5668	N-JFET	350	25	4	1500	7	5	400	2.5 dB @100 MHz	TO-92	2	VHF/UHF osc, mix, front-end amp
2N5669	N-JFET	350	25	6	2000	7	10	400	2.5 dB @100 MHz	TO-92	2	VHF/UHF osc, mix, front-end amp
J308	N-JFET	350	25	6.5	8000	7.5	60	1000	1.5 dB @100 MHz	TO-92	2	VHF/UHF osc, mix, front-end amp
J309	N-JFET	350	25	4	10,000	7.5	30	1000	1.5 dB @100 MHz	TO-92	2	VHF/UHF osc, mix, front-end amp
J310	N-JFET	350	25	6.5	8000	7.5	60	1000	1.5 dB @100 MHz	TO-92	2	VHF/UHF osc, mix, front-end amp
NE32684A	HJ-FET	165	2.0	−0.8	45,000	—	30	20 GHz	0.5 dB @12 GHz	84A		Low-noise amp

Notes:
[1]25°C.
For package shape, size and pin-connection information, see manufacturers' data sheets.

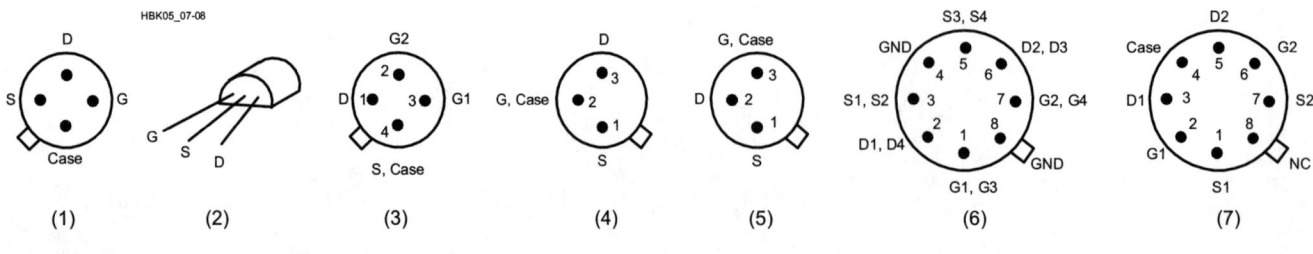

(1) (2) (3) (4) (5) (6) (7)

Low-Noise Transistors

Device	NF (dB)	F (MHz)	f_T (GHz)	I_C (mA)	Gain (dB)	F (MHz)	$V_{(BR)CEO}$ (V)	I_C (mA)	P_T (mW)	Case
MRF904	1.5	450	4	15	16	450	15	30	200	TO-206AF
MRF571	1.5	1000	8	50	12	1000	10	70	1000	Macro-X
MRF2369	1.5	1000	6	40	12	1000	15	70	750	Macro-X
MPS911	1.7	500	7	30	16.5	500	12	40	625	TO-226AA
MRF581A	1.8	500	5	75	15.5	500	15	200	2500	Macro-X
BFR91	1.9	500	5	30	16	500	12	35	180	Macro-T
BFR96	2	500	4.5	50	14.5	500	15	100	500	Macro-T
MPS571	2	500	6	50	14	500	10	80	625	TO-226AA
MRF581	2	500	5	75	15.5	500	18	200	2500	Macro-X
MRF901	2	1000	4.5	15	12	1000	15	30	375	Macro-X
MRF941	2.1	2000	8	15	12.5	2000	10	15	400	Macro-X
MRF951	2.1	2000	7.5	30	12.5	2000	10	100	1000	Macro-X
BFR90	2.4	500	5	14	18	500	15	30	180	Macro-T
MPS901	2.4	900	4.5	15	12	900	15	30	300	TO-226AA
MRF1001A	2.5	300	3	90	13.5	300	20	200	3000	TO-205AD
2N5031	2.5	450	1.6	5	14	450	10	20	200	TO-206AF
MRF4239A	2.5	500	5	90	14	500	12	400	3000	TO-205AD
BFW92A	2.7	500	4.5	10	16	500	15	35	180	Macro-T
MRF521*	2.8	1000	4.2	−50	11	1000	−10	−70	750	Macro-X
2N5109	3	200	1.5	50	11	216	20	400	2500	TO-205AD
2N4957*	3	450	1.6	−2	12	450	−30	−30	200	TO-206AF
MM4049*	3	500	5	−20	11.5	500	−10	−30	200	TO-206AF
2N5943	3.4	200	1.5	50	11.4	200	30	400	3500	TO-205AD
MRF586	4	500	1.5	90	9	500	17	200	2500	TO-205AD
2N5179	4.5	200	1.4	10	15	200	12	50	200	TO-206AF
2N2857	4.5	450	1.6	8	12.5	450	15	40	200	TO-206AF
2N6304	4.5	450	1.8	10	15	450	15	50	200	TO-206AF
MPS536*	4.5	500	5	−20	4.5	500	−10	−30	625	TO-226AA
MRF536*	4.5	1000	6	−20	10	1000	−10	−30	300	Macro-X

*denotes a PNP device

Complementary devices

NPN	PNP
2N2857	2N4957
MRF904	MM4049
MRF571	MRF521

For package shape, size and pin-connection information, see manufacturers' data sheets. Many retail suppliers and manufacturers offer data sheets on their Web sites.

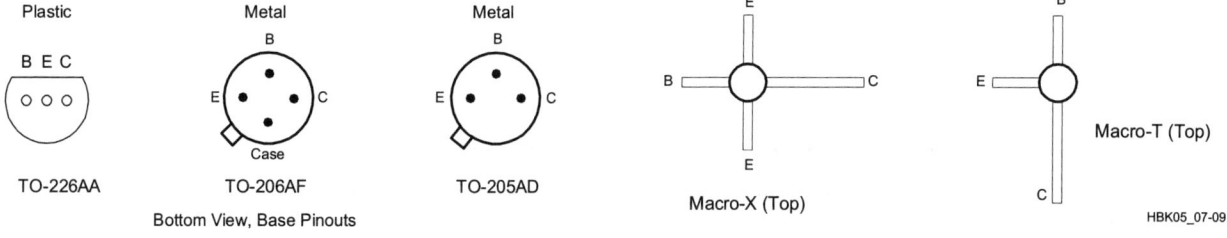

General-Purpose Transistors

Listed numerically by device

Device	Type	V_{CEO} Maximum Collector Emitter Voltage (V)	V_{CBO} Maximum Collector Base Voltage (V)	V_{EBO} Maximum Emitter Base Voltage (V)	I_C Maximum Collector Current (mA)	P_O Maximum Device Dissipation (W)	**Minimum DC Current Gain h_{FE}** $I_C = 0.1$ mA	**Minimum DC Current Gain h_{FE}** $I_C = 150$ mA	Current-Gain Bandwidth Product f_T (MHz)	Noise Figure NF Maximum (dB)	Base
2N918	NPN	15	30	3.0	50	0.2	20 (3 mA)	—	600	6.0	3
2N2102	NPN	65	120	7.0	1000	1.0	20	40	60	6.0	2
2N2218	NPN	30	60	5.0	800	0.8	20	40	250		2
2N2218A	NPN	40	75	6.0	800	0.8	20	40	250		2
2N2219	NPN	30	60	5.0	800	3.0	35	100	250		2
2N2219A	NPN	40	75	6.0	800	3.0	35	100	300	4.0	2
2N2222	NPN	30	60	5.0	800	1.2	35	100	250		2
2N2222A	NPN	40	75	6.0	800	1.2	35	100	200	4.0	2
2N2905	PNP	40	60	5.0	600	0.6	35	—	200		2
2N2905A	PNP	60	60	5.0	600	0.6	75	100	200		2
2N2907	PNP	40	60	5.0	600	0.4	35	—	200		2
2N2907A	PNP	60	60	5.0	600	0.4	75	100	200		2
2N3053	NPN	40	60	5.0	700	5.0	—	50	100		2
2N3053A	NPN	60	80	5.0	700	5.0	—	50	100		2
2N3563	NPN	15	30	2.0	50	0.6	20	—	800		1
2N3904	NPN	40	60	6.0	200	0.625	40	—	300	5.0	1
2N3906	PNP	40	40	5.0	200	0.625	60	—	250	4.0	1
2N4037	PNP	40	60	7.0	1000	5.0	—	50			2
2N4123	NPN	30	40	5.0	200	0.35	—	25 (50 mA)	250	6.0	1
2N4124	NPN	25	30	5.0	200	0.35	120 (2 mA)	60 (50 mA)	300	5.0	1
2N4125	PNP	30	30	4.0	200	0.625	50 (2 mA)	25 (50 mA)	200	5.0	1
2N4126	PNP	25	25	4.0	200	0.625	120 (2 mA)	60 (50 mA)	250	4.0	1
2N4401	NPN	40	60	6.0	600	0.625	20	100	250		1
2N4403	PNP	40	40	5.0	600	0.625	30	100	200		1
2N5320	NPN	75	100	7.0	2000	10.0	—	30 (1 A)			2
2N5415	PNP	200	200	4.0	1000	10.0	—	30 (50 mA)	15		2
MM4003	PNP	250	250	4.0	500	1.0	20 (10 mA)	—			2
MPSA55	PNP	60	60	4.0	500	0.625	—	50 (0.1 A)	50		1
MPS6531	NPN	40	60	5.0	600	0.625	60 (10 mA)	90 (0.1 A)			1
MPS6547	NPN	25	35	3.0	50	0.625	20 (2 mA)	—	600		1

Test conditions: IC = 20 mA dc; VCE = 20 V; f = 100 MHz

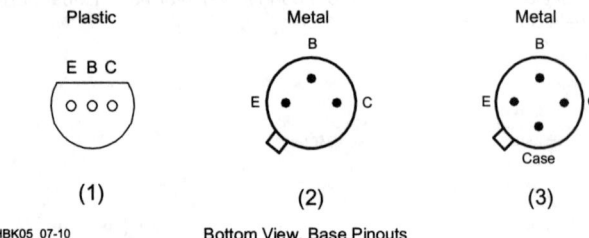

Bottom View, Base Pinouts

(1) Plastic — E B C
(2) Metal — B, E, C
(3) Metal — B, E, C, Case

RF Power Amplifier Modules

Listed by frequency

Device	Supply (V)	Frequency Range (MHz)	Output Power (W)	Power Gain (dB)	Package†	Mfr/ Notes
M57735	17	50-54	14	21	H3C	MI; SSB mobile
M57719N	17	142-163	14	18.4	H2	MI; FM mobile
S-AV17	16	144-148	60	21.7	5-53L	T, FM mobile
S-AV7	16	144-148	28	21.4	5-53H	T, FM mobile
MHW607-1	7.5	136-150	7	38.4	301K-02/3	MO; class C
BGY35	12.5	132-156	18	20.8	SOT132B	P
M67712	17	220-225	25	20	H3B	MI; SSB mobile
M57774	17	220-225	25	20	H2	MI; FM mobile
MHW720-1	12.5	400-440	20	21	700-04/1	MO; class C
MHW720-2	12.5	440-470	20	21	700-04/1	MO; class C
M57789	17	890-915	12	33.8	H3B	MI
MHW912	12.5	880-915	12	40.8	301R-01/1	MO; class AB
MHW820-3	12.5	870-950	18	17.1	301G-03/1	MO; class C

Manufacturer codes: MO = Motorola; MI = Mitsubishi; P = Philips; T = Toshiba.

†For package shape, size and pin-connection information, see manufacturers' data sheets. Many retail suppliers offer data sheets to buyers free of charge on request. Data books are available from many manufacturers and retailers.

Power FETs

Device	Type	VDSS min (V)	RDS(on) max (Ω)	ID max (A)	PD max (W)	Case†	Mfr
BS250P	P-channel	45	14	0.23	0.7	E-line	Z
IRFZ30	N-channel	50	0.050	30	75	TO-220	IR
MTP50N05E	N-channel	50	0.028	25	150	TO-220AB	M
IRFZ42	N-channel	50	0.035	50	150	TO-220	IR
2N7000	N-channel	60	5	0.20	0.4	E-line	Z
VN10LP	N-channel	60	7.5	0.27	0.625	E-line	Z
VN10KM	N-channel	60	5	0.3	1	TO-237	S
ZVN2106B	N-channel	60	2	1.2	5	TO-39	Z
IRF511	N-channel	60	0.6	2.5	20	TO-220AB	M
MTP2955E	P-channel	60	0.3	6	25	TO-220AB	M
IRF531	N-channel	60	0.180	14	75	TO-220AB	M
MTP23P06	P-channel	60	0.12	11.5	125	TO-220AB	M
IRFZ44	N-channel	60	0.028	50	150	TO-220	IR
IRF531	N-channel	80	0.160	14	79	TO-220	IR
ZVP3310A	P-channel	100	20	0.14	0.625	E-line	Z
ZVN2110B	N-channel	100	4	0.85	5	TO-39	Z
ZVP3310B	P-channel	100	20	0.3	5	TO-39	Z
IRF510	N-channel	100	0.6	2	20	TO-220AB	M
IRF520	N-channel	100	0.27	5	40	TO-220AB	M
IRF150	N-channel	100	0.055	40	150	TO-204AE	M
IRFP150	N-channel	100	0.055	40	180	TO-247	IR
ZVP1320A	P-channel	200	80	0.02	0.625	E-line	Z
ZVN0120B	N-channel	200	16	0.42	5	TO-39	Z
ZVP1320B	P-channel	200	80	0.1	5	TO-39	Z
IRF620	N-channel	200	0.800	5	40	TO-220AB	M
MTP6P20E	P-channel	200	1	3	75	TO-220AB	M
IRF220	N-channel	200	0.400	8	75	TO-220AB	M
IRF640	N-channel	200	0.18	10	125	TO-220AB	M

Manufacturers: IR = International Rectifier; M = Motorola; S = Siliconix; Z = Zetex.

†For package shape, size and pin-connection information, see manufacturers' data sheets. Many retail suppliers offer data sheets to buyers free of charge on request. Data books are available from many manufacturers and retailers.

Logic IC Families

Type	Propagation Delay for CL = 50 pF (ns) Typ	Max	Max Clock Frequency (MHz)	Power Dissipation (CL = 0) @ 1 MHz (mW/gate)	Output Current @ 0.5 V max (mA)	Input Current (Max mA)	Threshold Voltage (V)	Supply Voltage (V) Min	Typ	Max
CMOS										
74AC	3	5.1	125	0.5	24	0	V+/2	2	5 or 3.3	6
74ACT	3	5.1	125	0.5	24	0	1.4	4.5	5	5.5
74HC	9	18	30	0.5	8	0	V+/2	2	5	6
74HCT	9	18	30	0.5	8	0	1.4	4.5	5	5.5
4000B/74C (10 V)	30	60	5	1.2	1.3	0	V+/2	3	5 - 15	18
4000B/74C (5V)	50	90	2	3.3	0.5	0	V+/2	3	5 - 15	18
TTL										
74AS	2	4.5	105	8	20	0.5	1.5	4.5	5	5.5
74F	3.5	5	100	5.4	20	0.6	1.6	4.75	5	5.25
74ALS	4	11	34	1.3	8	0.1	1.4	4.5	5	5.5
74LS	10	15	25	2	8	0.4	1.1	4.75	5	5.25
ECL										
ECL III	1.0	1.5	500	60	—	—	−1.3	−5.19	−5.2	−5.21
ECL 100K	0.75	1.0	350	40	—	—	−1.32	−4.2	−4.5	−5.2
ECL100KH	1.0	1.5	250	25	—	—	−1.29	−4.9	−5.2	−5.5
ECL 10K	2.0	2.9	125	25	—	—	−1.3	−5.19	−5.2	−5.21
GaAs										
10G	0.3	0.32	2700	125	—	—	−1.3	−3.3	−3.4	−3.5
10G	0.3	0.32	2700	125	—	—	−1.3	−5.1	−5.2	−5.5

Source: Horowitz (W1HFA) and Hill, *The Art of Electronics—2nd edition,* page 570. © Cambridge University Press 1980, 1989. Reprinted with the permission of Cambridge University Press.

Three-Terminal Voltage Regulators

Listed numerically by device

Device	Description	Package	Voltage	Current (Amps)
317	Adj Pos	TO-205	+1.2 to +37	0.5
317	Adj Pos	TO-204, TO-220	+1.2 to +37	1.5
317L	Low Current Adj Pos	TO-205, TO-92	+1.2 to +37	0.1
317M	Med Current Adj Pos	TO-220	+1.2 to +37	0.5
338	Adj Pos	TO-3	+1.2 to +32	5.0
350	High Current Adj Pos	TO-204, TO-220	+1.2 to +33	3.0
337	Adj Neg	TO-205	−1.2 to −37	0.5
337	Adj Neg	TO-204, TO-220	−1.2 to −37	1.5
337M	Med Current Adj Neg	TO-220	−1.2 to −37	0.5
309		TO-205	+5	0.2
309		TO-204	+5	1.0
323		TO-204, TO-220	+5	3.0
140-XX	Fixed Pos	TO-204, TO-220	Note 1	1.0
340-XX		TO-204, TO-220		1.0
78XX		TO-204, TO-220		1.0
78LXX		TO-205, TO-92		0.1
78MXX		TO-220		0.5
78TXX		TO-204		3.0
79XX	Fixed Neg	TO-204, TO-220	Note 1	1.0
79LXX		TO-205, TO-92		0.1
79MXX		TO-220		0.5

Note 1—XX indicates the regulated voltage; this value may be anywhere from 1.2 V to 35 V. A 7815 is a positive 15-V regulator, and a 7924 is a negative 24-V regulator.

The regulator package may be denoted by an additional suffix, according to the following:

Package	Suffix
TO-204 (TO-3)	K
TO-220	T
TO-205 (TO-39)	H, G
TO-92	P, Z

For example, a 7812K is a positive 12-V regulator in a TO-204 package. An LM340T-5 is a positive 5-V regulator in a TO-220 package. In addition, different manufacturers use different prefixes. An LM7805 is equivalent to a mA7805 or MC7805.

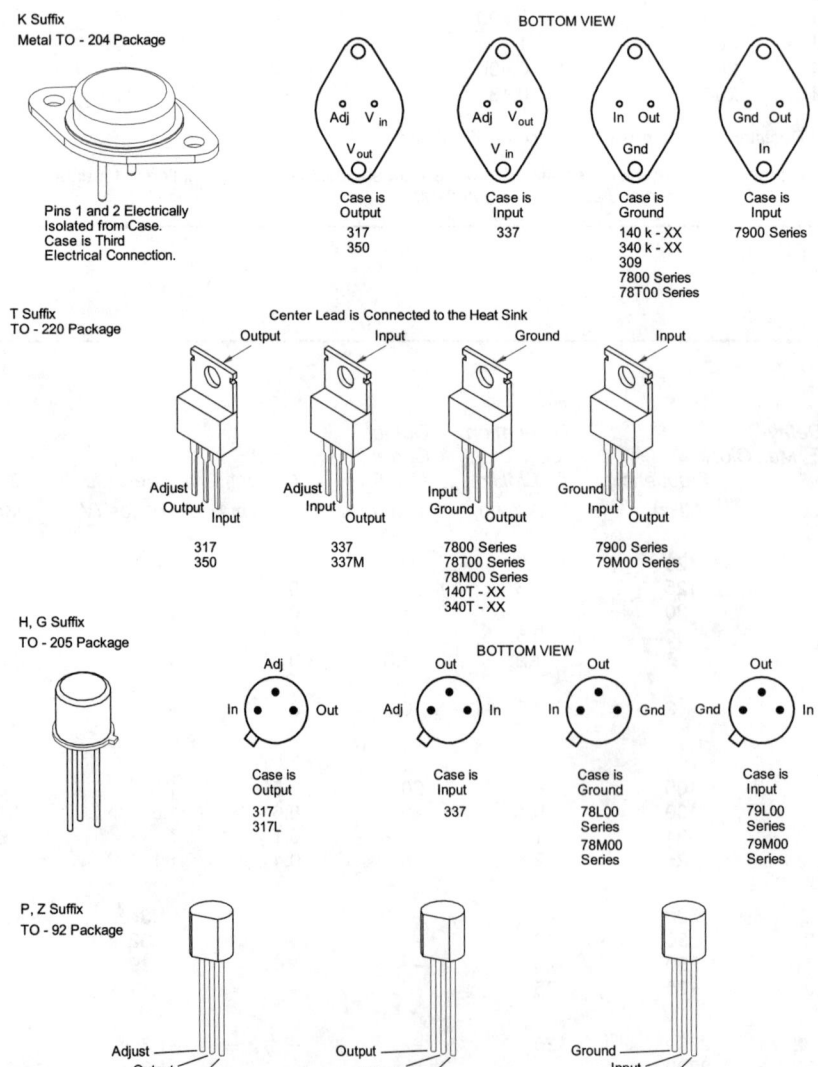

Coaxial Cable End Connectors

UHF Connectors

Military No.	Style	Cable RG- or Description
PL-259	Str (m)	8, 9, 11, 13, 63, 87, 149, 213, 214, 216, 225
UG-111	Str (m)	59, 62, 71, 140, 210
SO-239	Pnl (f)	Std, mica/phenolic insulation
UG-266	Blkhd (f)	Rear mount, pressurized, copolymer of styrene ins.

Adapters

PL-258	Str (f/f)	Polystyrene ins.
UG-224,363	Blkhd (f/f)	Polystyrene ins.
UG-646	Ang (f/m)	Polystyrene ins.
M-359A	Ang (m/f)	Polystyrene ins.
M-358	T (f/m/f)	Polystyrene ins.

Reducers

UG-175	55, 58, 141, 142 (except 55A)
UG-176	59, 62, 71, 140, 210

Family Characteristics:

All are nonweatherproof and have a nonconstant impedance. Frequency range: 0-500 MHz. Maximum voltage rating: 500 V (peak).

N Connectors

Military No.	Style	Cable RG-	Notes
UG-21	Str (m)	8, 9, 213, 214	50 Ω
UG-94A	Str (m)	11, 13, 149, 216	70 Ω
UG-536	Str (m)	58, 141, 142	50 Ω
UG-603	Str (m)	59, 62, 71, 140, 210	50 Ω
UG-23, B-E	Str (f)	8, 9, 87, 213, 214, 225	50 Ω
UG-602	Str (f)	59, 62, 71, 140, 210	—
UG-228B, D, E	Pnl (f)	8, 9, 87, 213, 214, 225	—
UG-1052	Pnl (f)	58, 141, 142	50 Ω
UG-593	Pnl (f)	59, 62, 71, 140, 210	50 Ω
UG-160A, B, D	Blkhd (f)	8, 9, 87, 213, 214, 225	50 Ω
UG-556	Blkhd (f)	58, 141, 142	50 Ω
UG-58, A	Pnl (f)		50 Ω
UG-997A	Ang (f)		50 Ω $^{11}/_{16}$"

Panel mount (f) with clearance above panel

M39012/04-	Blkhd (f)	Front mount hermetically sealed
UG-680	Blkhd (f)	Front mount pressurized

N Adapters

Military No.	Style	Notes
UG-29,A,B	Str (f/f)	50 Ω, TFE ins.
UG-57A.B	Str (m/m)	50 Ω, TFE ins.
UG-27A,B	Ang (f/m)	Mitre body
UG-212A	Ang (f/m)	Mitre body
UG-107A	T (f/m/f)	—
UG-28A	T (f/f/f)	—
UG-107B	T (f/m/f)	—

Family Characteristics:

N connectors with gaskets are weatherproof. RF leakage: −90 dB min @ 3 GHz. Temperature limits: TFE: −67° to 390°F (−55° to 199°C). Insertion loss 0.15 dB max @ 10 GHz. Copolymer of styrene: −67° to 185°F (−55° to 85°C). Frequency range: 0-11 GHz. Maximum voltage rating: 1500 V P-P. Dielectric withstanding voltage 2500 V RMS. SWR (MIL-C-39012 cable connectors) 1.3 max 0-11 GHz.

BNC Connectors

Military No.	Style	Cable RG-	Notes
UG-88C	Str (m)	55, 58, 141, 142, 223, 400	

Military No.	Style	Cable RG-	Notes
UG-959	Str (m)	8, 9	
UG-260,A	Str (m)	59, 62, 71, 140, 210	Rexolite ins.
UG-262	Pnl (f)	59, 62, 71, 140, 210	Rexolite ins.
UG-262A	Pnl (f)	59, 62, 71, 140, 210	nwx, Rexolite ins.
UG-291	Pnl (f)	55, 58, 141, 142, 223, 400	
UG-291A	Pnl (f)	55, 58, 141, 142, 223, 400	nwx
UG-624	Blkhd (f)	59, 62, 71, 140, 210	Front mount Rexolite ins.
UG-1094A	Blkhd		Standard
UG-625B	Receptacle		
UG-625			

BNC Adapters

Military No.	Style	Notes
UG-491,A	Str (m/m)	
UG-491B	Str (m/m)	Berylium, outer contact
UG-914	Str (f/f)	
UG-306	Ang (f/m)	
UG-306A,B	Ang (f/m)	Berylium outer contact
UG-414,A	Pnl (f/f)	# 3-56 tapped flange holes
UG-306	Ang (f/m)	
UG-306A,B	Ang (f/m)	Berylium outer contact
UG-274	T (f/m/f)	
UG-274A,B	T (f/m/f)	Berylium outer contact

Family Characteristics:

Z = 50 Ω. Frequency range: 0-4 GHz w/low reflection; usable to 11 GHz. Voltage rating: 500 V P-P. Dielectric withstanding voltage 500 V RMS. SWR: 1.3 max 0-4 GHz. RF leakage −55 dB min @ 3 GHz. Insertion loss: 0.2 dB max @ 3 GHz. Temperature limits: TFE: −67° to 390°F (−55° to 199°C); Rexolite insulators: −67° to 185°F (−55° to 85°C). "Nwx" = not weatherproof.

HN Connectors

Military No.	Style	Cable RG-	Notes
UG-59A	Str (m)	8, 9, 213, 214	
UG-1214	Str (f)	8, 9, 87, 213, 214, 225	Captivated contact
UG-60A	Str (f)	8, 9, 213, 214	Copolymer of styrene ins.
UG-1215	Pnl (f)	8, 9, 87, 213, 214, 225	Captivated contact
UG-560	Pnl (f)		
UG-496	Pnl (f)		
UG-212C	Ang (f/m)		Berylium outer contact

Family Characteristics:

Connector Styles: Str = straight; Pnl = panel; Ang = Angle; Blkhd = bulkhead. Z = 50 Ω. Frequency range = 0-4 GHz. Maximum voltage rating = 1500 V P-P. Dielectric withstanding voltage = 5000 V RMS SWR = 1.3. All HN series are weatherproof. Temperature limits: TFE: −67° to 390°F (−55° to 199°C); copolymer of styrene: −67° to 185°F (−55° to 85°C).

Cross-Family Adapters

Families	Description	Military No.
HN to BNC	HN-m/BNC-f	UG-309
N to BNC	N-m/BNC-f	UG-201,A
	N-f/BNC-m	UG-349,A
	N-m/BNC-m	UG-1034
N to UHF	N-m/UHF-f	UG-146
	N-f/UHF-m	UG-83,B
	N-m/UHF-m	UG-318
UHF to BNC	UHF-m/BNC-f	UG-273
	UHF-f/BNC-m	UG-255

Computer Connector Pinouts

(A) Parallel Port (DB 25 pin) Female

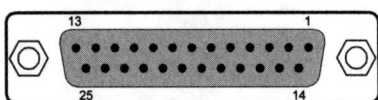

Pin	Signal	Pin	Signal
1	Strobe	10	Acknowledge
2	Data 0	11	Busy
3	Data 1	12	Paper Empty
4	Data 2	13	Select
5	Data 3	14	Auto Feed
6	Data 4	15	Error
7	Data 5	16	Initialize
8	Data 6	17	Select In
9	Data 7	18-25	GND

(B) Parallel Port (Centronics 36 pin) Female

Pin	Signal	Pin	Signal
1	Strobe	13	Select
2	Data 0	14	Auto Feed
3	Data 1	15	N/C (not connected)
4	Data 2	16	Signal GND
5	Data 3	17	Frame GND
6	Data 4	18	+5 V Out
7	Data 5	19-30	GND
8	Data 6	31	Reset
9	Data 7	32	Error
10	Acknowledge	33	External GND
11	Busy	34	N/C
12	Paper Empty	35	N/C
		36	Select In

(C) Serial Port (DB 9 pin) Male

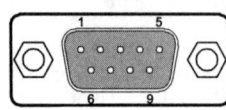

Pin	Signal
1	DCD (Data Carrier Detect)
2	RxD (Receive Data)
3	TxD (Transmit Data)
4	DTR (Data Terminal Ready)
5	GND (Signal Ground)
6	DSR (Data Set Ready)
7	RTS (Request To Send)
8	CTS (Clear To Send)
9	RI (Ring Indicator)

(D) Serial Port (DB 25 pin) Male

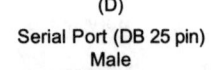
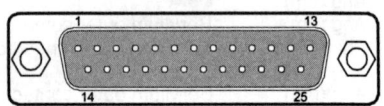

Pin	Signal	Pin	Signal
1	N/C (not connected)	20	DTR (Data Terminal Ready)
2	TxD (Transmit Data)	21	N/C
3	RxD (Receive Data)	22	RI (Ring Indicator)
4	RTS (Request To Send)	23	N/C
5	CTS (Clear To Send)	24	N/C
6	DSR (Data Set Ready)	25	N/C
7	GND (Signal Ground)		
8	DCD (Data Carrier Detect)		
9-19	N/C		

(E) Ethernet Connector (RJ45-8 pin) Female

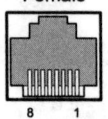

Pin	Signal
1	Output Transmit Data (+)
2	Output Transmit Data (-)
3	Input Receive Data (+)
4	N/C (not connected)
5	N/C
6	Input Receive Data (-)
7	N/C
8	N/C

(F) Ethernet Connector (RJ45-10 pin) Female

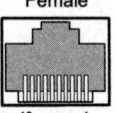

Pin	Signal
1	DCD (Data Carrier Detect)
2	DTR (Data Terminal Ready)
3	CTS (Clear To Send)
4	GND (Signal Ground)
5	RxD (Receive Data)
6	TxD (Transmit Data)
7	GND (Frame Ground)
8	RTS (Request To Send)
9	DSR (Data Set Ready)
10	RI (Ring Indicator)

(G) Mouse Port (DB 9 pin) Male

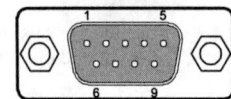

Pin	Signal
1	N/C (not connected)
2	Data
3	Clock
4	N/C
5	GND (Signal Ground)
6	N/C
7	RTS (12-9 V)
8	N/C
9	N/C

(H) Mouse Port (mini DIN 9 pin) Female

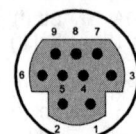

Pin	Signal
1	+5 V
2	X-A
3	X-B
4	Y-A
5	Y-B
6	Button 1
7	Button 2
8	Button 3
9	GND

(I) Game/Joystick Port (DB 15 pin) Female

Pin	Signal	Pin	Signal
1	+5 V	10	Button (B-1)
2	Button (A-1)	11	Position (B-X)
3	Position (A-X)	12	GND
4	GND	13	Position (B-Y)
5	GND	14	Button (B-2)
6	Position (A-Y)	15	+5 V
7	Button (A-2)		
8	+5 V		
9	+5 V		

(J) PC-AT Type Power Connector

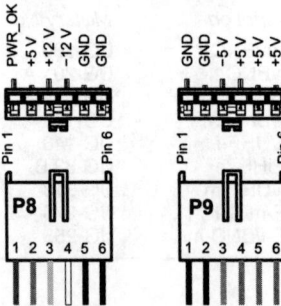

(K) PC-ATX Type Power Connector Viewed from Connector End

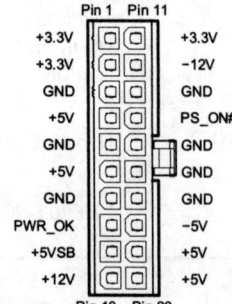

(L) Disk Drive, CD and Other Device Power Connector Viewed from Connector End

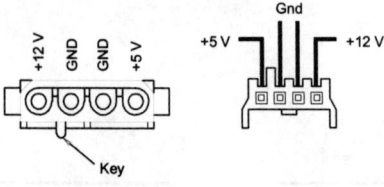

Note: All figures not drawn to same scale.

FEEDBACK

Please use this form to give us your comments on this book and what you'd like to see in future editions, or e-mail us at **pubsfdbk@arrl.org** (publications feedback). If you use e-mail, please include your name, call, e-mail address and the book title, edition and printing in the body of your message. Also indicate whether or not you are an ARRL member.

Where did you purchase this book?
☐ From ARRL directly   ☐ From an ARRL dealer

Is there a dealer who carries ARRL publications within:
☐ 5 miles   ☐ 15 miles   ☐ 30 miles   of your location?   ☐ Not sure.

License class:
☐ Novice   ☐ Technician   ☐ Technician with code   ☐ General   ☐ Advanced   ☐ Amateur Extra

Name _____   ARRL member? ☐ Yes ☐ No
                                                              Call Sign _____

Daytime Phone (   ) _____   Age _____

Address _____

City, State/Province, ZIP/Postal Code _____

If licensed, how long? _____   e-mail address: _____

Other hobbies _____

Occupation _____

For ARRL use only		HOR EXP
Edition	1 2 3 4 5 6 7 8 9 10 11 12	
Printing	1 2 3 4 5 6 7 8 9 10 11 12	

From _____
_____
_____

Please affix postage. Post Office will not deliver without postage.

EDITOR, HANDS-ON RADIO EXPERIMENTS, VOL 2
ARRL—THE NATIONAL ASSOCIATION FOR AMATEUR RADIO
225 MAIN STREET
NEWINGTON CT 06111-1494

— — — — — — — — — — — — — — please fold and tape — — — — — — — — — — — — — — —